AF674942

ABOUT THE AUTHOR

Dr. N. Ghosh Hajra heading Darjeeling Tea Research and Development Centre, Tea Board of India as Project Manager. He did his Master degree from the University of Calcutta, and a Doctorate from Burdwan University. He has been the recipient of CSIR Post Doctoral Scholarship. Nikhil Ghosh Hajra was a visiting scientist in Belgium at the invitation of the University of Antwerpen. Dr. Ghosh Hajra has been Invited Speaker at several International conferences including those held at Bordeau, France and Brussels, Belgium. He is associated with different professional bodies including Academy of Plant Sciences, India.

Dr. Ghosh Hajra has authored 66 research papers, monographs in professional journals of International repute and one book besides a large number of popular articles. His work and publications attracted awards and gold medal.

Tea Cultivation : Comprehensive Treatise

N. Ghosh Hajra

International Book Distributing Company
Lucknow, U.P

Published by
INTERNATIONAL BOOK DISTRIBUTING COMPANY
(Publishing Division)
Chaman Studio Building, 2nd Floor,
Charbagh, Lucknow - 226 004 (India)
Tel. : Off. : 450004, 450007, 459058 Res. : 343274, 342829, Fax : 0522-458629
E-Mail : ibdco@sancharnet.in

First Edition 2001

ISBN 81-85860-57-2

Composed and Designed at :
Panacea Computers
33, Nehru Road, Sadar Cantt.
Lucknow - 226 002 U.P.
Tel. : 0522-481164, 480546
E-mail : prasgupt@hotmail.com

Printed at :
Army Printing Press
33, Nehru Road, Sadar Cantt
Lucknow - 226 002
Tel. : 0522- 481164, 480546
E-mail : armyprintingpress@rediffmail.com

Preface

Tea has come to occupy an important place among plantation crops of the world and it is known world over as one of the healthy beverages. Tea plantation a wonderful agro-asset and does not cause any imbalance in the eco-system; rather it really contributes a lot in maintaining ecological balance. The lush green tea plantation retains the soil cover, checks soil erosion and also conserves the soil moisture. Tea is cultivated in 31 countries, scattered from 45^0 north to 33^0 south of the equator. Because of the long history and the large-scale production, trade and consumption, tea occupies an important place in the world agricultural economy.

More than forty research institutions in the world are engaged in research on different aspects of tea. The information generated in most of the institutions not easily accessible to the researchers, progressive planters and others. I felt necessary to review this plethora of information on different facets of tea cultivation and placed between covers before the readers. Many related information was not possible to include due to space problem and therefore discretion had to be exercised in their choice. However, the present book provides adequate information emanated from current and classical research on tea backed by discussions and opens up new vistas of discoveries for researchers as well as serve ready reckoner for the vigilant tea growers, and offers alternate solutions to persistent problems for policy planners. The tea estate managers can seek alternatives to choose scientifically tested ideas to decide on future strategies to increase their production and improvement in product quality.

Processing of tea fall outside the scope of the book.

Every effort has been made to ensure the accuracy of the contents of the book. The views offered in the present book is personal and not those of the Tea Board of India. However, I am pleased to hear of any errors/ loopholes which have escaped my notice and also welcome suggestion(s) for incorporation in the next edition

N. Ghosh Hajra

Dedicated to

My wife

Sudeshna

# Acknowledgements

I wish to express my sincere thanks and gratefulness to Shri N. K. Das, IAS, Chairman, Tea Board and Shri B. Banerjee, IAS, Deputy Chairman, Tea Board for providing me the opportunity and inspiration for publication of this book.

I am indebted to the tea planters of Darjeeling hills, friends in tea industry the world over for their encouraging support in a number of ways during the preparation of this book.

It is a special pleasure to acknowledge the constant encouragement provided by my colleagues of Darjeeling Tea Research and Development Centre and my friends Shri D. Chowdhuri, Deputy Director, Rubber Research Institute of India, Guwahati and Dr. G. Sircar, In-Charge, New Alluvial Zone, BCKVV, Nadia.

I would also like to acknowledge the help taken in writing this book from various reports, journals and books available at different libraries and say thanks to the authors for their dedicated work.

I feel obliged and extremely thankful to Shri Suneel Gomber, International Book Distributing Co., for taking endless pain during its preparation, printing and publication.

Finally, I am grateful to my father K. K. Ghosh Hajra, mother J. Ghosh Hajra and wife for their active support and keen interest right from the time when this work was contemplated.

Nikhil Ghosh Hajra

Contents

Chapter 1

History of tea cultivation

1.1 Introduction

Tea has been domesticated and grown for nearly two centuries but the place of its origin and dispersal is still a matter of speculation. However, the natural home of the tea plant is considered to be within the fan shaped area included between the Naga, Manipuri and Lushai Hills along the Assam-Burma frontier in the west, through to China, probably as far as the Che-Kiang (Zhejiang) province in the east, and from this line generally south through the hills of Burma and Thailand into Vietnam (Harler, 1966). A recent study, based on the geological history of eastern Yunan and the distribution of *Camellia* species in that area indicated that the long narrow region of Wenshan and Honghe, located between 22° 40′-24° 10′ N and 103° 10′-105° 20′ E was the centre of origin of the tea plant (Yu, 1986).

In China, tea has been used as medicinal decoction during 4th century and later in 7th century tea became common beverage but no wild genotype of tea has been found in China. The word tea is derived from T'e of the Chinese Fukien dialect and tea is known as ch'a in Cantanese language. The wild types recorded from time to time in southern China are possibly relics from cultivation in the unrecorded past (Eden, 1976). However, the current distribution patterns of different varieties of tea suggest that the probable centre of origin of tea is near the source of Irrawaddy River and from there it dispersed to south-eastern China, Indonesia and Assam. Tea varieties specifically these three principal regions have characteristic biological features, but it is rarely that only one particular type is to be found in a particular locality. The cohabitation of more than one type in a locality suggests intermixing between the types is indeed a common feature suggesting a common genetic base for all varieties of tea (Banerjee, 1993).

According to Kingdon-Ward (1950), tea however, probably originated further north of the Irrawaddy basin, and that the three prime locations are basically the secondary centre of origin. Tea varieties due to their distinct and specific morphological features were even considered as distinct subspecies (Watt, 1889), and identified as *bohea, stricta, viridis* and *lasio-calyx*. On the basis of morphological and biological features, tea was grouped as *assamica, sinensis* and *cambodia*.

1.2 Tea producing countries

From the main centres of its primary origin in south-east-Asia, tea has spread far and wide into tropical and sub-tropical areas and adopted broad characteristics correspond to regions of tropical rain forest, tropical Savannah and summer rain areas. At present tea is grown in more than thirty countries of which a dozen could be regarded as major tea growing countries. The area, productivity and export of tea of some countries are given in Table 1.1.

1.2.1 Tea in India

North-east India

The East India Company has initiated searching of tea in Assam as an alternate source of tea which was then supplied to U.K. mainly from China. Lord William Bentinck, the Governor-General of India appointed the Tea Committee in 1834 to investigate and recommend suitable areas for tea cultivation in India.

The Tea Committee sent G. J. Gordon to China in order to collect tea seed as well as tea makers and those familiar with the cultivation of tea. Gordon left Calcutta on the sailing ship *Water Witch* in June 1834.

In 1823, Major Robert Bruce while visiting as member of trading mission discovered tea plants growing wild in some hills near Rangpur (now Sibsagar), the then capital of Assam (Ukers, 1935). Assam was then Burmese occupation and in 1824 war broke out with Burmese. Robert Bruce died in the same year. His younger brother Mr. Charles Alexander

Table 1.1 : Area, production, productivity and export of tea in different countries during 1998.

	Country	Area (ha)	Production ('000 Kg)	Export ('000 Kg)	Productivity (kg/ha)
***Asia**					
	** Bangladesh	49246	55700	22215	1131
	** China	1103000	665030	217152	602
	** CIS (USSR)	NA	10000	-	-
	** India	436344	870410	210338	1994
	Iran	35000	60000	25000	1714
	** Indonesia	156681	166120	70000	1060
	** Japan	51200	82600	752	1613
	Malaysia	3053	6300	800	2063
	Myanmear	58700	56000	40	954
	Nepal	3600	2900	15	806
	** Sri Lanka	195000	280000	265305	1435
	Taiwan	220659	22640	2482	974
	Thailand	6400	7928	270	1238
	** Turkey	76752	110000	2000	1433
	Vietnam	71000	42000	16500	591
Africa					
	Burundi	7950	6669	5784	839
	** Kenya	118650	294160	263023	2479
	Malawi	18798	40360	41000	2147
	Mauritius	1109	1480	194	1334
	Mozambique	15900	1600	500	1006
	Republic of Congo	9000	2500	2000	277
	Rwanda	11000	14850	5000	1350
	South Africa	6327	10840	-	1713
	Tanzania	20925	24333	22218	1162
	Uganda	20570	26425	23350	1285
	Zimbabwe	6858	17750	10800	2588
South America					
	Argentina	39000	50000	41239	1282
	Brazil	6000	8000	3208	1333
	Ecuador	NA	1900	1200	
	Peru	2000	2400	200	1200
Australia					
	Papua New-Guinea	4000	6000	5000	1500
	Queenslands	NA	1250	-	NA

* Including parts of Europe; ** Major tea producing countries; NA - Not available
Source: Tea Board India, Calcutta; ConceptGUI, Inc. and Report of different Organizations.

Bruce was appointed in 1824 to command a division of Gumboats in upper Assam. During the progress of hostalities with Burma in 1826 he brought down certain plants and seeds which were identified as true tea camellia of commerce. But there is dispute in this regard.

According to Baildon (1877), the tea plant of Assam was discovered by local Assamese nobleman, Dewan Maniram Dutta Barua, who later worked in the Assam company for some time. It might be that Dewan Maniram brought the plant to the notice of Robert Bruce during his visit to Rangpur in 1923. Further, the role of Sungpho tribe of Assam in bringing local plant to the notice of the outside world can not be ignored. It was a Sungpho Chief who supplied the seeds and plants to C. A. Bruce. This clearly shows that the Sungphos were familiar with the tea plant (Barua, 1989).

The Tea Committee reported to the Government that the tea plant was indigenous to Assam, especially in its upper reaches and about the possibility of its commercial cultivation in India. By that time it was too late to recall G. J. Gordon from China and any way a consignment of some 80,000 seeds was on its way to India. These seeds reached Calcutta in January of 1835 and nurseries were raised in the Government Botanical Gardens in Calcutta. From this consignment of China tea seeds, the resultant young plants of which 20,000 were allocated to the hill districts of Kumaon in north India; 20,000 to upper Assam on the north-east frontier and the remaining 20,000 plants to the Nilgiri hills of south India. Even after discovery of tea plants in Assam, the seeds were imported from China. The experimental planting carried out at Saikhown near Sadiya in Assam where many plants died. The surviving plants were again replanted to a new site near Chabua about 25 kilometres east of the present Dibrugarh town.

In 1835, a scientific commission was also constituted with Dr. Nathaniel Wallich, Dr. W. Griffith and Dr. J. Mclell to prepare a report on Indian indigenous tea and to advise on the most suitable sites for establishing experimental tea gardens. Dr. N. Wallich favoured the Himalayan region while others suggested upper Assam as the site for experimental planting where "wild" tea grown under natural condition. Commercial plantings were started with local Assam tea and seed originated in China from 1836 onwards. The experimental tea samples were sent to Calcutta in 1836. Later an invoice of eight chests of Assam tea were sent to London in 1838 and the consignment was auctioned in London on January 10, 1839. Different corporate tea companies in the middle of the nineteenth century were organised to operate in Assam and else where. Griffiths (1967) and earlier by Ukers (1935) provide a detailed history of the growth and development of tea in India.

The high prices fetched by the tea sent in London in 1838-1839 at once attracted the attention of enterprising businessmen in Calcutta as well as in London. In 1839, the first private owned tea company the Assam Company was formed in London by the Board of Control and Court of Directors of the East India Company. In March 1840, practically all the East India Company's plantations, except those at Chabua, Dinijoy and two other small gardens, were handed over to the Assam Company. Later in 1852, the company emerged as the first successful tea company in the world and performance of the Assam Company attracted others to invest on tea. Later, through the pioneering efforts of local entrepreneurs mostly Assamese, the plantation began to spread (Table 1.2) and flourish in other parts of the state. The tea areas of Assam are shown in Figure 1.1.

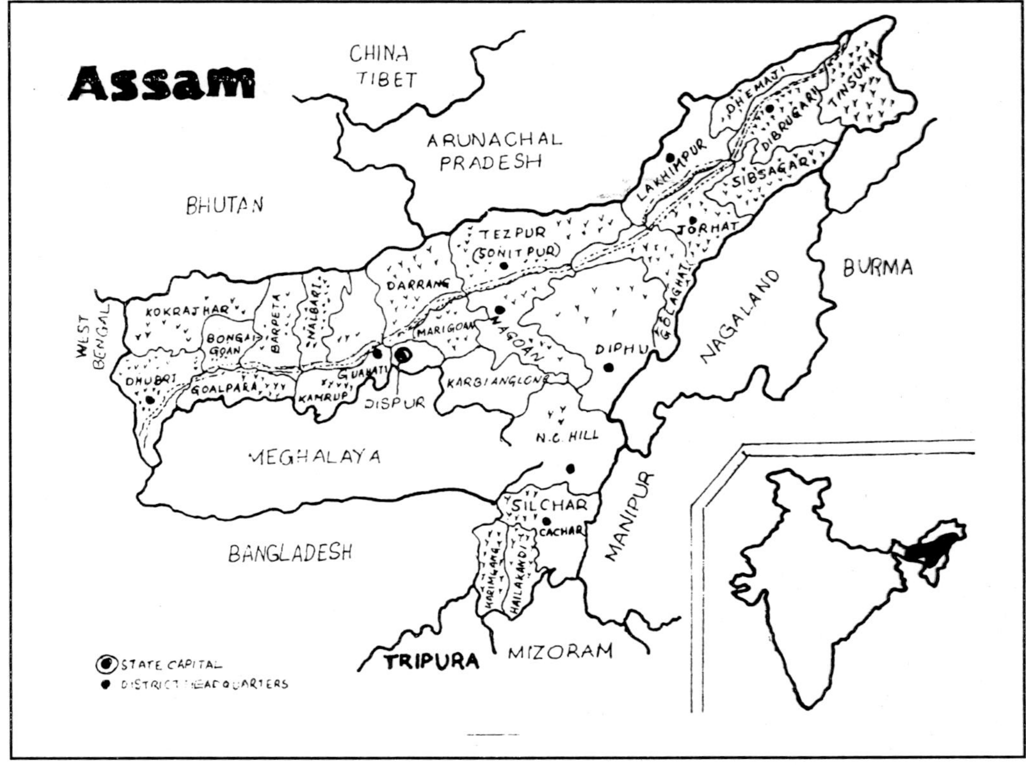

Fig. 1.1: Tea growing areas in Assam.

Table 1.2 : Trend in change of area and production of tea in India.

States	1951	1961	1971	1981	1991	1994	1996	1998
Assam								
*Area	155674	162367	182325	203038	233284	227120	227240	228260
**Production	150370	182311	223665	305130	400554	400732	424864	425430
West Bengal								
Area	79580	82705	88409	93971	101922	99967	102550	103950
Production	74158	86258	104087	128259	155216	158825	165375	169946
Tripura								
Area	4773	5055	5444	5333	5898	5938	5980	6010
Production	1873	2689	2960	3219	5154	5827	6657	6548
Bihar								
Area	1644	534	460	460	22	76	76	76
Production	1098	55	41	1	100	136	140	150
Uttar Pradesh								
Area	2605	2084	1818	1804	903	914	1070	1070
Production	901	804	690	542	520	338	390	400
Himachal Pradesh								
Area ***	4317	4183	4183	4183	2063	2063	2312	2312
Production	1125	1188	888	607	1184	1314	1320	1370
Sikkim								
Area	-	-	-	192	171	172	172	172
Production	-	-	-	32	95	96	98	104
Orissa								
Area	-	-	-	-	214	219	219	219
Production	-	-	-	-	10	46	47	85
Arunachal Pradesh, Manipur & Nagaland								
Area	-	-	-	85	705	1731	2105	2160
Production	-	-	-	-	100	655	735	1633
Tamil Nadu								
Area	33375	32723	34646	37037	38634	48854	49050	49142
Production	25225	39768	57531	71597	119937	117520	113755	130594
Kerala								
Area	33203	39784	37271	35589	34678	36817	36798	36821
Production	25775	39545	42729	47631	66803	63127	61962	69637
Karnataka								
Area	1669	1794	1870	1901	1976	2095	2105	2105
Production	874	1779	2877	3409	4526	4293	4691	5415

* Area in hectare; ** Production in thousand kg; *** Area registered with Tea Board.
Source: Tea Board India, Calcutta; Report of different Organizations.

Tripura

Maharaja Birendra Kishore Manikya was pioneer in introducing tea cultivation in Tripura state. The available information reveals that Ranibari was the first tea estate established as early as in 1917. Early entrepreneurs to ven:ure out on tea were mostly Bengalis who earn royal patronage including free gifts of land for the purpose. But Tripura tea industry had suffered from the very beginning owing to inadequate technical knowledge on appropriate field practices resulting in low yield and slender economic health through the decades (Anonymous, 1991). The tea growing areas in Tripura shown in Figure 1.2.

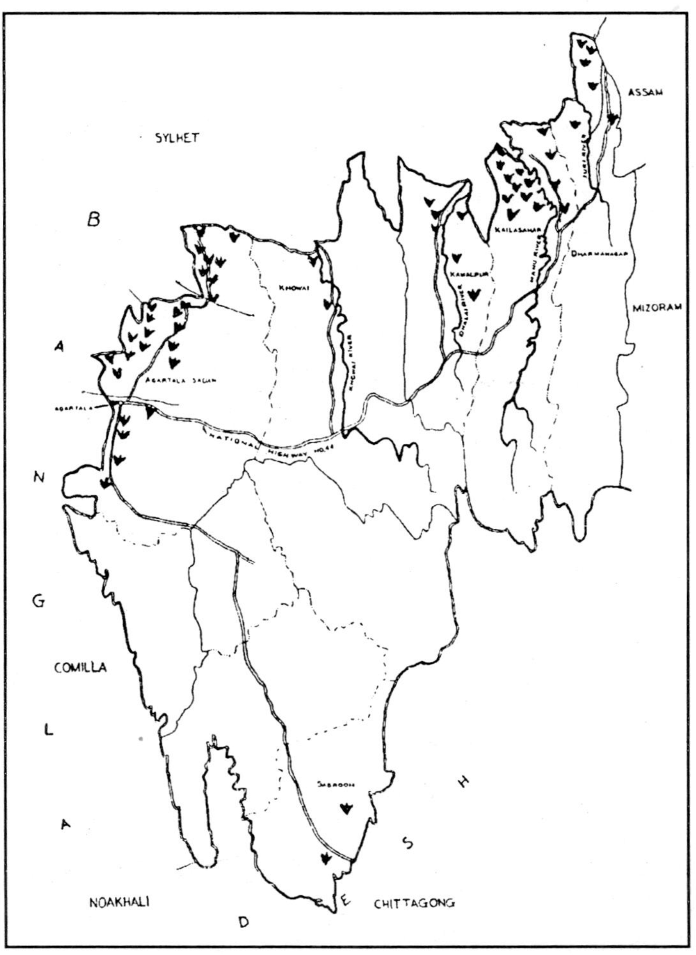

Fig. 1.2: Tea growing areas in Tripura [Source: Anonymous (1991)].

Dooars

The tea industry in Dooars did not begin commercially before 1874-75. In J. F. Grunning's *Gazetteer of the district Jalpaiguri* (1911) it has been recorded that the first tea garden established in the district was Gazaldubi in the year 1874. This garden was established by Dr. Brougham, who had started Dootoriah garden in Darjeeling in 1859, employing as his Manager, Richard Haughton, who is described in the Gazetteer as the pioneer of tea in Jalpaiguri district. In 1875, two other gardens viz. Fulbari and Bagrakote were established. Fulbari was opened out by late Mr. Pillans and was owned by Col. Money and Bagrakote was opened out by late Mr. North and owned by Mr. W. S. Cresswel (Ghosh, 1970). Mr. Cresswell, in fact, started many more gardens in the Dooars and subsequently launched the renowned Broking Firm of M/S W. S. Cresswell & Company.

In 1877, the first Indian, Munshi Rahim Baksh, a Peshkar (clerk) of the Deputy Commissioner, received Jaldhaka Grant on 17.8.1877 for the purpose of cultivation of tea for 728 acres. In 1877, Baintbarrie, Dam Dim, Kumlai and Washabarrie tea estates were started. At this time one Mr. Shillingford and his family came into the picture and were responsible for starting many gardens. In 1878, Kalabari Grant for 800 acres was taken by Mr. Johnson on behalf of Land Mortgage Bank on 9.3.1878. Subsequently the grant was passed on to the hands of Dr. Nilratan Sircar, the then eminent physician of Bengal and Smt. Sarojini Roy. In 1878, Good Hope, Ranicherra, Manabarrie, Balabarrie Altadanga, Money hope, Chael and Patabarrie of Lish River Grants were taken (Ghosh, 1970). On 19.9.1878, Babu Kali Mohan Roy and Smt. Durgabati Sen somehow got the Altadanga Grant for 310 acres which was later on passed to Munshi Rahim Baksh. But the first Indian started tea gardens in 1879. It was Mugalkata tea garden, the owner being the Jalpaiguri Tea Company Ltd (1879), a registered Joint Stock Company. Bhagwan Chandra Bose, father of the illustrious scientist Sir J. C. Bose was posted as a Deputy Magistrate in the district and he took a good deal of initiative in the formation of tea gardens by the Indians. In 1881, altogether ten grants were issued but apparently new gardens Oodlabari, Bytagool, Hahiapatha and Neora Nuddy were started. In 1882, Bibi Meherunnessa and Bibi Gulabjan are the first Indian ladies took Chengmari Taluk Grant for 777.82 acres. The efforts, however, failed and the grant was transferred to J. Anderson of the Carron Tea Company Ltd. The promoters of Jalpaiguri Tea Co. Ltd., established

Northern Bengal Tea Corporation in Jalpaiguri in 1882 and Needeem and Carron (Mujnai Grant) tea estates were started. After a quiet spell in 1883 when Toonbarrie was only started lot of activities had taken place in 1884. In 1884, Prasanna Kumar Das and Chandra Kanta Das had taken Youngtong Grant and Chalauni Grant to start two tea gardens but they failed and subsequently transferred the Grants to the Europeans. Beharilal Ganguli, Mohim Chandra Roy and Jagat Chandra Roy took a Grant of 758 acres but failing to establish tea garden relinquished the Grant. In this time Shillingford family took number of Grants in the Mal, Chalsa and Nagrakata area. Seventeen gardens were established in this year. Duncan Brothers also started their activities in tea and they established Meenglass during 1884, Hope in 1885 and Jiti and Chalauni in 1886. Two small gardens viz. Ellenbarrie and Manabarrie had opened out earlier were taken over by Duncan Brothers. In 1885, though sixteen Grants were issued of which one was resumed (Ghosh, 1970) but seven gardens were started. In 1886, Bibi Rahimannessa wife of Munshi Rahim Buksh, started Mal Nuddy tea garden of 329 acres of land. In this year, Binode Behari Dutta took a grant for 600 acres but was unable to start a garden and ultimately relinquished the same. In 1887, Patharijhora and Glencoe tea estates were started. Glencoe T. E. later a year passed on to the New Glencoe Tea Co. Ltd. In 1888, Munshi Rahim Buksh took Kujih Grant of 468 acres in Chengmari Taluk but he failed and the Grant passed on to Carron Tea Co. Ltd. The Sahas of Amla Sadarpur (Nadia district) established Totapara and Ambari tea estates in 1892 and 1908 respectively.

After 1880 the cultivation of tea extended rapidly in the track between Teesta and Diana rivers. The first tea garden opened out in the district was planted with china jat which was for a long time considered to be only suitable. This was later superseded by hybrids from Assam indigenous and china variety which gave higher yield was found more profitable.

Eleven companies under Indian ownership formed during the period from 1879 to 1910 (Table 1.7). The Dooars Planters' Association, the first Association of the European planters was started in 1880 which later became a branch of Indian Tea Association with their headquarters at Binaguri. This was followed by formation of the Indian Tea Planters' Association in January 1919 with the exclusive objective of looking after the interest of the Indian tea planters.

Darjeeling

It has been mentioned in the *Imperial Gazetteer of India* (1908, P172) that the introduction of tea "into Darjeeling is due to Captain James, who persuaded Government to obtain seed from China, which he distributed among residents of the district to experiment with". A very important event was the posting of Dr. A. Campbell as Superintendent of Darjeeling in 1838 who was pleading with the Government all along for establishment of tea plantation either directly by Government or by the native residents in Darjeeling region. He pioneered the experimental growth of tea in the hills in early 1841 by planting few seeds from the China stock around his bungalow which is now known as Beechwood, 2134 meters above sea level. The first experimental trial at Darjeeling was made in 1841 with a few seeds grown in the Kumaon plantations from China stock and it was quite successful. When W. B. Jackson came on tour to Darjeeling in 1854 noticed one tea plantation but Hunter reported in 1876 the existence of two or three gardens in 1853 in Darjeeling. But he had not mentioned their names and location. However, the real date of commencement of the tea industry may be taken at 1856-57. The tea gardens viz. Alubari, Pandam and Steinthal opened in 1856 in Darjeeling district (*Imperial Gazetteer*, 1908, P172). Mr. Stolke planted Steinthal while the Barnes brothers planted at Mineral Springs, Bannockburn, and Soom. Captain Samler was the first planter of Darjeeling who in 1856 started the Alubari tea garden under the management of Kurseong and Darjeeling Tea Company. The largest tea concerned in the district was Darjeeling Company Limited which owned four gardens viz. Ambootia, Ging, Tukda and Phoobshering established between 1860 and 1864. The headquarters of the said company was in London. Its local management was vested in the hands of the Superintendent with five European assistants. In spite of monthly wage as high as Rs. 5 for men, 3 for women and Rs 2 for children were paid there was a shortage of workers. These people were encouraged to settle down in the company's garden permanently by assigning on them small plots of land for cultivation of cereal crops such as maize, millet etc. The entire labour force in the gardens were Gorkhas. Dr. Brougham started Dootoriah garden in 1859. Lebong Tea company opened Tukvar and Badamtam tea gardens before 1864. Singell and Makaibari tea estates were opened in this period. Singell tea estate was planted by Mr. James White who later turned his attention to the plains and started first garden at Chamta near Sukna in 1862.

The tea industry in Darjeeling district during the four years following 1870 continued to develop at an even greater pace than before, and by 1874, the number of tea gardens had increased from 56 to 113 (Ghosh, 1987). In 1874, there were 129 Europeans employed as managers or assistant managers of tea gardens in Darjeeling, and there were 1373 natives in posts of trust or authority employed under the Europeans (Hunter, 1876). In 1891, there were 177 tea gardens but thereafter the number decreased. The reason may be the tea estates after 1891 were more organised under big limited concerns so that total area under tea although increased, the number of gardens decreased. However, the growth was by no means a smooth sailing. There were many difficulties like communication bottlenecks, administrative problems, shortage of provisions and manpower.

After independence, the British planters who owned about 90 % of the estates in Darjeeling hills started selling their holdings to local entrepreneurs and by the mid-1950 a large number of estates had changed hands. There are around seventy tea estates functioning at present in Darjeeling hills.

Despite above, a few tea estates were also opened during 1862 in Ranchi, a plateau region of Chotanagpur of Bihar state. But it could not succeed due to various reasons. After a lapse of 120 years in 1892, a small tea plantation company, i.e., Sonapur Tea Estate started tea cultivation with only 2 hectares in Sonapur–Panwara village of Pothia block in Kishanganj district. Subsequently the cultivation spreads in adjoining blocks viz. Pothia, Thakurganj and Kishanganj and it has covered nearly 1500 hectares.

North-west India

Himachal Pradesh

The available records show that Dr. Jameson, then Superintendent, Botanical Gardens, N. W. Provinces of India visited Kangra district of the earstwhile Punjab in 1849 and recommended that the lower slopes of the Dhauladhar range lying between 9000 to 1400 meters with an average rainfall 152 cm was ideal for tea cultivation. The first commercial plantation was established at Holta near Palampur in 1852, at an elevation of 1260 meters above sea level. Thereafter, the tea industry flourished and expanded its area in Kangra district and state. Britishers had started a number of tea gardens in mid-1850s. The seeds for planting were largely obtained from China and a small quantity of China hybrid seeds and plants from Kumaon hills of Uttar Pradesh. R. A. Ballard, the then Honorary Secretary of Kangra Tea Planters' Association, wrote in 1892: "There are altogether 34 gardens owned by Europeans, varying in size 10 to 612 acres under tea, of these 34 gardens three are owned by large companies".

However, natural disaster earthquake struck the industry in 1905 and destroyed the organised tea industry. Most of the factories were razed to the ground. The area under tea in Himachal Pradesh was 4000 ha in 1950, declined to 2300 ha in nineties due to poor economic returns. Only 4 per cent of total tea growers were fully dependent on tea for their livelihood. More than 1100 ha fall under neglected and abandoned category for various reasons. The corporate sector of the industry abandoned the plantation which were taken over by the labourers or bought by the local traders.

With the change of management, fragmentation of land holdings started. Consequently the productivity and quality both declined due to lack of knowledge and inputs. The productivity came down to 1/6th of the national average. In early 60's the Reserve Bank of India appointed a Committee which recommended a series of measures to rehabilate the fast depleting industry of the valley.

Four Cooperative Tea Factories at Bir (1964, capacity 100000 kg), Palampur (1980, capacity 500000 kg), Baijnath (1981, capacity 200000 kg) and Sidhbari (1984, capacity 200000 kg) were established from the period from 1964 to 1984. The industries department of Government of Himachal Pradesh extended the benefits of inputs subsidies to small growers. The Kangra Cooperative Marketing Society was established to market Kangra teas in the Amritsar auctions. The PFA act was amended to incorporate new standards of Kangra Tea. The "Agmark" was introduced to protect the fair name of Kangra tea.

As bench mark survey conducted during 1993 and it was seen of the total 1660 tea holdings of 2063 hectare, 81% were smaller than 1 hectare and contribute only 6% of the total production. Of the total holdings, only 0.5% holdings were totally dependent on income from tea and about 60.3% planters were getting no income from tea thus indicating abandoned state of their tea gardens (Gupta, 1995).

At the moment, out of 2063 ha area under tea, more or less 40% is either abandoned or neglected. Of the remaining 1300 hectare now, 40% or 520 hectare has achieved optimum level of production while 60% or 780 hectare is still averaging 650 ha^{-1} (Anonymous, 1996).

The Himachal Pradesh Government has initiated steps for bringing more area under tea in non-traditional areas falling in different districts. There are twelve districts in Himachal Pradesh, out of which some districts have been surveyed and found suitable for tea cultivation. Chamba district is one of these districts where 3200 ha have been identified on the basis of soil and agroclimatic conditions as potential tea growing area of the district which is comprised of vallies like Sihunta, Chowari, Banikhet and Chamba.

Dehradun valley in Uttar Pradesh state had some plantations in a climate like Kangra but on more flat soil. These are in a bad shape. Efforts are on to revive the tea industry.

South India

In 1832, one Dr. Christi was experimented for growing tea in Nilgiris and some of the tea plants were distributed for performance trial to various parts of the Nilgiri hills (Muraleedharan, 1991). The tea plants sent from Calcutta Botanical Gardens, which were grown from the seeds brought from China, were planted in these hills. Most of the plants sent to Nilgiri hills died but those that were planted at the experimental farm Ketti, near Ootacamund fared best. But Weatherstone (1992) reported that he was unsuccessful in finding any original China plants where he made search around Ketti during 1985. The earliest record of commercial planting was in 1875 in Peermade in Kerala (Muraleedharan, 1991). In 1859, field scale efforts to plant tea were made in Thiashola and Dunsandle estates in Nilgiris (Arunachalam, 1995). In 1878, the development of Kanan Devan hills exclusively with tea was a landmark in the history of tea cultivation in south India. The planting of tea started in Central Travancore in 1875, Wynaad in 1889 and Anmallais (Coimbatore District) in 1897. In 1893, the area under tea was estimated to be about 1200 ha in Nilgiris, 100 ha in the Wynaad, 130 ha in the High range, Kerala and 2000 ha in the rest of Travancore (Barua, 1989). The tea growing tracts of south India (Fig. 1.3) extending along the Western Ghats vary in their elevation from 300 to 2300 m above MSL. The main tea growing states in south India are Tamil Nadu and Kerala. There are number of small tea growers (around 47500) concentration in south India.

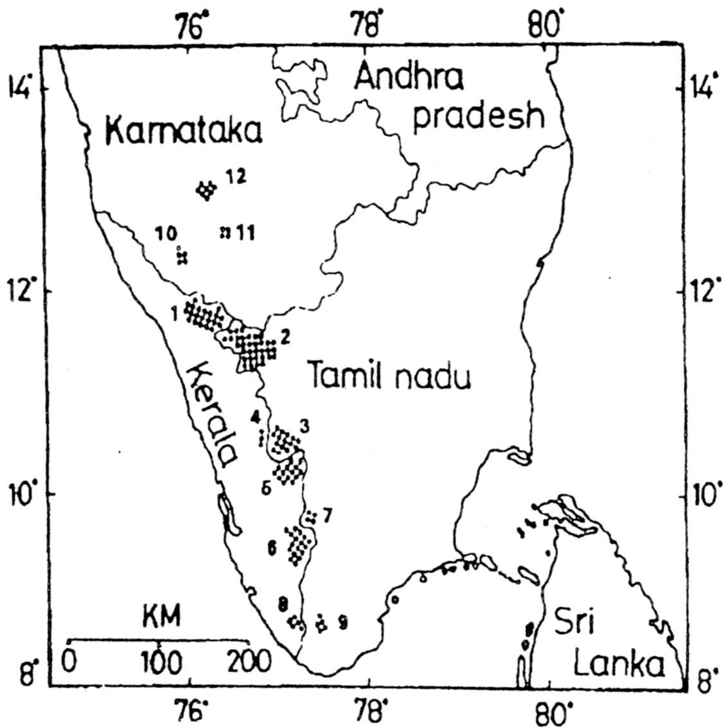

Fig. 1.3: Tea growing areas in south India. 1. Wynad; 2. Nilgiris; 3. Anamallais (Coimbatore); 4. Nelliampathy (Palghat); 5. High Ranges (Idukki); 6. Vandiperiyar & Peermade (Idikki); 7. High Wavys (Madurai); 8. Trivandrum; 9. Singampatty (Tirunelvelli); 10. Coorg; 11. Hassan and 12. Chikmagalur [Source: Muraleedharan (1991)].

The China *jat* and their hybrids were initially used in south Indian plantation but gradual change over to the Assam hybrids took place over the years mainly in estates at lower elevations. The vegetative clones are at present being used in new planting.

1.2.2 Growth of tea in other countries

China

The history of tea cultivation in China can be traced back to three thousand years ago. As early as Cheng-Huang period of Xi-Zhon dynasty in 1100 BC, it was recorded in the monograph of "Shang-Shu, Gu-Ming that tea had been used in the "offer sacrifices to Gods or ancestors". "Tea

Scripture" written by Luyu of Tang dynasty around 758 AC was the first monograph on tea in the world. After Xi period (around 130 AC), the understanding on tea was more detailed. As per available record in "Tea Scripture", the area under tea in Tang dynasty was distributed in 13 provinces of China. The tea production in Yuan dynasty was around 100000 tonnes and increased to 150000 tonnes in Qing dynasty (Chen and Lu, 1993). In China, tea is a kind of traditional export commodity. China by exporting tea has established trade relationship with over 30 countries in the Middle-east, Southeast Asia, East and West Europe.

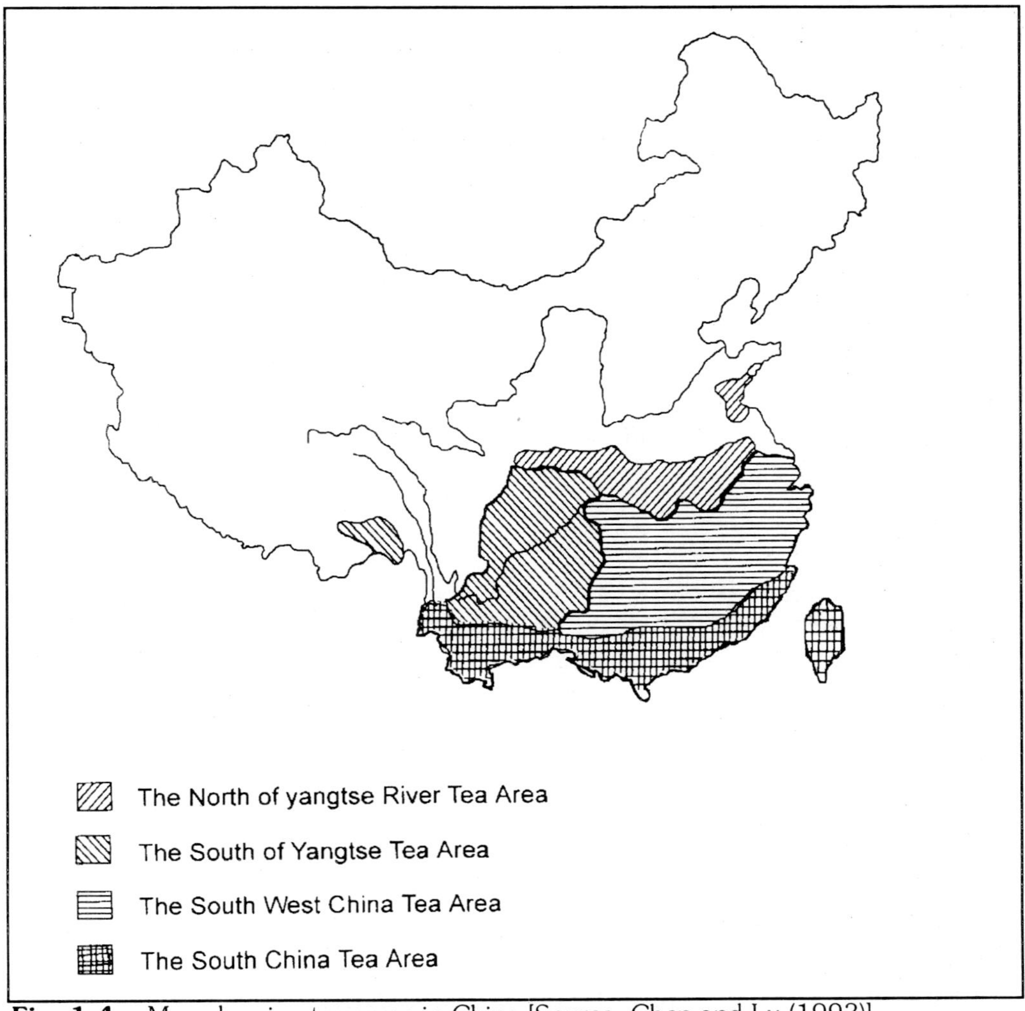

Fig. 1.4 : Map showing tea areas in China [Source: Chen and Lu (1993)].

The tea plantations in China are located in between 94 ^{0}E to 122 ^{0}E longitude, and from 18 ^{0}N to 37 ^{0}N latitude, on elevations up to 2600 m. But the areas for economic cultivation are concentrated in the region between 102 ^{0}E eastwards and 32 ^{0}N southwards. The climate of tea areas belongs to the humid monsoon climate of the east coast type. In the subtropics of China, the climate is not as dry as the general subtropics: it is moist, has a lot of precipitation and high temperature in the summer (Huang, 1989). The tea producing areas are distributed in 19 provinces viz. Anhui, Jiangxi, Zhejiang, Jiangsu, Hunan, Sichuan, Guangxi, Yunnan, Guizhou, Guangdong, Hainan, Hubei, Taiwan, Fujian, Shanxi, Shandong, Gansu, Henan and Tibet. The four major tea producing provinces are Zhejiang, Hunan, Sichuan and Anhui together account for 60% of the national output (Chen and Lu, 1993). The tea areas are divided into 4 parts namely the south China tea area, the southwest China tea area, the south of Yangtse River tea area and north of Yangtse River tea area (Fig. 1.4). China for a long time held a pre-eminent position in tea production and export. But after the ravages of World War II, the tea area had shrunk to 154600 ha in 1949 and production down to 41 million kg, averaging an yield of only 268 kg/ha. The rates at which tea productivity in China has been lately increasing and was very impressive. The target proposed various kinds of teas are produced in China. According to processing methods and technologies, Chinese teas are basically divided into six categories. These are green tea, black tea, oolong tea, white tea, yellow tea and black-black (post-fesmented) tea. Of these, green tea forms the main bulk and black tea accounts for about 20 percent of total production. Oolong tea is a semi-fermented tea and mainly made in Fugian, Guangdong and Taiwan provinces. White tea, a light fermented tea primarily made in Fujian province. Black-black tea is a post fermented tea and consumed mainly by the people living in remote border areas in China.

Japan

In Japan, tea was introduced from China in the early part of the eighth century by the Buddhist priests. According to the paleography, the history of Japanese tea dates from the early nineth century. It has achieved unique development in the period of 1200 years. In the cultural aspects of tea, "Sado" or the art of tea ceremony was completed in the sixteenth century and it has been exerted a great influence not only upon the habits of tea drinking but also the spiritual life of the Japanese.

Tea is grown in Japan mostly in mountainous areas, and about 60% of tea plantations are located on the hill sides and plateaus where other crops are difficult to grow. In Japan, frost is the most damaging climatic hazard from April to early May when shoots are about to be plucked. The plantations could be devastated by just one frost. Tea areas are concentrated mostly in the south western half of the Japanese Archipelago from the Kanto district to the southern tip of Kyushu Island. Five prefectures (Shizuoka, Kagoshima, Mie, Saitama and Kyoto) in Japan plant 69 % of the tea area and Shizuoka prefecture dominates with 41.4 %. Figure 1.5 shows the tea

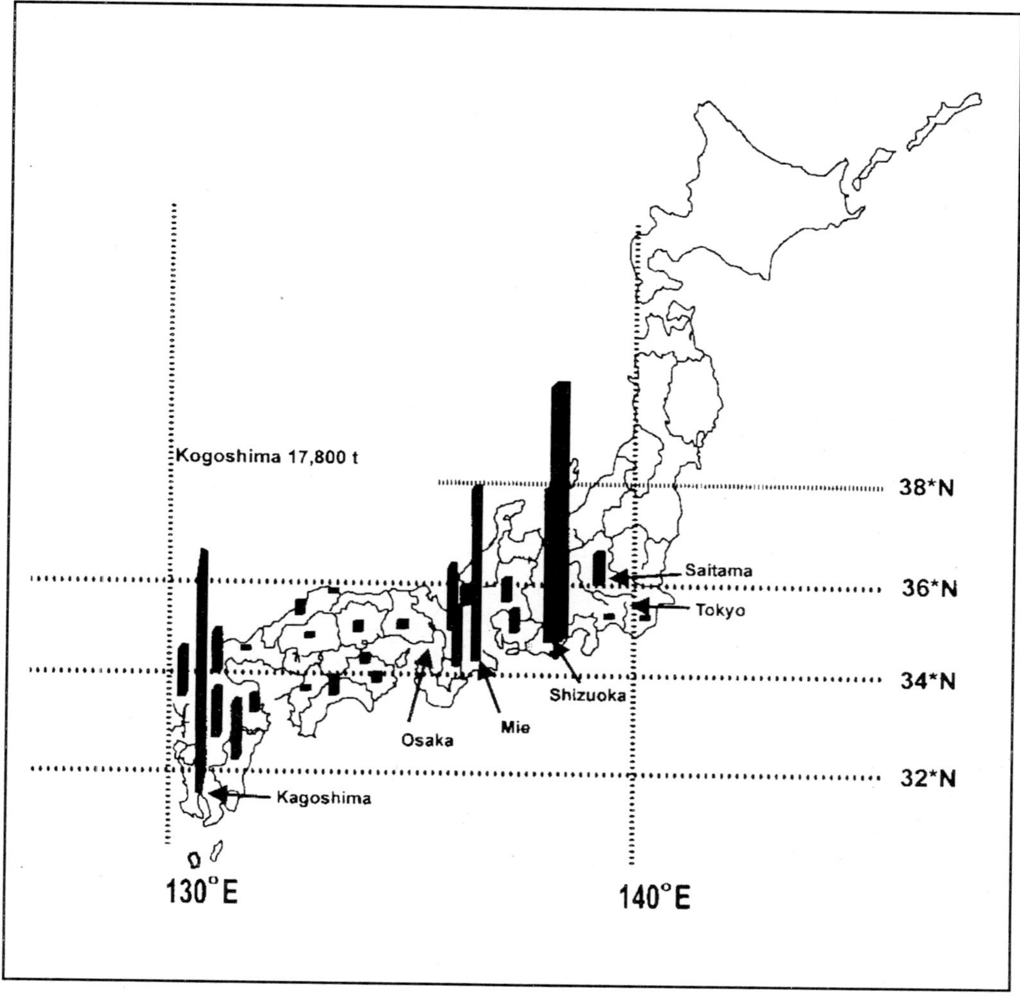

Fig. 1.5 : Main tea producing regions in Japan [Source: Kodomari and Hara (1999)].

areas in Japan. Black tea production was practically ceased after 1971. Japanese green tea is classified into six groups viz. gyokuro, kabusecha, tencha, sencha, tamaryokucha and bancha. Until 1960, 90 % of tea plants were seedlings. Since then clonal planting has spread to 80 % of the tea area which is dominated (85 %) by clone Yabukita (Kodomari and Hara, 1999). In Japan, tea plant usually sprouts four times in a year, and two or three of them are utilised for plucking. The best quality of green tea is made from the first crop (April-May), which shows higher content of amino acids and lower content of catechins.

Sri Lanka

Dr. N. Wallich of Calcutta Botanic Gardens sent Assam tea seeds to Sri Lanka (then Ceylon) in December 1839 and seeds were received at the Royal Botanic Gardens at Peradeniya near Kandy in 1839. Thereafter, the Government of Sri Lanka received 200 plants in 1841 which were planted in Nuwara Eliya. Two further consignments of plants were received in early 1841 and 1842 from China and Assam, and were planted in Rothschild estate at Udapusselawa and Pennyland estate, Dolasbage respectively (Gnanapragasam, 1999). Planting of tea in Sri Lanka in a serious way began in the decade of the 1880s. Tea in Sri Lanka was essentially a replacement of coffee which was fast dying out because of attack of the dreaded coffee rust (Hemileia vastatrix). Sivapalan (1991) analysed the tea scenario of Sri Lanka in successive eras. Most of the areas are situated above 900 m. Tea in Sri Lanka is produced under varied ecological regimes and for easy reference these are grouped on the basis of elevation into three distinctive broad categories of low (below 600 m), medium (between 601 to 1200 m) and high elevation (above 1200 m) tea areas. The entire production of tea is in the form of black tea and almost of which (98%) is manufactured by orthodox method. Plucking continues throughout the year and black tea is the main product. Currently, Sri Lanka is the leading exporter of tea. The tea districts of Sri Lanka are shown in Figure 1.6.

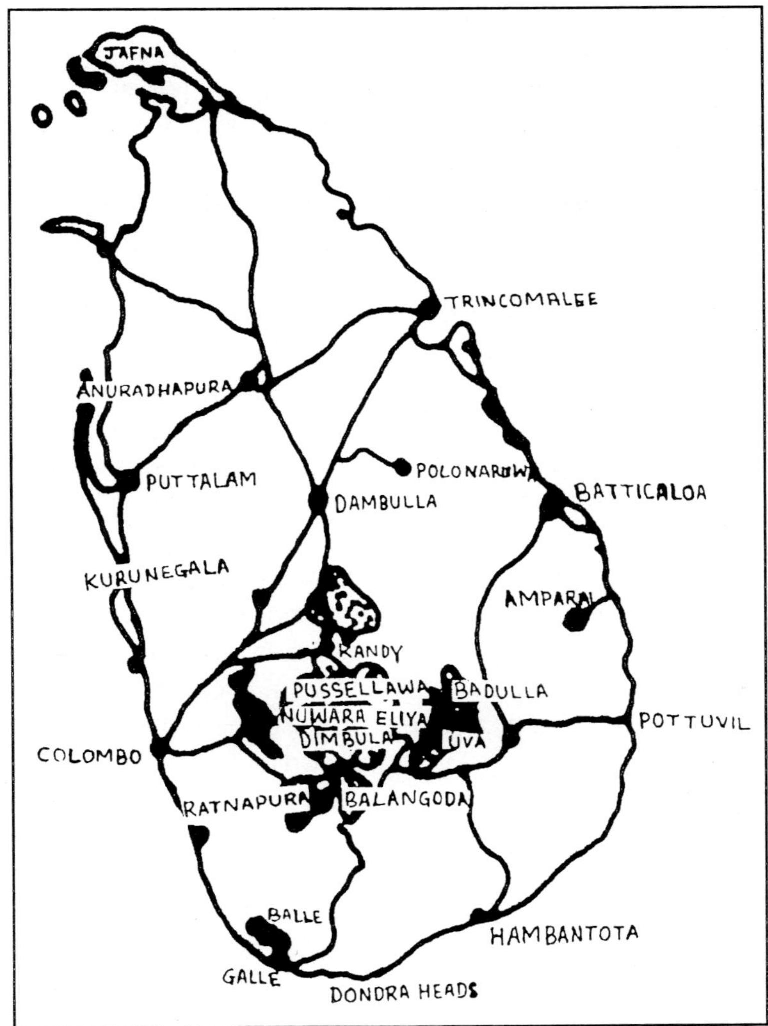

Fig. 1.6 : Tea districts of Sri Lanka

Indonesia

It is a big maritime country consisting of thousands of islands. There are five big islands viz. Sumatera, Sulawesi, Kalimantan, Irian Jaya and Java. In addition, there are two group of islands Maluku and Nusa Tenggava. Tea grows mostly in Sumatera and Java. More than 80 % of the tea areas are located in western Java. New plantations are coming up in the southern part of Sulawesi island. Tea seeds were imported from Japan in 1826 through von Siebold, a Dutch representative in Japan and planted in the Bogor (Buitenzorg) Botanical Gardens. The experimental plot was started in Wanayasa, Purwakarta (Arfin and Semangun, 1999). Seeds were also collected from China, though commercial planting was taken up from 1878 using Assam seeds collected from Sri Lanka by R. E. Kerkhoven. Before World War II tea plantings in Indonesia covered an area of 138000 ha consisting of 75000 ha under smallholders and 63000 ha under large estates. Further, Indonesia was capable to export 83000 tons or 18.5 % of the total world tea export. Many large tea plantations within twenty years after World War II have been abandoned and total tea area was reduced by nearly 50 percent. During the war, when Japan occupied Indonesia for about two and half years, more than half of the tea areas mainly the lower elevations were either uprooted or destroyed badly. Even after independence in 1945, the tea plantations remained neglected for about two decades. The tea companies owned by the Dutch were nationalised in 1958. It was only from 1965 vigorous steps were taken for revival of tea in Indonesia and restoring its past glory. About 32 percent of the plantations belong to state estate, 23 percent are private large tea estate and rest 45 percent are under smallholders. Indonesia produce black, green and Jasmine -scented tea. Most of the green tea is mixed with Jasmine flowers and reprocessed into aromatic (Jasmine) tea that is widely consumed in the domestic market.

Taiwan

According to the history of Taiwan by Mr. Lien Ya-Tang, one Mr. Ko Chao who brought few tea plants to Taiwan from Wu-i, Fukien Province, China and planted in north Taiwan during the Chia-Cheng period (1796 - 1820). In fact, it was the beginning of tea cultivation in Taiwan. However, commercial tea cultivation has been started in northern Taiwan using seeds from tea plants brought by Chinese mainlanders from Fukien Province about two hundred years back. In 1865, the first export of 82 metric tons of Taiwan oolong tea was officially recorded (Chiu, 1991). Later John Dodd, an English businessman had exported 121 metric tons oolong tea to United States. In 1881, Wu Fu-Lao, a tea trader from Fukien came to Taiwan to produce Paochung tea and Taiwan started to export Paochung tea (Chiu, 1991). Thereafter, in 1885, Liu Ming-Chuan the first Chinese Governor of Taiwan appointed by Ching dynasty who started to develop the tea industry systematically and total tea export in 1893 increased to more than 9800 metric tons.

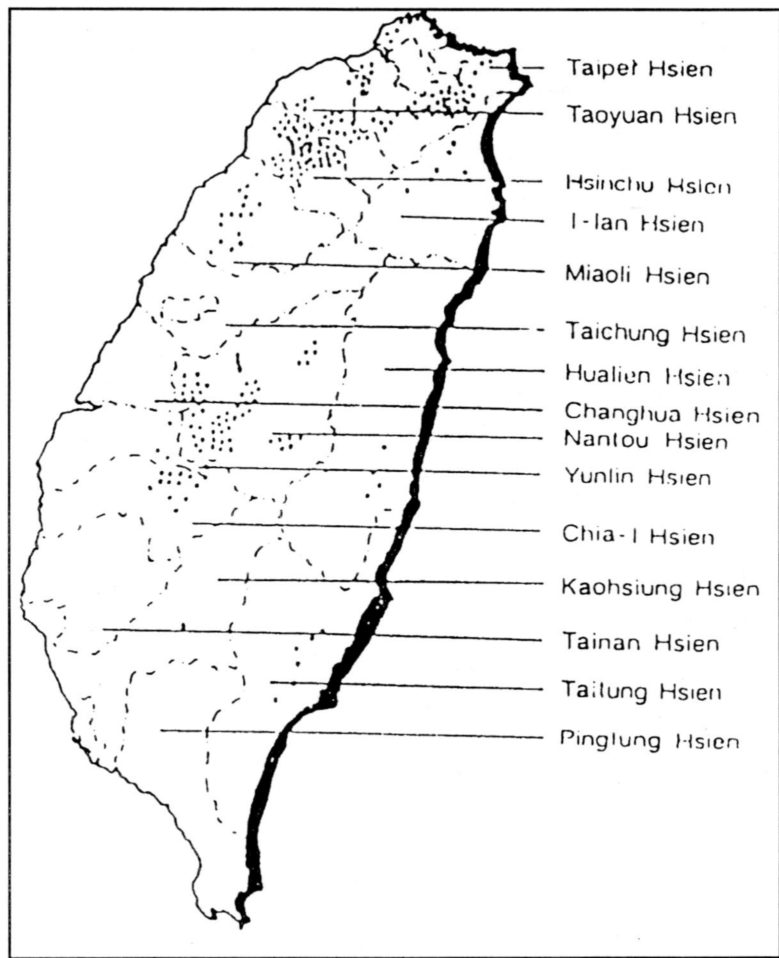

Fig. 1.7: Distribution of tea gardens in Taiwan in 1990 [Source: Chiu (1991)].

Tea is grown on table land and broken foothills on the northern part of the island between the mountain range and the alluvial plains bordering the north-west coast at elevations ranging from 80 m to 300 m above sea level. The tea plantations were found in Taipei, Taoyuan, Hsinchu, Miaoli, Nantou, Chiayi, Yunlin, Yilan, Hwalien and Taifung countries (Fig. 1.7).

Bangladesh

In Bangladesh, the main tea growing areas lie to the east of the Ganges-Jamuna flood plains in the hill areas bordering India's Cachar tea growing districts. 93 percent of the tea areas are located in the north-east of Sylhet zone in the six valleys and the rest in the south-east of Chittagong zone and the hill tracts. The introduction of tea in Chittagong dates from 1839–1840. Mr. Sconce, the then collector of the district, first received some tea seeds from Assam and some plants of China variety from Calcutta Botanic Garden and planted them. Another gentleman, Mr. Hogue laid out plantation of an acre or two at Kodala. The oldest tea garden established in 1854 by the British was Malnichera in Sylhet and its commercial production began shortly thereafter in 1857. The area under tea comprises some 48000 ha on 158 estates, of which 115 estates produce 92 percent of the tea. Out of 158 estates at present, three are run by the Tea Board of Bangladesh of which one is run on an experimental basis; there are thirteen tea estates functioning under one public limited company; fiftysix estates are private limited ones; twentyfive tea estates owned by Sterling companys and rest are proprietary concerns. This apart, six tea estates are abandoned and thirtyseven gardens are declared sick. Tea bushes remain dormant during winter season restricting plucking to about nine months in a year. Almost the entire production is CTC tea. A very small amount of green tea is produced for the export market. Chittagong is the port city where weekly auctions are held for export and local sales. It is also the port of shipment. In the economy of Bangladesh, tea is considered a major export commodity and is a foreign exchange earner. Half of the produce is exported annually.

Nepal

Innumerable mysteries surround the origins of tea in Nepal. Historians, however, believe that Nepal began growing tea in its eastern regions around the same time as tea estates were begun in neighbouring Darjeeling. Indeed, some surviving bushes date back to Nepal's mid-19th century ruler, Jang Bahadur Rana, but much of the Government's newer estates were planted a mere 20 years ago under a British developmental project (Masty, 1998). Today, there are two types of tea grown in two places in Nepal. There is high grown tea from the eastern hills near Darjeeling cultivated in the Government estates as well as outgrowers. Though it represents a small quantity but it is of high quality tea exported to Germany fetches premium prices. Then there are plantations in Terai, the flat lands and jungles leading from India up to the foothills of the Himalayas. Terai produces CTC tea which is consumed domestically. The private tea estates including small holdings are located near Government owned plantations in Terai and the eastern hills are earning profits. However, upto date expertise on cultural practices and manufacturing processes are needed throughout the sector.

Col. Gaj Raj Singh Thapa (the then district administrator) had established a tea garden in Ilam (altitude 1060 m) in 1863 followed by another in nearby Soktim in terai at an altitude of 500 m in 1865. Kanyam tea garden was established in 1971. The other tea estates are Tokla, Burne, Baradasi and Chilinkot were established in 1972,1978,1976 and 1983 respectively comprising of 1500 ha tea. Tokla and Burnie planted with polyclonal and bioclonal tea.

The Government owned tea estates which are under Nepal Tea Development Corporation (NTDC) features two high-grown orthodox producing tea estates viz. Ilam, Kanyam in the Jhapa region of eastern Nepal and five CTC producing estates in the Terai. NTDC was established in 1966 and cultivated 951 ha of tea till date. Despite, 1706 and 889.4 ha are under private tea estates and small holders respectively.

Vietnam

Vietnam has a longstanding tradition and customs of growing and drinking tea. There are century old giant tea trees found in high mountainous areas such as Suoi Gian (Yen Bai province), Thong nguyen Cao bo (Ha Gian province) which retains primitive characteristics and original qualities (Tien, 1993). The tea production in Vietnam took commercial shape in 1918. However, tea industry was at a standstill for a long time due to colonial aggression and war for national independence. In fact, frequent and prolonged wars have made severe damage to the tea industry and had only been healed after peace was restored during 1975. Since 1980, tea production has been consolidated and strengthened. A national economic and technical managing body viz. Union of Vietnam Tea Farming-Industry Enterprises (UTE) under the Ministry for Agriculture and Food Industry came into being in 1980. This was a turning point ushering in a new important growth stage of Vietnamese tea industry in all field.

Thailand

Tea areas in Thailand are concentrated on mountain slopes, covering 6400 ha. Tea, especially *assamica* type has been grown for centuries at wide spacing and kept unpruned. Plucking is done at long intervals and made into pickled tea called miang which is consumed solely in the north.

Chinese traders have introduced primary green and semifermented tea factories. The manufactured tea is reprocessed in secondary green tea factories located mostly in and around Chiang Mai province. The product is mainly sold locally to traders in Bangkok. Small quantity is exported to Japan and the USA (Noppakoonwong and Nillavesana, 1999). The northern mountainous region of Thailand where tea is grown extends over nine provinces (Fig. 1.8). Rainfall is concentrated during the period from May to October with an average annual rainfall in the range of 1100–1700 mm. The main constraints of tea growing in the northern hills include the severe six months dry period, remoteness, restricted market etc.

Iran

The history of the habit of tea drinking in Iran dates back to the end of the 15th century A.D. Before introduction of tea in Iran most of the Iranians used to drink coffee, but due to difficulties mainly in transportation and easy access to China through the "silk road" induced people to switch over from coffee to tea. Since then people have become habituated to drink tea. Due to uncertainty in trade relation with China some Iranians took interest in cultivating tea in Iran. In the year 1882 Haji Mohammad Hossein Esfahani brought some tea seeds from India and cultivated in Iran but he was unsuccessful. Later in 1899, Mr. Mohammad Mirza Kashefelsaltane, the then Iranian Consulate General in India met Mr. Thompson the Director General of tea plantations in Kangra Valley and decided to cultivate tea in Iran. About 3000 tea seedlings were taken from Kangra, Himachal Pradesh, north India to Iran. The seedlings were planted on low hillocks of Chahr Hanessar of Lahijan in Gillan province in 1900 (Hosseini, 1993). This was proved to be successful. The farmers took active part in tea cultivation and tea plantation spread over the Gillan and Mazandaran provinces. The first tea garden was established in Tunekabun area in Lahijan in 1901. Tea areas of Iran are situated only northern part of the country i.e. on the southern shore of the Caspian Sea between 49^{0} -42" to 50^{0} -8" east longitude and 37^{0}–25" to 36^{0}-61" north latitude. They are scattered in an area which is 203 km long and 90 km wide, surrounded by the Alborz Mountain Range. Of these holdings, 85 % are less than one hectare in area. There are 130 tea factories (10 factories under Government and rest private) with a capacity of handling 2429 tons of green leaf regularly. The tea areas are located only in the northern part of the country i. e. on the southern shore of the Caspian Sea scattered in an area of 203 km long and

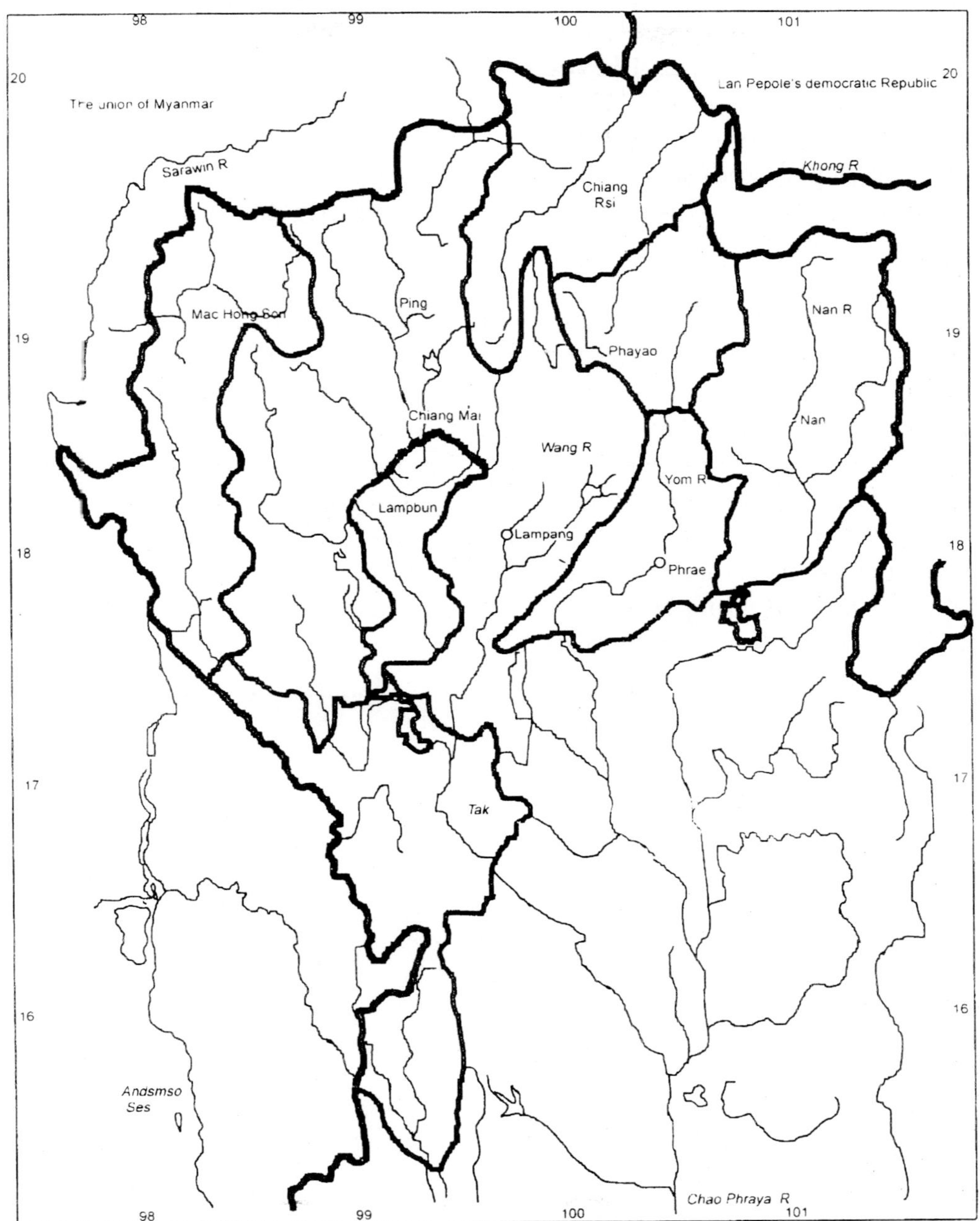

Fig. 1.8 : Map showing Northern mountainous provinces of Thailand where tea is grown [Source: Noppakoonwong and Nillavesana (1999)].

90 km wide. The tea areas experiencing severe cold during the period from January to March and temperature may fall -5^{0}C with frequent snowfall. Thereafter, the weather between April and July gets hot and dry and temperature rises up to 35^{0}C. Actual period of production of green leaf is limited to 6 months starting from May to October. The rainfall during June, July and August is not sufficient to induce new shoots particularly in drought-prone areas. During this period supplementary irrigation is advised.

Turkey

The habit of tea drinking in Turkey began in the early sixteenth century. Tea was first planted in 1888 around Bursa but it was not successful mainly because of ignorance in cultivation technology and environmental situation (Vanli, 1991). Cultivation of tea started in 1939-1940 dominated by China hybrid seeds obtained from neighbouring Georgia in the former Soviet Union. Tea planting commenced in Rize contiguous to the Russian border on the Black Sea coast and gradually extended through Trabzon, Giresun to Ordu over a distance of nearly 300 km in the strip of land which lies between the Iranian plateau and the Caspian Sea (Fig. 1.9). Black tea production was first started in Rize in 1947.

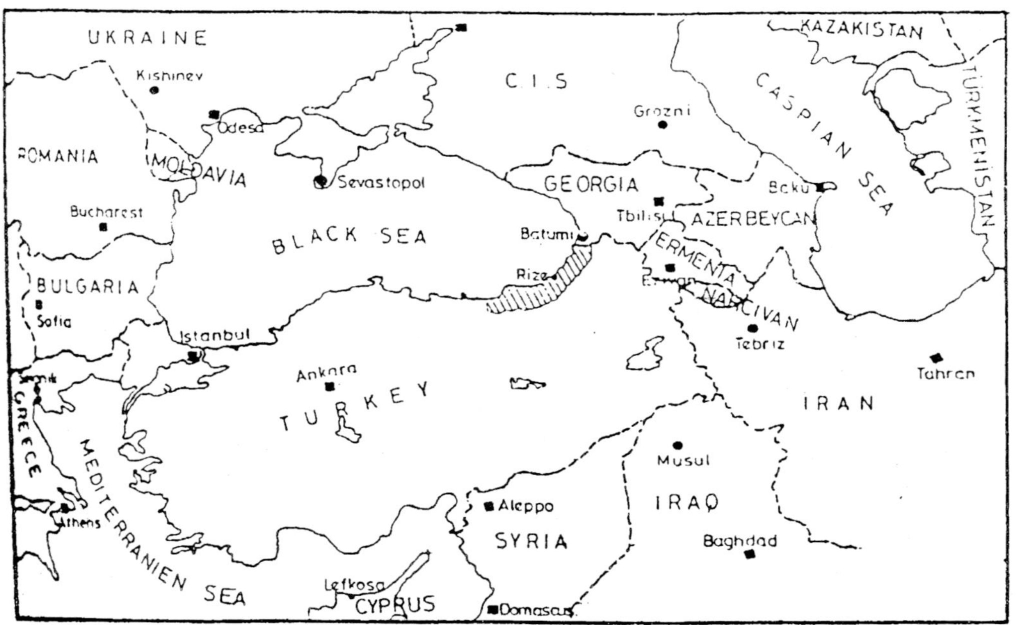

Fig. 1.9: Area under tea plantation in Turkey.

The law 1497 adopted on 6th December, 1971 has put an end to the monopoly of the state and established the Tea Corporation of Turkey, a state-owned enterprise having an autonomous management was entrusted with a task of ensuring production, processing and marketing of tea. The Corporation maintained its existence, modelled after the law 440 on state Economic Enterprises, until 19th December, 1984 when the law 3092 entered into force. Its 13 years of existence made its imprint in the tea industry in Turkey and paved the way for considerable developments in the field (Vanli, 1991). Turkey experiences very low temperature, snow and frost during winter. The plucking seasons, yield distribution in different months follows same pattern as in the U.S.S.R. Tea growing areas have showed rapid growth and increased from 27880 ha to 90575 ha in 1990 and later it was decreased to 76609 ha in 1995. In 1995, the overall production was 120,000 tons, of which 80,000 were processed by the state sector, 40,000 tons by the private sector. In other years, state production has reached 110,000 tons, resulting in a substantial surplus for export.

USSR (CIS)

Cultivation of tea started towards the end of the last century with the seeds transported from China. Initially tea was planted on the eastern shore of Black Sea in the Republic of Georgia. Later it was extended to the Republic of Azerbaijan and the Russian Republic of Krasnodar region (Fig. 1.10).

Fig. 1.10 : Tea growing areas in former USSR (marked).

Initially the development was very slow and by 1905 only 400 ha were under tea. There were only two tea processing factories available near Batumi till 1915. The intensive development of tea plantations and true tea industry started only at the beginning of thirties. Planting is done on flat terraces and gentle slopes of low hills up to an elevation of 700 m above sea level but maximum tea areas are located below 300 m elevation (Gokhale, 1958). The area under tea comprising of 67000 ha in Georgia, 13000 ha in Azerbaijan and 1800 ha in Krasnodar territory from the middle of eighties. During 1989, the annual harvest in Georgia was 451200 tons, in Azerbaijan 34000 tons and in Krasnodar territory 6800 tons. From this green leaf, the total production of processed tea was 122700 tons (71700 tons of black tea and 51000 tons of green tea) and besides 1500 tons of soluble black tea and 1200 tons of liquid tea concentrate for making various beverages. 150 factories roughly were in operation in order to process this amount of raw material (Zaprometov and Jinjolia, 1991). However, the conditions of tea plantations during last two decades became worse. The tea areas due to high latitude experience very low temperature during winter. In Georgia, the minimum temperature at times drops to -8^{0}C while in Krasnodar and Azerbaijan the minimum temperature of -12^{0}C is occasionally faced. During a part of winter frost is very common and sometimes tea bushes for 2-3 weeks remain covered with snow. China and its hybrids are used as planting material. Plucking operation is partly done by hand or machine and it lasts from May to October. Black, green and brick teas are manufactured in USSR.

Malawi

In 1878 Jonathan Duncan, a gardener at the Church of Scotland Mission in Blantyre, obtained seeds from the botanic garden in Edinburgh but these seeds failed. Another batch of seeds was brought by Dr. Elmslie in 1886 from Edinburgh, as well as a few from Kew Gardens. Mr Duncan tried again and obtained just two bushes which were planted in the grounds of the Mission in Blantyre. One of these bushes became the ancestor of the original tea plantations in Mulanje and vegetatively propagated bushes from this original bush are still growing in Tea Research Foundation (Whittle, 1999). During 1891, Mr. Henry Brown procured few seeds from Mr. Duncan and half of it were planted in his estate (Thornwood) and other half on a neighbouring estate (Lauderdale) owned by John Moir. Thereafter, tea estates from 1908 onwards were opened in the Mulanje (700 m to 1000 m elevation) and Cholo (900 m to 1300 m elevation) districts in the south (Fig. 1.11). The first commercially planted tea in Malawi was China hybrid in character. Expansion of tea cultivation then spread over to the northern district of Nkhata Bay in 1960. In 1997, Malawi produce 44 million kg black tea from an area close to 19 thousand hectares. Although the total area under tea has increased by some 3000 ha since the 1970s, this has almost entirely been due to the development of the smallholder sector (Whittle, 1997). Initially planting was done with China hybrids but later Assam jats were introduced. Though tea bushes do not face any winter dormancy in Malawi but production drops drastically during cold and dry months (August to November). The maximum temperature during dry months occasionally rises up to 36^{0}C or more while in winter it drops to 7^{0}C or little lower. Malawi has made a tremendous effort to improve its quality, and its primary clonal grades obtain good prices.

South Africa

Early records show that tea was popular in southern Africa during 1670s and probably much earlier (Paterson, 1995). It is known that tea was included in the cargoes of Portuguese ships sailing round the Cape in the 1550s, but the first record of tea drinking does not appear until 1676. In Africa, tea was first grown south of the Limpopo river in the Durban Botanical Gardens, Natal, in 1850. The plants were brought from Kew Gardens, London. In 1867, serious attempt to grow tea commercially was pioneered in Natal by James Liege Hulett, James Brickhill and W. B. lyle (Paterson, 1995). Assam jat seeds were brought to Durban from Calcutta and distributed among these three members. Mr. J. L. Hulett planted out 4000 plants on the Kearsney estate near Stanger in November, 1877, and one year later, 1200 of these plants were survived. The first South African tea was manufactured in 1881/82. The Natal tea industry moved ahead steadily and in 1886, Natal produced 80000 pounds (equivalent to 36287 kg) of black tea and consumed locally. In 1887, Natal tea was exhibited in London at the time of Queen Victoria's jubilee. By 1889, Natal had about a dozen estates with 1090 acres (equivalent to 441 ha) under tea. At the turn of the century, cultivation was extended to Kwazulu/Natal with a production of 472 tons. But the industry faced a heavy blow during 1911 when indenture of labour from India was abolished and production by 1949 virtually had come to a halt because of high cost of labour and packaging. In 1963/64 Sapekoe was established by the Industrial Development

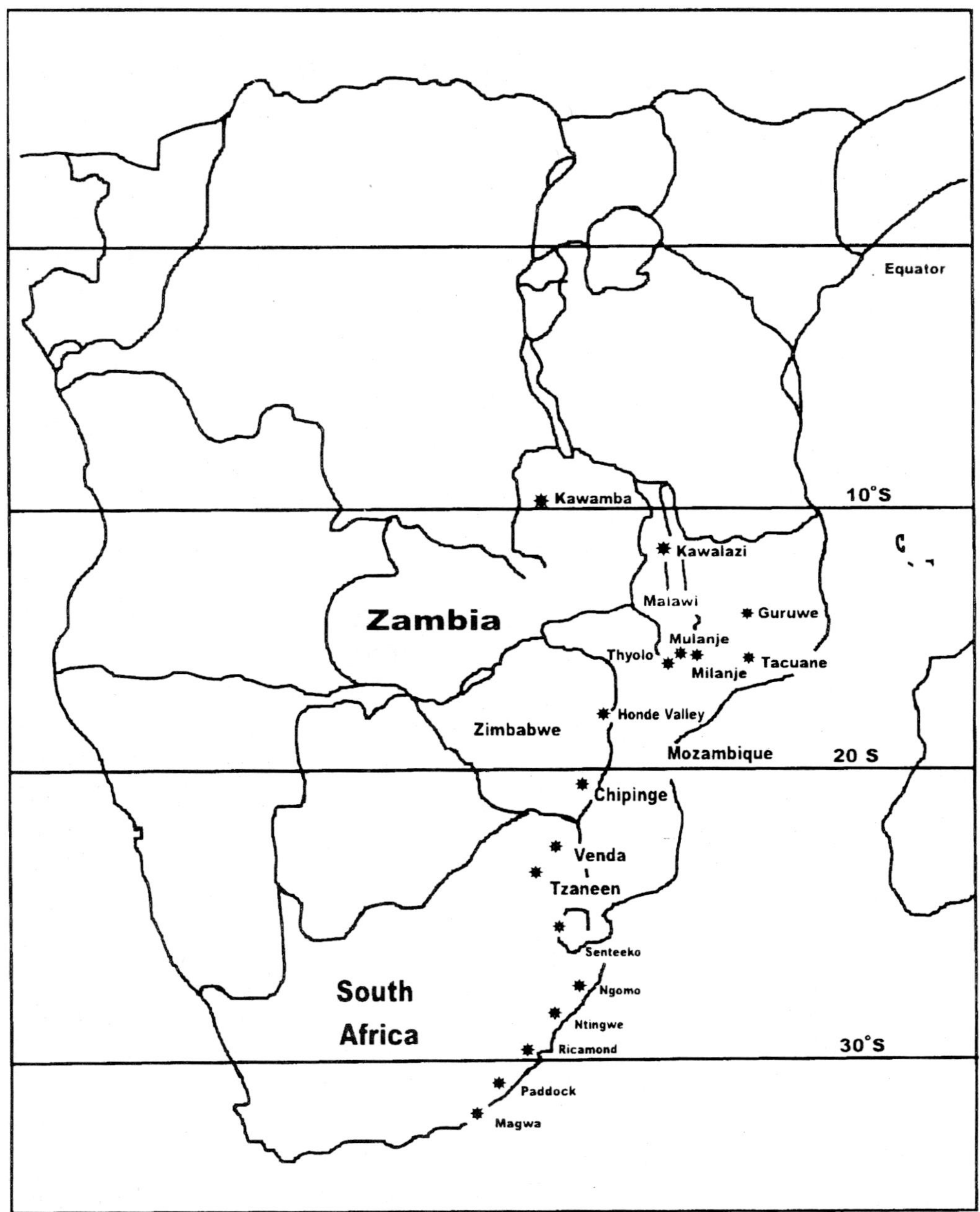

Fig. 1.11 : Tea growing areas in Central and Southern Africa [Source: Whittle (1999)].

Corporation and prospective tea growing areas were considered to be along the line of the Drakensberg Mountains in the eastern Transvaal and in parts of Natal. Sapekoe started its first two estates near Tzaneen at Grenshoek Farm and Middlekop Farm. Thereafter, new tea estates have been established in Venda (Fig. 1.11) and the Levubu area near Louis Trichardt. Further, the Mamedi River estate has also been developed. New development also started in central Zululand near Ntingwe. In 1963, work pertaining to the establishment of tea nursery began at Grenshoek (Paterson, 1995). Growing of tea also began in the Transkei and covered 2000 ha under tea. In Transvaal, Natal and Zululand, tea is planted above 900 m while average elevation of tea areas in Transkei is 300m. In percentage terms, the industry in South Africa has the largest clonal planting, all most all using TRF (Tea Research Foundation) clones. South Africa currently has over 36% of its area under clonal tea (Whittle, 1997). Plucking operation is concentrated during November to March. Severe drought have affected production in past several years. Tea areas experience periods of high day and low night temperature despite strong winds.

Zimbabwe

In 1927, tea was first planted in Rhodesia at New Year's Gift estate in the Melsetter (now Chimanimani) area of Chipinge district, with an initial area of 4 hectares. Expansion of the industry was relatively slow till 1960. The Chipinge district (Fig. 1.11) contains Ratelshoek, Jersey, Zona, Avontuur and New Year's Gift estates owned by the Tanganda group, and Southdown, Rosscommon and Clearwater estates, owned by Southdown Holdings. The Honde Valley in Inyanga district, the other area where tea is also grown (Whittle, 1999). Attempts have recently been made to develop smallholder sector in Chipinge and Inyanga districts.

Kenya

Tea was introduced in Kenya in 1903. Commercial tea growing however did not start until 1920s, when extensive clearing of agricultural land in the white highlands (European settled areas) began. There was an interruption of expanding tea industry during the Second World War but after the war the expansion of the industry continued in Limuru, Kericho, Nandi Hills and Sotik. Production of tea in Kenya was virtually confined to estates until the 1960s. Then came the special Crops Development Authority (set up in 1960) to develop smallholders' tea production. The Kenyan

Government established the Kenyan Tea Development Authority (KTDA) in 1964 to develop tea among the smallholders and also to extend cultivation to all suitable areas. At present 66 percent of the tea areas in Kenya is owned by smallholders and there are over 200000 individual growers and the holding is on an average 1 acre or slightly more than half hectare. The small holders' contribution to Kenya's tea has moved from 4% in 1965 to close to 55% in 1990 (Othieno, 1993). The principal tea growing districts of Kenya are located in an area lying between Mount Kenya and Lake Victoria on either side of the Rift Valley. The tea estates are located on either side of the equator. The main tea growing districts are Kericho in the west followed by Nandi, Sotik, and Kissii around Kericho (Fig. 1.12). Planting is done at an elevational range of 1600m to 2250m. Kenya produces only black tea. About 90% is of CTC manufacture. Between 8- 15 percent is used for local consumption and rest is exported. The industry remained dominated by large commercial houses such as Brooke Bond (now Unilever group of companies), African Highlands Produce (Finlays), George Williamsons and Eastern Produce Africa Limited companies.

In terms of bulk exports, Kenya is the major world player, exporting 89% of its 1998 production. Kenya tea producers today have realised good margins of profit simply because the country has offered a reliable source of supply of good medium teas, while the traditional sources of India and

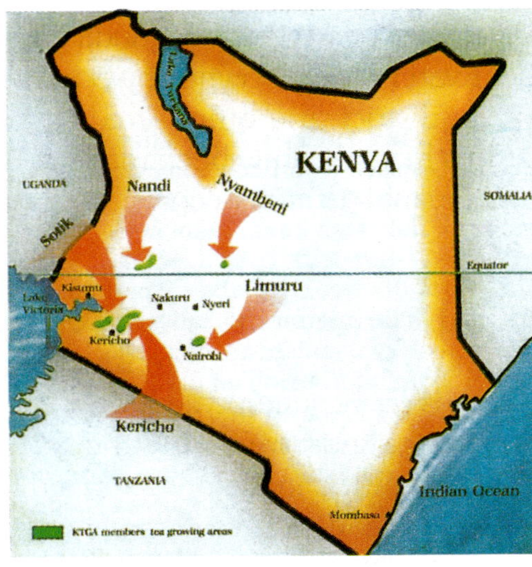

Fig. 1.12 : Kenya's position in Africa and its tea growing areas (marked).

Sri Lanka have not, and when the market has one of its periodic dramatic rises, as is happening now, tea producers prosper.

Tanzania

Although experimental planting of tea in Tanzania was made in 1904 by German settlers in the Usambara mountains at the Agricultural Research Station at Amani and at the Kyimbila Mission in the Rungwe district near Tukuyu in the southern highlands, commercial production did not begin until 1926 (Carr *et al.*,1988). The first estate, known as Ambangulu, was developed by Mr. C. Gee in the Usambaras in 1926. At about this time Germans who had been forced to leave Tanganyika during World War I began to return under the British mandate to the southern highlands, particularly to Mufindi and Tukuyu. A land development survey commission in 1929 recommended that tea should replace coffee and following the appointment of tea officer, free seed was distributed to interested settlers during the period from 1930 to 1934. A small tea factory was opened in Mufindi in 1930 (Moffet, 1958). In 1932, Dr. H. Mann, tea expert from India, identified 20000 ha of land suitable for tea in the Usambaras and the southern highlands. In order to maintained prices, the planted area was officially restricted to 1174 ha, a limit which held good until 1938. Planting began on a large scale in Usambara mountains (900 m to1400 m elevation) as well as in Mufindi and Rungwe (2100 m or little higher elevation). Later on the tea cultivation was extended to Bukoba district (1200 m) on the western shore of Lake Victoria which is 2^0 south of the equator. However, World War II then intervened; many German settlers were interned at the outbreak of hostilities, and their eatates and farms were taken over by the British Custodian of Enemy Property (Carr *et al.*, 1988). In September 1940, the Tanganyika Tea Company Ltd (a subsidiary of Brooke Bond Ltd) leased all 'enemy' tea estates of Tanganyika for five years from the Custodian of Enemy Property. In 1942, Tanganyika Tea Company Ltd was permitted to plant 400 ha tea in Mufindi (Fig. 1.13) and built a factory (Lugoda). The same company during 1950 took over the leases permanently and began to cover more areas with tea. Later, Tanzania Tea Authority (TTA) was established in 1968 and has its headquarters in Dares Salaam. TTA manages a number of estates and operates six tea processing factories despite supervision of marketing of tea.

Uganda

In 1900, tea plants obtained from Kew and Edinburgh Botanical

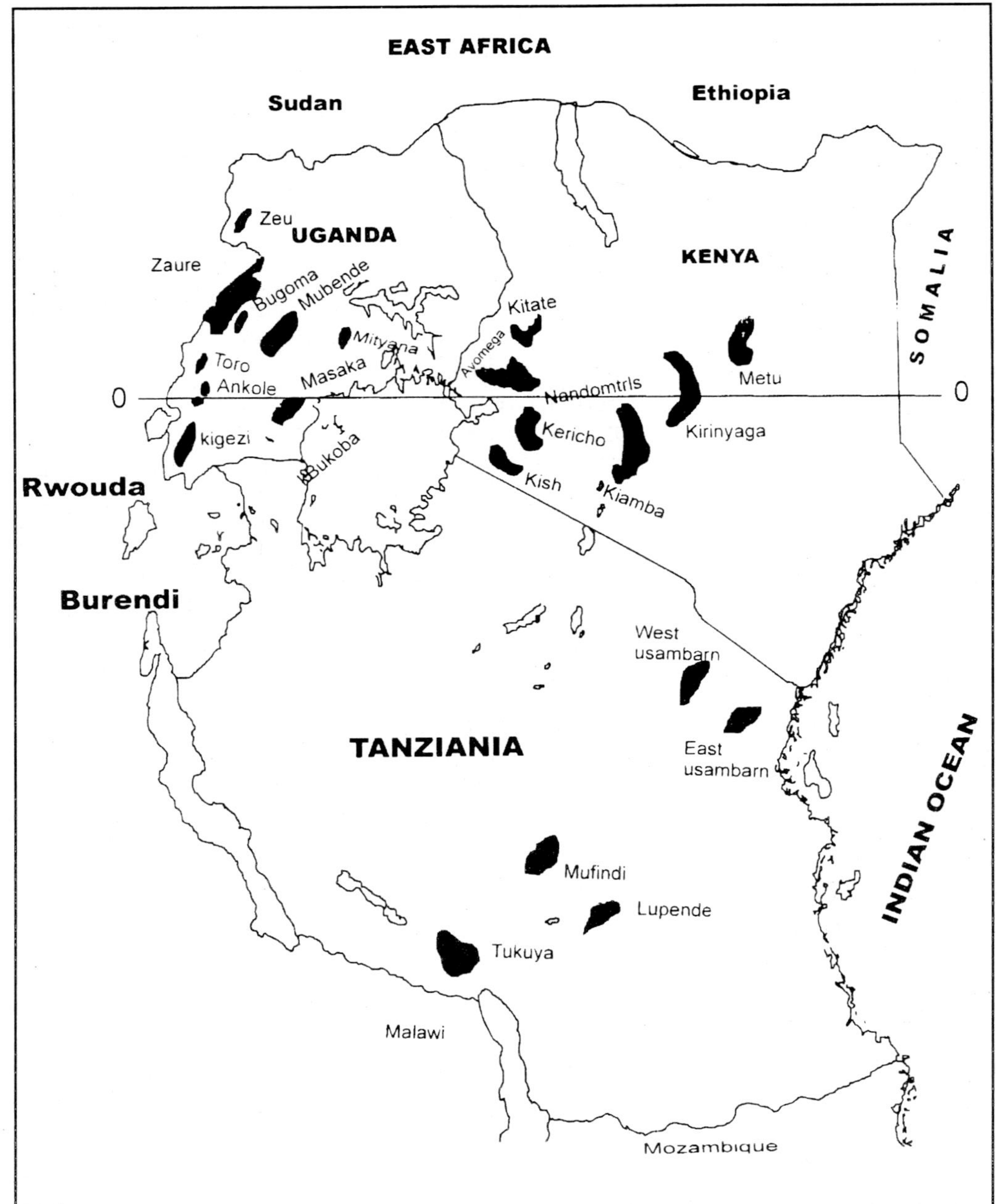

Fig. 1.13 : Tea growing areas of East Africa (marked) [Source: Owuor (1990)].

Gardens were planted in the Botanical Gardens in Entebbe, Uganda. Later in 1909, seeds were imported from Assam and were grown in Kampala

Government plantations. Seeds of these plants were used for starting commercial plantations in 1931. More land was subsequently brought under tea in Mengo, Toro, Masaka, Ankole, Mityana, Bunyoro and Kigezi districts (Fig. 1.13). These tea plantations are located on either side of the equator to the north and west of Lake Victoria. Until 1926 however, only one estate namely Kakonde was functioning on a large scale. Tea is grown on an average between 1300 m to 1800 m elevation. Plucking of green leaf continues throughout the year. Though a pronounced period of drought is not experienced but there is usually a deficit of water in the soil during drier months (June-August and December-February).

During 1960's there was considerable development of the tea industry in Uganda, with a production in over 30 factories, and in 1972, Uganda was the second largest tea producer in Africa after Kenya, with a crop of 23 million kg. Unfortunately, over the following decade, the situation deteriorated under the Presidentship of Idi Amin, reaching its nadir in 1980, during this period vast areas of tea had been abandoned and only a few tea factories had somehow been kept going. The official yield was 1.5 million kg but the quality was appalling. Slowly rehabilitation started. Some emergency funding was committed to the work and seven factories were rehabilitated. Of which four of them were outgrowers and three belongs to Agricultural Enterprises Ltd. Further, joint venture operations with Mitchell Cotts was initiated by taking over vast areas of abandoned tea and rehabilitating three factories. The fund obtained from International Finance Corporation. The Uganda Tea Corporation rehabilitated their one factory and built a new factory at Kasaku (Bunston, 1994). The overall situation improved when President Museveni took charge in January, 1986.

Uganda is capable of producing good medium coloury tea. The only real investment, with new factories, has come from the local entrepreneur, Mzee Karmali, who bought estates in western Uganda (which used to be owned by European private interests) and had developed a real attachment to Uganda. Other investments have come from the Commonwealth Development Corporation, which, with James Finlay, has taken over the old estates of Agricultural Enterprises Ltd. Other investments in tea have been from former Asian owners of estates to rehabilitate their factories, or from ill-conceived plans to rehabilitate the factories of Uganda Tea Growers' Corporation which still have to realise their full potential (Kitching, 1998). UTGC have four factories handling outgrowers' leaf in the west of Uganda-

Mabale and Mpanga near Fort Portal, Igara near Busheni in Ankole district and Kayonza in Kigezi. Mitchell Cotts, pioneer tea harvester in Uganda has three factories viz.Kiamara and Toro/Kahuna near Fort Portal and Mityana.

Republic of Congo (Zaire)

Mountainous Kivu district, west of Lake Kivu, are the main tea growing areas of Republic of Congo. The pattern of crop growth, temperature and precipitation do not differ from those in Uganda. Area and production in 1994 were 9 thousand hectare and 3 million kg respectively.

Mauritius

In 1760, Father Galloys introduced tea in Mauritius. Sir Robert Farquhar, Governor of Mauritius, encouraged tea cultivation on commercial scale in 19th century. 190 hectares of plantations were established by the end of the century. At the onset of World War II an area of 850 ha under tea was planted. After World War II, the Government started to encourage the growth of tea industry and small quantity of tea was first exported in 1948. In 1955, the Government launched the Tea Small Holding Scheme and the Tea Project Planters Scheme. By 1975, additional 1234 ha of tea was planted. Almost 90% of the total area is under small holders having plot size less than 1 ha (Ramlogun, 1993).

Mauritius looked toward diversifying the economy of the island and finding a crop that is able to resist cyclones. Development plans allowed for production to reach 2740 tons by 1971. Today, only one estate is in production, the rest having been uprooted and replaced by sugarcane, the main agricultural export of the island (Kitching, 1998). The production of Mauritius during 1985 was 8.1 million kg and it has come down to 1.7 million kg during 1997.

Mozambique

Tea cultivation in Mozambique began in the early twenties in Milange district which is adjacent to the Mulanje district of Malawi. Tea planting later extended to the other districts viz. Gurue, Socone, Tacuane etc. It is considered that there is immense scope of expansion of tea cultivation in Mozambique. The tea areas located in between 15^0 to 17^0 S latitudes and 300 to 1000 meters. The pattern of plucking and temperature are more or less same as in Malawi. Black tea is the main product obtained from Assam type bushes.

Argentina

The first seed introduced in the country was the Assam variety called Betjan. In 1923 an immigrant, perhaps from Austria, introduced tea in the Misiones Province. According to the official records, the Ministry of Agriculture distributed tea seeds brought from Ceylon (now Sri Lanka) to the farmers of Misiones, Chaco, Formosa and northern Corrientes during 1925. Tea plantations were established in Tres Capones, Loreto, La Mered and Puerto Bomberg in between 1923 and 1930 (Cirelli, 1999). Geographical boundaries of tea areas range from 26^0 to 31^0 south parallels and 150 to 390 m altitude (Fig. 1.14). The distribution of rainfall is irregular and the total varies in between 1500 and 2500 mm a year, and the sunshine ranges from 10 to 12 hours a day. There are 8434 producers with plantations up to 10 h, 443 up to 50 h and 49 producers with more than 50 h. Three principal varieties can be distinguished in Argentina tea plantations viz. *China* (25 %), *assamica* (17 %) and *cambodiensis* (798 %) (Cirelli, 1999). The plucking takes place from October to April (6 months). Today tea harvesting is fully mechanized.

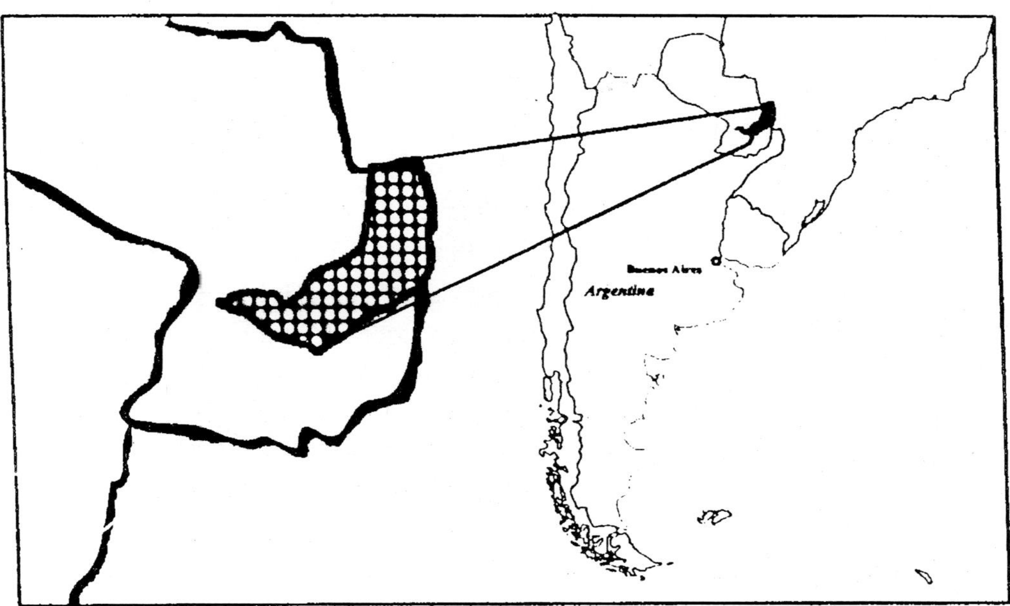

Fig. 1.14 : Tea cultivation area of Argentina (Misiones and N. E. Corrientes) [Source: Cirelli (1999)].

Peru

The beginning of the tea industry in Peru dates back to 1913, when seeds were brought at the suggestion of Dr. Benjamin de la Torre Mar, a surgeon appointed by the province of La Convencion (Cavero, 1997). The Peruvian Consul to Yokohama (Japan) sent first shipment of one hundred twenty pounds tea seeds to Peru which had been sown by Dr. de la Torre at his Huyro farm. Unfortunately he died in 1915. In 1928, the Government showed interest on the cultivation of tea and issued a Supreme Resolution authorising to contract experts from Ceylon (now Sri Lanka) for advise on tea cultivation. It was estimated that 10000 ha (represents 0.13% of the total area of the country) spread over 3500 ha in Cusco (Huyapata and Santa Teresa in the province of La Convencion), 3000 ha in Huanuco (located in the province of Leoncio Prado in natural woodland), 1000 ha in Ayacucho and others 2500 ha are suitable for tea cultivation. In the beginning, the green leaf production sites were located in the Departments of Ayacucho, Cusco, Huanuco, Junin, Madre de Dios and Puno on the edge of the forest of the eastern side of the Andes mountains situated between the tropical and subtropical latitude where the annual rainfall is between 3000 mm and 5000 mm and average temperature is between 19°C and 24°C. Tea growing in Peru largely depends on availability of rain water during the period from October to July.

In 1936, first plantation was started in Huanuco. By 1941, 800000 plants were established in the Official Colonization Centre of Tingo Maria in Huanuco and a black tea processing factory was set up in Hermilio Valdizan (Huanuco). In 1943, another factory was made in Huyro (Cusco).

In 1977, the Foreign Ministry had signed an agreement with Holland which gave rise to the REINTEP project (project for Full Recovery of Tea in Peru). This committed the Dutch to provide technical and financial assistance over a period of four years in the high tea production areas. Further, a black tea processing factory with a capacity of manufacturing 60 tons of green leaf per day was built in Amaybamba. This is the largest factory at present in Peru.

Australia

Sir Joseph Banks established a few tea bushes in Melbourne's Botanical Gardens in 1824, but Herbert Frederick Cutten (1855-1920) planted tea first on a large scale and is acknowledged as the father of the tea industry in

Australia. He was one of his four brothers who migrated from India. The family tea estate was established in 1884 at Bingil Bay, near Innisfail on the north-east coast of the state of Queensland's. The tea estate was not however a success and it was destroyed by a cyclone and an ensuing tidal wave. Some plantings were trialled on Queensland's highest mountain, Mt Bartle Frere (1711m) in 1936, but it was not until some 23 years later, in 1959, that tea became a driving force. Dr. Allan Maruff, a doctor from Calcutta started growing tea near Innisfail with seeds collected from Bingil Bay. In fact his dedication and enthusiasm led to the establishment of the Nerada plantation near Innisfail. Recently the Australian Agricultural Department and individuals had shown interests for cultivation of clonal plants for Japanese green tea market and number of Yabukita and Sencha clonal plants were lifted from Japan in 1991 and subsequently multiplied for large scale planting in 1994/95 (Ford, 1994).The tea areas are located approximately at an elevation of 160 m and receives insufficient rains for growing tea which falls during January to June. Irrigation is required during dry months.

Papua New Guinea

In the early 1900s, some Assam jat tea seeds were brought from Malaysia and cultivated in the Brisbane Botanical Gardens. Much later, in 1930s, the Australian Administration of Papua New Guinea, Department of Agriculture, brought some of these seeds to an emerging experimental station (as such establishments were called in those days) to a place named Garina in the Morobo Province of Papua New Guinea. Later in the early 1950s, the Department of Agriculture build a small tea processing factory for manufacturing Garina tea for local consumption (Emslie, 1998). In the Western Highlands, the first recorded tea was planted by Bobby Gibbs, a coffee plantation settler, in his Tremarne plantation. One Mr. Bucknell who had a mixed farm called Moitemp on the North Wahgi swamp had also planted tea but both the areas were not developed into a commercial tea estates. However, their seed bearers provided planting materials later in the Highlands.

One Australian family headed by Ivor Manton and his son developed Warrawau, Tigi and Kuri tea estates. All of these were planted with Assam jat tea seeds that had been grown at Garina. Manton was followed by David Bennet at Kurumul estate in 1962-63, and then by W. R. Carpenter & Co.

estates in 1964, whose estates at Kudjip, Kindeng, and Aviamp were planted with Garina and Moitemp seeds.

Large tea estates viz. Pugam and Minjigina opened by Dick Hagon and Pioneer Concrete respectively in the late 1960s and early 1970s. Both the tea estates are located on the upper Mount Hagen ranges at an elevation over 6200 feet. Thereafter, one Mr. John Colman planted part of his Madan estate with tea and ANG Corporation developed tea on their Bunum Wo estate at the same time as Carpenters. Paraguai, a small tea estate was also opened at this time by three Australian partners on the border of the Western and Southern Highlands (Emslie, 1998). Clonal tea varieties developed at Kuk Research Station by John Morgan were also planted in mid-1970s under a project funded by the World Bank in the Southern Highlands. Unfortunately all of them have been abandoned. To date, only two large groups viz. W. R. Carpenters with 1800 ha and Warren Plantations with 1000 ha and one privately owned tea estate in the Western Highlands are surviving (half of originally planted) in Papua New Guinea. There are six factories, four of which are owned by Carpenters and two by Warrens, are operating at present. The main market for their teas are Australia, U.S.A., U.K., South Africa and Malaysia.

References

Anonymous, 1991. *Techno-economic survey of Tripura tea industry (a reappraisal).* Tea Board of India, Calcutta, 1-130.

Anonymous, 1996. *Evaluation report and recommendations of Technical Expert Committee on tea industry of Himachal Pradesh*. Tea board of India, Calcutta, 1-27.

Arfin, M. S. and Semangun, H. H. 1999. Tea industry in Indonesia. In *Global Advances in Tea Science* (ed. N. K. Jain). Aravali Books International (P) Ltd., 65 – 82.

Arunachalam, K. 1995. *A hand book on Indian tea*. Arunachalam Associates (Plantation Consultants), Wellington Bazar, The Nilgiris, p. 271.

Baildon, S. 1877. *Tea in Assam*. Calcutta.

Banerjee, B. 1993. *Tea production and processing*. Oxford & IBH publishing Co. Pvt. Ltd., New Delhi, p. 336.

Barua, D. N. 1989. *Science and practice in tea culture*. Tea Research Association, Calcutta - Jorhat, p. 493.

Bunston, M. 1994.Tea production in Uganda. *Tea International*, **4** : 17-18.

Carr, M. K. V., Stephens William and Congdon, T. C. E. 1988. Tea in Tanzania. *Outlook on Agriculture*, 17(1): 18-22.

Cavero, Y. E. D. 1997. Tea in Peru. *Tea & Coffee Trade Journal*, **169(9)**: 92-101.

Chen, Zongmao and Lu, Zhenhuri. 1993. Tea industry in China. *Proc. Intl. Symp. on Tea Science & Human Health*. TRA, Calcutta, January 11-14, 1993, India. 22-30.

Chiu, Tsai-Fua. 1991. The present status on tea cultivation, manufacture and utilization in Taiwan. *Proc. Intl. Symp. on Tea Science*. Shizuoka, August 26-29, Japan. 34-44.

Cirelli, A. F. 1999. Tea industry in Argentina. In *Global Advances in Tea Science* (ed. N. K. Jain). Aravali Books International (P) Ltd., New Delhi, 209 – 214.

Eden, T. 1976. *Tea*. Longman, London, p.236.

Emslie, M. 1998. Tea in Papua New Guinea. *Tea & Coffee Trade Journal*, 170 (4): 130-136.

Ford, R. 1994. Madura tea estates, Australia. *Tea International*, **5**: 8-11.

Ghosh, B. C. 1970. Development of tea industry in the district of Jalpaiguri. *Jalpaiguri District Centenary Souvenir*. p.238.

Ghosh T. K. 1987. *Tea gardens of West Bengal*. B. R. Publishing Corp. Pvt. Ltd., New Delhi. pp.262.

Gnanapragasam, N. C. 1999. Sri Lanka tea industry. In *Global Advances in Tea Science* (ed. N. K. Jain). Aravali Books International (P) Ltd., New Delhi, 103 – 120.

Gokhale, N. G. 1958. Report on a tour of the U.S.S.R. and seven east European countries. *Memo 25, ITA, TES*, pp.30.

Griffith, P. 1967. *The history of Indian tea*. Weidenfield and Nicolson, London. p. 730.

Gupta, A. K. 1995. Tea in Himachal Pradesh. *Proc. 32 UPASI Sci. Conf.*, Sept. 11, 1995, 25 - 34.

Harler, C. R. 1966. *Tea growing*. Oxford University Press, London. p. 162.

Hosseini, S. J. 1993. Tea industry in Iran. *Proc. Intl. Symp. on Tea Science and Human Health*, TRA, Calcutta, January 11-14, 1993. 48-51.

Huang, S. Metecrology of the tea plants in China: A review. *Agric. Forest Meteorol*, 47: 19-30.

Hunter, W. W. 1876. *A statistical account of Bengal. vol. x*, London. pp.164.

Kingdon-Ward, F. H. 1950. Does wild tea exist ? *Nature*, **165**: 297 - 299.

Kitching, S. 1998. A retrospective of east African tea production. *Tea & Coffee Trade Journal*, 170(3): 68-73.

Kodomari, S. and Hara, Y. 1999. Tea in Japan. In *Global Advances in Tea Science* (ed. N. K. Jain). Aravali Books International (P) Ltd., New Delhi, 89 – 102.

Masty, S. J. 1998. Tea-time on the roof of world. *Tea & Coffee Trade Journal*, 170 (1): 82-85.

Moffet, J. P. (ed.). 1958. *Handbook of Tanganyika* (2nd edition). Government of Tanganyika, Dar es Salaam.

Muraleedharan, N. 1991. *Pest management in tea*. UPASI Tea Research Institute, Valparai, Coimbatore, Tamil Nadu, p. 130.

Noppakoonwong, U. and Nillavesana, S. 1999. Tea production in Thailand. In *Global Advances in Tea Science* (ed. N. K. Jain). Aravali Books International (P) Ltd., New Delhi, 131 – 142.

Othieno, C. O. 1993. Tea industry in Kenya. *Proc. Intl. Symp. on Tea Science and Human Health*. TRA, Calcutta, January 11-14, India, 60-70.

Owuor, P. O. 1999. Tea in east Africa – Kenya, Uganda and Tanzania. In *Global Advances in Tea Science* (ed. N. K. Jain). Aravali Books International (P) Ltd., New Delhi, 89 – 102.

Paterson, B. 1995. Letter from South Africa. *Tea International*, **3(8)**: 2-8.

Ramlogun, H. G. S. 1993. Tea industry in Mauritius. *Proc. Intl. Symp. on Tea Science and Human Health*. TRA, Calcutta, January 11-14, India. 71-76.

Sivapalan, P. 1991. Current status of tea production in Sri Lanka. In *World Tea. Intl. Symp. On Tea Science*, August 26 – 29, 1991, Shizuoka, Japan, pp. 89 – 96.

Tien, D. M. 1993. Tea industry in Vietnam. *Proc. Intl. Symp. on Tea Science and Human Health.* TRA, Calcutta, January11-14, India. 103-106.

Ukers, W. H. 1935. All about tea. *Tea and Coffee Trade Journal Co.*, New York, USA.

Vanli, H. 1991. Tea production and consumption in Turkey. *Proc. Intl. Symp. on Tea Science*. Shizuoka, August 26-29, Japan, 97-104.

Watt, G. 1889. *Dictionary of economic products of India*. Govt. of India.

Weatherstone, J. 1992. Historical introduction. In *Tea Cultivation and Consumption* (eds. K. C. Willson and M. N. Clifford), Chapman & Hall, 1-23.

Whittle, A. H. 1997. Clonal tea in Southern Africa. *Tea & Coffee Trade Journal*, **169(12)**: 62 - 66.

Whittle, A. H. 1999. Tea industry in central and south Africa. In *Global Advances in Tea Science* (ed. N. K. Jain). Aravali Books International (P) Ltd., New Delhi, 149 – 170.

Yu, F. 1986. Discussion on the originating place and the originating centre of tea plant (in Chinese). *J. Tea Sci.,* **6**: 1-8.

Zaprometov, M. N. and Jinjolia, R. 1991. Tea plant in USSR: some remarks concerning cultivation, science and industry. *Proc. Intl. Symp. on Tea Science*. Shizuoka, August 26 - 29, 1991, Japan, 105-107.

Chapter 2

Botanical characteristics

2.1 Introduction

Tea is economically most important of all *Camellia* species. The genus *Camellia* belongs to the family Theaceae, tribe Gordonieae. This genus accommodates as many as 82 species (Sealy, 1958) and over 200 species (Zhijian *et al*., 1988). The tea plant was originally included in the genus *Thea* by Linnaeus (1752). *Camellia*, where all non-teas were included was considered to be a 'section' under genus *Thea* (Roberts *et al*., 1958). Subsequently, the genus *Thea* and *Camellia* are merged. Hence, *Thea sinensis* L., the original name of the tea plant has now become *Camellia sinensis* (L.) O. Kuntze. Tea is essentially an outbreeding crop and its races are highly infertile. As a result of enormous natural hybridization not only between the tea species but also with *Camellias* and large-scale dispersal of tea seeds in south-east Asia over the centuries, resulted in extreme heterogeneity in the existing seedling population. Tea taxonomy is very complicated, and hybridization early in its commercial history has left an almost total absence of pure plants (Willson, 1999).

2.2 Classification

Wight (1962) on the basis of foliar, reproductive and growth features assigned specific ranks to China and Assam varieties. The nomenclature *Camellia sinensis* (L.) was retained for *C. sinensis* var. *sinensis* (L.), but Weight (1962) named *C. sinensis* var. *assamica* as *C. assamica*. A third

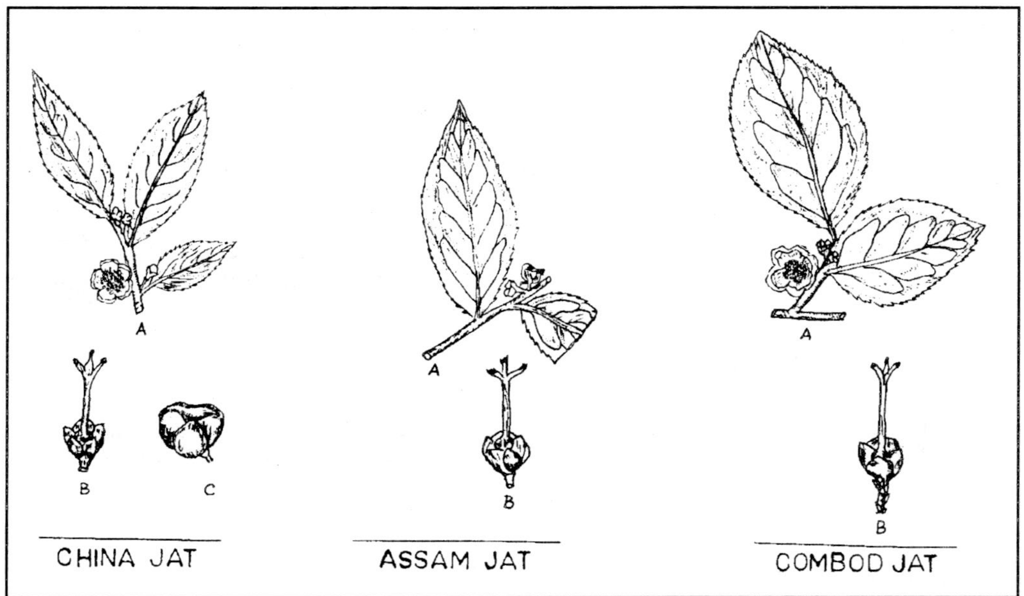

Fig. 2.1: Outline of three races of tea. A. Flowering shoot; B. Pistil; C. Fruit [(Source : Bezbaruah (1976)].

form described as the southern form of tea (Roberts *et al.*, 1958) or Cambod race and resembles to Planchon's *Thea lasiocalyx* (Barua, 1963), was proposed by Wight (1962) to be a sub-species of *C. assamica* and named *C. assamica* subsp *lasiocalyx* (Planchon ex Watt). The tea varieties have developed into a complex taxonomic group of related plants. A fourth group called 'hybrid variety' with mixed taxonomic characters occupies the largest tea area under cultivation through seed propagation (Wight, 1958). Although major phenotypic groups had a common genetic base, and contrasting characters might have their origins to one or other varieties, the variation in the wide range of intermediaries can be greatly sorted out by applying principles of numerical taxonomy. A summary of the foliar, floral characteristics and growth features that differentiate the three main races of tea are presented in Table 2.1 and Figure 2.1.

2.3 Botanical composition of jats

Distinctive types of tea based on foliar characters and the district or the garden growing the seed is given to the strain and this is known as the 'jat'of the tea. The term 'jat' in tea nomenclature is considered close to 'variety'. The multi-stemmed China type has two jats — '*Low-jat*' and '*High jat*'. The former 'jat' has narrow leaves, profusely indented margin and flat, rather dull-coloured laminae, whereas later has comparatively larger leaves, wider in proportion with blunter tips, fewer serrations and a crinkled surface (Subramanium, 1995). The single-stemmed Assam type has *Burma*, *Lushai*, *Manipuri*, *light-leaved Assam* and the *dark-leaved Assam jats*. *Burma* jat comes from Myanmar hills and has spiny, crinkled and slightly

Table 2.1 : Habitat and morphological characteristics of three principal taxa of cultivated tea.

Characteristics	Taxa		
	China	Cambod	Assam
Habitat	Shrubs	Under-trees	Trees
Leaves size (cm)	10-14.5 x 4-6.0	12.5-15.5 x 4.5-6.5	13.0-16 x 4.5 -6.0
Leaf angle to branch (degrees)	50	50 - 70	70
Internodal length (mm)	15 ± 0.5	45 ± 1.0	65 ± 1.5
Individual leaf area (mm^2)	120 ± 45	382 ± 124	1554 ± 352
Leaf area index	8.54	5.25	3.65
Leaf length/ breadth ratio	2.5	2.3	2.4
Laminar angle to nearest 5^0	100 ± 10	120 ± 5	190 ± 20
Shape	Elliptic-oblong or obovate-oblong	Elliptic-oblong	Elliptic to obla-anceate
Upper surface	Slightly to predo-minantly rugose, upfolded to flat	Rugose to smooth, upfolded	Smooth, flat to slightly upfolded
Colour (young leaves)	Dark-green	Light-green, occasi-onally pigmented with anthocyanin	Light-green to dark green
Vestiture on lower surface	Lamina sparsely appressed and pub-escent, midrib den-sely pubescent	Lamina glabrous, mid-rib sparsely pubescent when young	Lamina nearly glabrous to spar-sely villous, mid-rib sparsely appr-

			essed, pubescent to densely villous
Margin	Obtusely serrulate to double serrate, flat to wavy	Serrulate, flat to slightly recurved	Obtusely serrulate to bluntly serrate or denticulate flat
Apex	Acute to shortly accuminate, acumen obtuse or retuse with a caducous mucron, downturned	Shortly accuminate, acumen obtuse, rarely retuse, mucronate, downturned	Shortly to gradually acuminate, acumen obtuse or retuse, slightly downturned or straight
Base posture	Cuneate to rounded	Acute to rounded	Obtuse to rounded
leaf posture	Well upturned	Slightly upturned to horizontal	Horizontal to errect
Flower			
Sepals	Five, glabrous to silvery villous within, ciliate along margin	Five, glabrous to sparsely puberulous, along margin	Five, glabrous to silvery villous, ciliate along margin
Petals	Six to eight, obovate-oblong to sub-orbicular	Six to seven, obovate to sub-orbicular	Six to eight, obovate to sub orbicular glabrous
Ovary	Globular, sparsely hirsute to densely silvery villous	Globular, glabrous to densely or sparsely pilose	Globular to obovoid, sparsely pilose to silvery villous
Style	Free upto half its length, arms three to four, geniculate, spreading horizontally	Free upto one third its length or shortly split, arms three ascending	United to a greater part, arms three, spreading horizontally or ascendingly spreading

Source: Mohanan and Sharma (1981); Banerjee (1987)

errect leaves and has dark-green, older leaves having a bluish tinge. *Lushai* is form of Lushai hills and has pendent flabby leaves. *Manipuri* is a hardy and high yielder of quality tea. The leaves are dark-green, shiny, rather narrow and have a semi-upright leafpose. It is softer than *Burma* jat. The *light-leaved Assam* is found in the Naga hills and has a semi-upright, glossy, crinkled, light green leaves. The *Lushai* and the *light-leaved Assam* are generally not planted in Assam because the former is very delicate and the later is poor yielder and has a balsam-like odour. *Dark-leaved Assam* has soft, glossy, crinkled and slightly pendent leaves (Subramanium, 1995).

2.4 Biology of the plant

2.4.1 *Vegetative structure*

Sealy (1958) used characters of leaf, floral morphology and growth habit in assigning taxonomic categories within *Camellia*. Though in preliminary differentiation of taxa vegetative characters have been found useful but due to plasticity reproductive features are found more reliable than vegetative characteristics (Wight, 1959). Under natural conditions, the tea plant is an evergreen tree and wildly grows into a medium size tree (Fig. 2.2). Under cultivation it is pruned down and trained as low spreading bush to ensure that a maximum crop of young shoots can be obtained and

Fig. 2.2 : Full-grown plants of *Camellia assamica*.

maintained. The root system of a tea bush can be divided into main roots, subsidiary roots and feeder roots. The lateral roots give rise to a surface mat of feeding roots which lack root hairs when mature. Root system varies in tea bushes depending on genetic make up and soil environment. Starch is stored in roots.

2.4.2 *Reproductive structure*

Tea flowers (Fig. 2.3) appear in the axils of scale leaves, either solitary or in clusters, with short pedicels, fragrant; calyx persistent with 5-7 sepals; petals 5-7, white, waxy, obovate, concave; stamens numerous with 2-celled yellow anthers; petal and outer stamens united for a short distance at base

Fig. 2.3 : Tea flower.

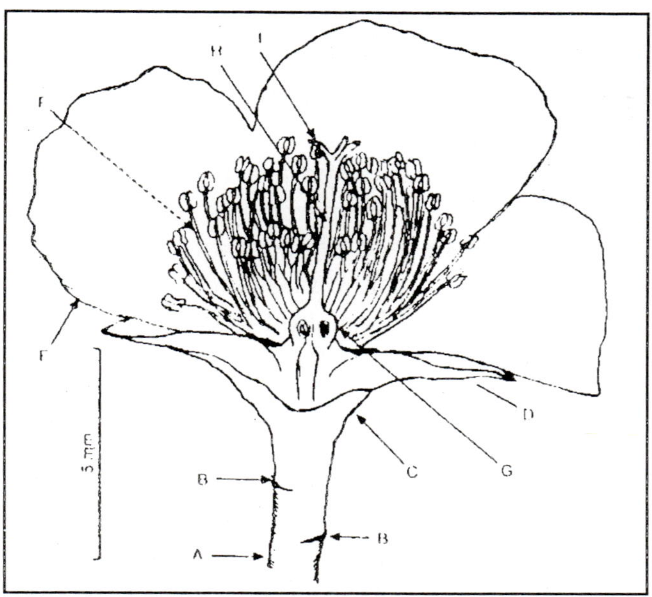

Fig. 2.4: Section through tea flower. A. Pedicel; B. Scars; C. Receptacle; D. Sepal; E. Petal; F. Stamen; G. Ovary; H. Column; I. Style arms [(Source: Barua (1993)].

(Fig. 2 4); ovary superior, hairy, with 4-6 ovules per carpel; styles short, free to almost completely fused, stigmatic lobs 3-5.

Fruits (Fig. 2.5) are capsule thick walled, brownish green, 3-lobed

Fig. 2.5 : Tea plant bearing fruits.

and usually 3-celled, becoming slightly rough at maturity, dehiscing during the dry period by splitting from apex into 3 valves.

2.4.3 *Growth pattern*

Unlike other perennials, tea is unparallel because only vegetative parts–the two leaves and the bud–are commercially exploited. Tea is maintained as a low bush in a continuous phase of vegetative growth. The term 'flush' describes the growth by the apical bud between two successive phases of dormancy, which in effect means that a bud passes through alternate phases of growth and dormancy (Wight and Barua, 1955). The phase of resting period or dormancy of the shoot is *banjhi* period and the apical bud is then termed as the *banjhi* bud. The bud in *banjhi* period is very small, blunt without the usual closely serrated margin and is closed usually by two bud scales or cataphylls and this is called the *janams* (Fig. 2.6). The position of

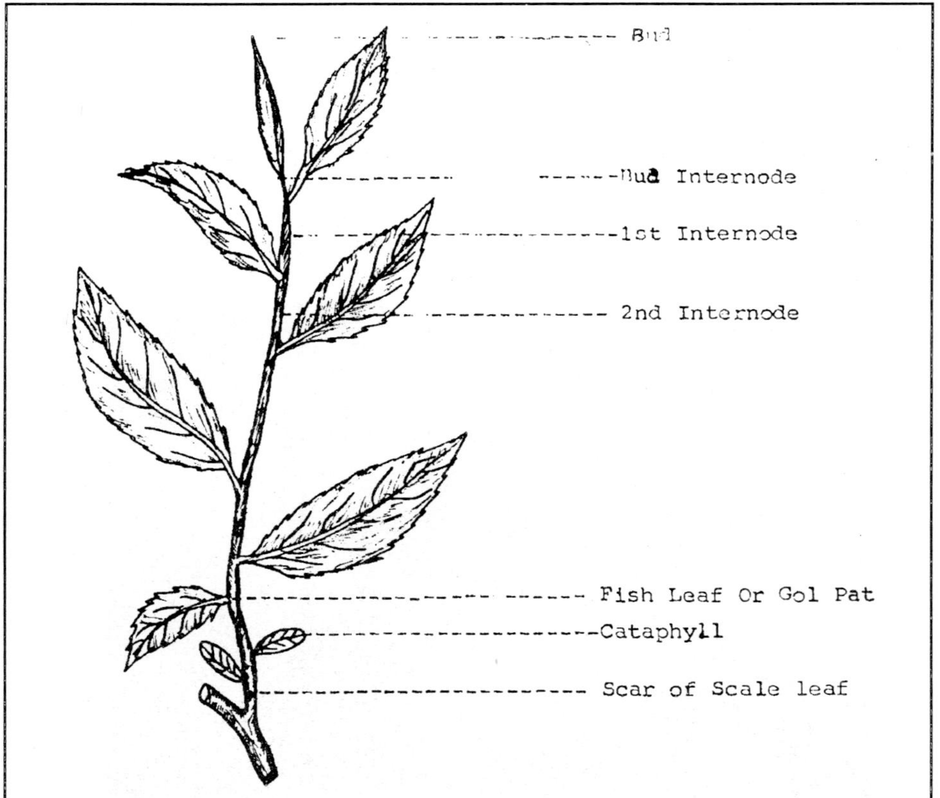

Fig. 2.6: A typical tea shoot development.

janam or fish leaf on the shoot is of great importance in considering the standard of plucking. The tip of a new leaf then become visible which is bigger than the *janam* leaf but smaller than normal leaf . Sometimes this may be partly serrated and lack an apex. Such a leaf is called as gol-pat or fish leaf. As the bud continues to unfurl, normal leaves are produced which grow in size to that of mature flush leaves with elongated internodes. After the successive development of normal leaves, the shoot again goes *banjhi*, thereby completing one phase of production, followed by further ones of the same type, and the process recurs. This periodicity between *banjhi* and productive phase of shoot is independent of plucking and during a single phase, the typical growth consists of two scale leaves, a fish leaf and four flush leaves.

2.4.4 *Winter dormancy*

Tea cultivated near equator flushes throughout the year but further from the equator at latitudes beyond about 16^0, bushes remain completely dormant during the winter season. Harvesting of the crop stops during the winter months from November onwards and does not start until the following months of March (Assam, Dooars, Kangra region in Himachal Pradesh and Darjeeling). The problem of winter dormancy in tea not yet resolved. Low temperature and day length apparently interact in inducing dormancy but their respective roles are not clear. This has also been discussed in section 6.6.

2.5 The non-commercial *Camellias*

In addition to the three principal tea producing taxa, there are number of species with similar characteristics which do not produce the necessary chemicals but may be involved in hybridization. Most important are *C. irrawadiensis* Barua (Wilson's *Camellia)* and *C. taliensis* (W. W. Smith) Melchoir (Forest's *Camellia*). *C. irrawadiensis* which was originally found in Upper Burma in 1917 by L. O. Wilson at an altitude of 7600 ft. is characterised by the presence of numerous pellucid punctations on leaves. Leaves are very glossy, completely glabrous, soft, supple and brick red in colour on both surface. Sclereids are thick, elongated slightly branched with an unrestricted lumen extending the whole length (Barua and Wight, 1958). 4-5 styles divide distally into four or five parallel errect limbs which are acutely bent to form horizontal arms. *C. irrawadiensis* makes a product visually like tea but contains no caffeine and spurious as an infusion. *C.*

taliensis externally closely resembles *C. assamica* except size of floral parts. Flowers are 5-6 cm in diameter and fully expanded petals (Sealy, 1948). Styles are divided into five distinct horizontal arms at the distal end. *C. taliensis* also fails to produce commercially acceptable tea. *C. lutescens* Dyer, has been occasionally reported from south India (Sharma and Venkataramani, 1974) characterised by three very short free styles and non acceptable tea. The other characteristics of *C. lutescens* are elliptic-oblong tapering or rounded below caudate-acuminate membranous leaves, styles short, stigmas recurved, capsules drupaceous when young. A few other taxonomically important non-tea *camellia* spp. are *C. japonica* L., *C. kissi* Wall, *C. pitardii* var. *yunnanica* Cohen-Stuart, *C. sasanqua* Thunb., *C. reticulata* Lindl., C. caudata Wall. These species of *Camellia* L. seem to have been involved in the evolution of the present day commercial tea population (Wood and Barua, 1958; Wight, 1959; 1962; Bezbaruah and Gogoi, 1972; Sharma and Venkataramani, 1974).

2.6 Hybrid differentiation

2.6.1 *Anatomical features of tea leaf*

A transverse section of tea leaf (Fig. 2.7) shows three groups of tissues viz. epidermis, mesophyll and veins. The epidermis is a single layer cell forming a thin skin which covers the entire leaf surface. Its main function is to protect the internal tissues. Over epidermis there is a very thin waxy layer called cuticle which protects the epidermis from excessive loss of moisture, mechanical injury and against the entry of parasites (Dutta, 1978). Though the clones with thicker cuticle will show positive affinity towards

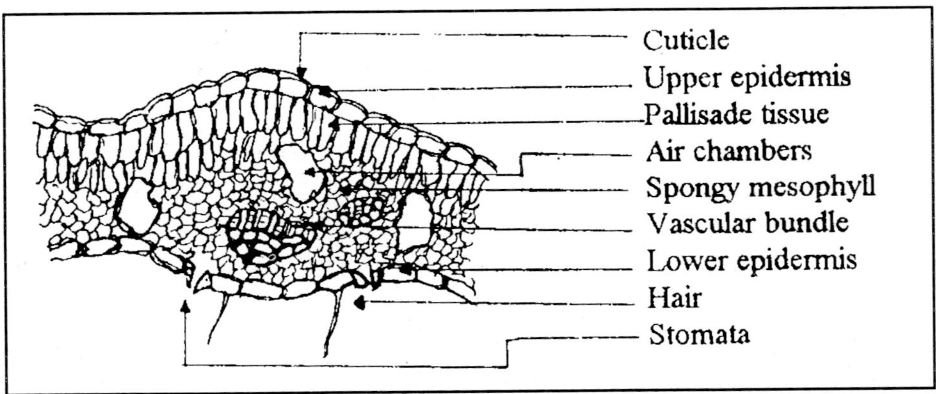

Fig. 2.7 : Cross section of a tea leaf of China hybrid.

disease resistance and drought tolerance, it might impose problems during manufacturing. Withering in such cases will be delayed leading to very slow moisture removal resulting in higher energy input and extended optimum fermentation time. The epidermis contributes 7 to 16 percent of the total leaf thickness of the different clones (Balasubramanian and Muraleedharan, 2000). Like most of the dicotyledonous plants, the mesophyll or ground tissue of the tea leaves is differentiated into palisade and spongy parenchyma. The palisade layer is made out of one or two narrowly elongated cells arranged in stakes as seen in transition. The palisade parenchyma in China hybrids usually consists of two sometimes three layers of cells but it has one or rarely two layers in Assam and Cambod series. In between palisade and spongy mesophylls numerous air spaces are present. All these cells have chloroplasts which are responsible for manufacturing of plant food. Underneath the spongy mesophyll of irregular shape and lower epidermis are present. The stomata with prominent guard cells are distributed in the abaxial surface of the leaf and the remaining portion is covered by a cuticle. Internal cavities of the leaf where these stomata are located are called air-chamber. The stomatal density varies with the clones. Trichomes or pubescence distributed on the lower surface of the leaf are one of the characteristics of tea leaf (Eden, 1976). The trichomes shed gradually with the leaf development and their presence on the bud is a sign of good leaf and have been used as one of the quality indices of some kind of tea (Liang *et al*., 1993). This grows out from the epidermis of fresh leaves at angles ranging from 45° to 75 ° (Wu, 1964) and attached to the lower epidermis through an expanded joint whose wall mainly consist of fibre. The outer surface of the joint is unevenly covered by waxy substances and the internal surface is smooth. The dense trichome on tea buds and tender leaves may protect the leaf by diminishing evapotranspiration and thus help the buds and leaves to conserve moisture (Yan, 1990). Amma (1986) considered distribution, size and lumen length of the leaf hair to identify Japanese taxa. The Assam series characterised by the shortest hair, while Japanese polyploids have the longest hair, the Darjeeling Chinary series have hair longer (Fig. 2.8) than Assam series but shorter than the Japanese polyploids (Amma, 1986). The veins are vascular bundles which branch out from the vascular bundles of the mid-rib into the leaf blades. These veins extend through the mesophyll tissues.

Barua and Dutta (1959) have usefully utilized the variation in the distribution and morphology of sclereids in differentiating tea taxa

Fig. 2.8: Scanning electron micrograph of the trichome hairs on the undersurface of the leaf of China hybrid.

(Fig. 2.9). Sclereids, also known as stone cells give strength, rigidity, flexibility and elasticity to the plant body and thus enables it to withstand strains. Normally there are no living contents inside the cavity of sclereids. This cavity in fact called as lumen. The anatomical variations of the three races of tea are presented in Table 2.2.

Table 2.2 : Distinguishing anatomical characters of the three races of tea.

China	Cambod	Assam
1. Sclereids absent or rare	Sclereids numerous	Sclereids numerous
2. Sclereids slender, almost without spicules	Sclereids stout with a few spicules	Sclereids stout with numerous spicules
3. Lumen of sclereids almost completely closed	Lumen of sclereids of irregular width, closed in places	Lumen of sclereids of irregular width but not closed
4. Palisade parenchyma usually consists of two, occasionally three layers of cells	Palisade parenchyma has one or rarely two layers	Palisade parenchyma has one or rarely two layers

Source: Barua (1963; 1989).

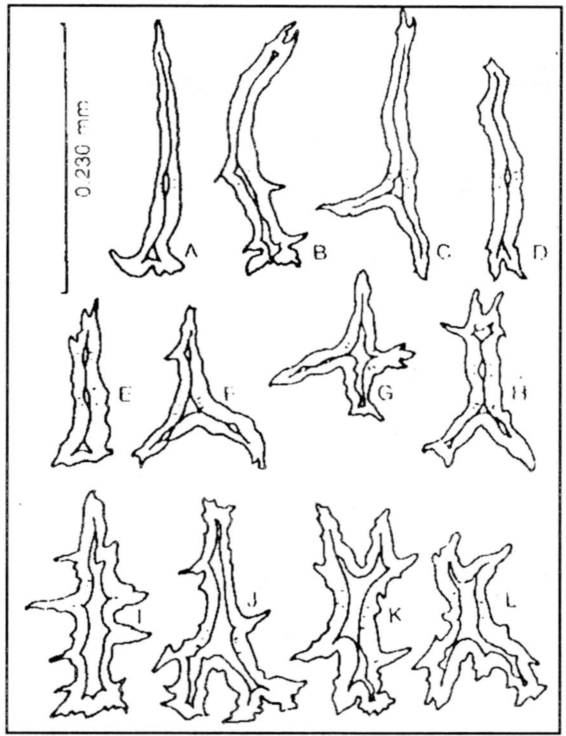

Fig. 2.9: Sclereids in tea. A – D. China plant; E – H. Assam plant; I – J. Cambod plant [Source: Barua (1963)].

2.6.2 *Biochemical differences*

The genetic differences between hybrids are reflected in biochemical composition of their leaves. Chromatographic analysis shows that triglycosides of quercetin and kaemferol are generally found in Chinary hybrids but these are absent or present only in traces in Assam hybrids (Banerjee, 1992).

Higher ratio of volatile compounds observed in Chinary hybrids compared to Assam hybrids (Mohanta and Singh, 1990). The content of monoterpene alcohol is found to be about five times higher in Chinary hybrids. Takeo (1981) demonstrated that *assamica* cultivars contained high levels of linalool while *sinensis* cultivars produced mainly geraniol and intermediate characteristics showed in the hybrids of *assamica* and *sinensis* varieties. The *assamica* varieties were shown to have terpene index, i.e. the ratio of the gas chromatographic peak areas due to linalool to those due to linalool plus geraniol, close to a value of one while *sinensis* varieties have terpene indices close to xero. However, there is need to be developed additional chemotaxonomic criteria for hierarchical classification of *Camellia*.

2.6.3 *Cytotaxonomy*

A comparable chromosomal structure and numbers appear to maintain in most species of *Camellia* suggests a monophyletic origin of all tea species. Even in some wild *Camellia* (*C. caudata*, *C. kissi* and *C. irrawadiensis*) the diploid chromosome number is the same (2n = 30) as in cultivated varieties. In Japanese taxa, more deviations from normal chromosome number are known. *C. sasanqua* is a hexaploid (2n = 90) and *C. sinensis* var. *macrophylla* a triploid (2n = 45). *C. rosaeflora* which is a native of Sri Lanka is also a triploid (Karasawa, 1935). A natural triploid from China has also been reported (Zhan *et al.*, 1987).

Generally, leaf size, dry weight of leaf and rooting ability of triploids and tetraploids are higher than those in diploids. Pentaploids and aneuploids are poor rooters, and have smaller leaves than those of diploids, triploids and tetraploids (Banerjee, 1992). Pollen variability and fertility of triploids are usually poor than tetraploids but less than the diploids (Visser, 1969). Sarma (1985) obtained twenty triploid progenies by crossing TV1, TV3 and TV7 with tetraploid stocks. Marked variations in the contents of theaflavins (TF), thearubigins (TR), caffeine and tannin were observed in the hybrids. Caffeine was found higher in the parents, while tannin varied with the ploidy levels. TF and TR were inversely correlated with the ploidy level. It was assumed that the inheritance from the male parents might be responsible for TF and TR variations. However, cytological investigations in tea have been restricted mostly to determine the number of chromosome (Bezbaruah, 1971), rather than the determination of cytological basis of species.

References

Amma, S. 1986. Identification of tea clones *Camellia sinensis* (L.) O. Kuntze by pubescence on the undersurface of young leaves. *In* : *Development of New Technology for Identification and Classification of Tree Crops and Ornamentals* (eds K. Kitamura, T. Akihama, H. Kukimaru *et al.*), Fruit Research Station, Ministry of Agriculture, Forestry and Fisheries, Japan, pp. 19-24.

Balasubramanian, S. and Muraleedharan, N. 2000. Anatomical variations in the leaf of the tea clones. *Planters' Chronicle*, 96(2): 85–86.

Banerjee, B. 1987. Can leaf aspect affect herbivory? A case study with tea. *Ecology*, **68:** 839-843.

Banerjee, B. 1992. Botanical classification of tea. *In* : *Tea Cultivation to Consumption* (eds K. C. Willson and M. N. Clifford), Chapman & Hall, London, pp. 25-51.

Barua, D. N. 1989. *Science and practice in tea culture.* Tea Research Association, Calcutta, p. 493.

Barua, P. K. 1963. Classification of the tea plant. *Two and a Bud*, **10:** 3-11.

Barua, P. K. and Dutta, A. C. 1959. Leaf sclereids in the taxonomy of *Thea camellias* II. *Camellia sinensis* L. *Phytomorphology.*, **9:** 372-382.

Barua, P. K. and Wight, W. 1958. Leaf sclereids in taxonomy of *Thea Camellias* I. Wilson's and related *Camellias*. *Phytomorphology*, **8:** 257 - 264.

Bezbaruah, H. P. 1971. Cytological investigations in the family Theaceae - I. Chromosome numbers in some *Camellia* species and allied genera. *Caryologia*, **24:** 421-426.

Bezbaruah, H. P. 1976. The tea varieties in cultivation — an appraisal. *Two and a Bud*, **23:** 13-19.

Bezbaruah, H. P. and Gogoi, S. C. 1972. An interspecific hybrid between tea (*Camellia sinensis* L.) and *C. japonica* L. *Proc. Indian Acad. Sci.*, **76B:** 219-220.

Dutta, A. C. 1978. Cross-section of a typical tea leaf. *Two and a Bud*, **25(2):** 11.

Eden, T. 1976. *Tea*. Longman, London, p. 236.

Karasawa, K. 1935. On the somatic chromosome number of triploid *Thea*. *Japanese J. Genet.*, **11:** 320.

Liang, Y., Pan, Gensheng and Xu, Xinqing. 1993. Effect of Maofeng tea processing on leaf trichomes of tea (*Camellia sinensis* L.). *J. Sci. Food Agric.*, **62:** 35-39.

Linnaeus, C. 1752. *Species Plantarum Ed I.*

Mahanta P. K. and Singh, R. 1990. Flavour components of Assam and Darjeeling tea in relation to agropractices and processing. *Assam Review & Tea News*, **79:** 39-43.

Mohanan, M. Sharma, V. S. 1981. Morphology and systematics of some tea (*Camellia* spp.) cultivars. *Proc. PLACROSYM IV*, 3 - 5 December, 1981, Mysore, India, pp. 391-400.

Roberts, E. A. H., Wight, W. and Wood, D. J. 1958. Paper chromatography as an aid to the identification of *Thea camellia*. *New Phytologist*, **57:** 221-225.

Sarma, P. C. 1985. Cytogenetical and morphological studies on some triploid tea (*Camellia sinensis* L.) produced from some triploid-diploid crosses. *Ph.D Thesis*, Guwahati University, Assam, India.

Sealy, J. 1948. *Camellia taliensis* Curts. *Curts' Botanical Magazine*, **164:** 9684.

Sealy, J. 1958. A revision of the genus *Camellia*. *Royal Horticultural Society*, London

Sharma, V. S. and Venkataramani, K. S. 1974. The tea complex. I. Taxonomy of tea clones. *Proc. Indian Acad. Sci.*, **80B:** 178-187.

Subramaniam. B. 1995. *Tea in India*. Publication & Information Directorate and Wiley Eastern Limited, New Delhi, p. 123.

Takeo, T. 1981. Variation in amounts of linalool and geraniol produced in tea shoots by mechanical injury. *Phytochemistry*, **20:** 2149-2151.

Visser, T. 1969. Tea *Camellia sinensis* (L.) O. Kuntze. *In* : *Outlines of Perenial Crop Breeding in the Tropics* (eds F. P. Ferwerda and F. Wit), H. Veenman & Zonen, N. V., Wageningen, pp.459-493.

Wight, W. 1958. The agrotype concept in tea taxonomy. *Nature*, **181**: 893-895.

Wight, W. 1959. Nomenclature and classification of the tea plant. *Nature*, **183:** 1726–1728.

Wight, W. 1962. Tea classification revised. *Current Sci.*, **31:** 298-299.

Wight, W. and Barua, D. N. 1955. The nature of dormancy in the tea plant. *J. Expt. Bot.*, **6:** 1 - 5.

Willson, K. C. 1999. *Coffee, Cocoa and Tea*. CABI Publishing, Wallingford, UK, p. 300.

Wood, D. J. and Barua, P. K. 1958. Species hybrids of tea. *Nature*, **181:** 1674-1675.

Wu, C. T. 1964. Studies on heredity variation and morphology of pubescence on young tea shoots. *J. Agric. Assn. China*, **47:** 1-22.

Yan, X. 1990. Morphology and structure of tea and its quality inspection. Agricultural Publishing House, Beijing, People's Republic of China, pp. 90-100.

Zhan, Z., Ke, Nanjin and Chen, B. 1987. The cytology of tea clonal cultivar Fujian Shufxian and their infertile mechanism. *Proc. Int. Tea Quality- Human Health Symp.*, China, pp. 46.

Zhijian, Gu, Lifang, Xia and Lishan, X. 1988. Report on the chromosome number of some species of *Camellia* in China. *Acta Botanica Yunnanica*, **10(3):** 291-296.

Chapter 3

Plant improvement

3.1 Introduction

The tea plant is predominantly cross pollinator and highly heterogenous. It was observed that tea sets better with pollen from another bush, the average set of the plant with its own pollen being about one quarter of that obtained by cross pollination (Wight and Barua, 1939). Further, selfing results in smaller seeds with reduced germinability or no seeds at all (Mamedov, 1961; Sebastiampillai, 1963). Breeding in the strict sense and evolution of pure line races in the polymorphic species *Camellia sinensis* are time consuming and beset with practical difficulties. Fortunately, a large variation in several important and desirable characters occurs from bush to bush in the existing tea populations. Exploitation of such plant variability by careful selection for high yield, excellence of cup characters of made tea, fair resistance to drought and some important pests and diseases and adaptations for different environment appear to offer a ready means of developing improved planting material in this perennial crop. Other objectives of breeding vary according to the requirements of different regions. The achievements of all desirable objectives simultaneously will not be practicable but some of these can be realised at a time through the selection of novel plants and their multiplication as vegetative clones or by breeding improved strains of seeds.

3.2 Selection procedure

Selection of mother bush from the existing population is most important for production of improved seed and clonal varieties or hybrids. The fact that a tea bush in a field appears vigorous and apparently healthy does not mean much because it may produce huge crop of inferior tea, or a less amount of crop of superior tea. The bush population even in a small field of tea shows great variation in regard to growth habit, inherent quality and yield. Wight (1939) observed that about 10 percent of the bushes in a commercial plantation in north eastern region produce only 2 percent of the total crop and about 0.5 percent bushes produce as much as or more than 300 percent of the yield of an average bush. It was seen in Sri Lanka that the pluckable shoots on about 1500 randomly selected bushes in a field varied from 20 to 100 in number and only about 5 percent of the bushes gave counts above the mean (Bezbaruah, 1975). In another study at Tocklai, India it was observed that crop yield of individual bushes in a field may vary upto 500 percent between the lowest and highest yielding bushes, and that approximately 67 percent of these variations may be due to environment and about 33 percent was genetic (Bezbaruah, 1975). Selection of bushes, therefore, is really a hard task. Wight (1956) stated that the desirable attributes for high crop yield and perhaps quality may not likely to occur in more than one out of every 40000 or even more bushes. Further, tea commercially accepted if produced from the principal tea-producing taxa, *Camellia sinensis* (L.) O. Kuntze (China type), *C. assamica* (Masters) Wight (Assam type) and *C. assamica* ssp. *lasiocalyx* (Planch. ex Watt) Wight (Cambod type) and their allies thereof (Wight and Barua, 1957). But tea produced from *C. irrawadiensis* Barua and *C. taliensis* (W. W. Smith) Melchoir, although looks like tea, not acceptable commercially as tea. Therefore, selection process has had to be restricted within those plants which ultimately produce tea acceptable for consumption (Banerjee, 1988). Although yield comparison of the mother bushes is the recommended procedure of screening out the high yielders for final selection of clones but it was observed that all the high yielding mother bushes may not be genetically high yielders and significant correlation may not exist between the individual yield of mother bushes and that of their vegetative offsprings (Visser, 1969).

3.3 Selection criteria for yield

The protracted nature of breeding programmes makes it essential to use the correct selection criteria. The aim of the tea breeding programme is to develop high yielding tea per unit area of bush surface with acceptable quality under different agroclimatic conditions. Clones have been selected from tea fields in production, from seedling nurseries and from special plantings of selected seeds. Ideally, selection should be possible at the ortet or nursery stage. Studies of morphological and physiological characteristics of clones, in relation to the components of yield should aid selection efficiency (Squire, 1985) and so reduce the proportion of poor clones involved in the clone trial. In Assam, where high irradiance and leaf temperatures can occur, a more vertical leaf pose to reduce overheating has been shown to lead to higher yields (Barua, 1961; Hadfield, 1968). Other parameters apparently correlated with high yield include the thickness of palisade tissues (Pochet and Flemal, 1982), and in Malawi, leaf phosphorus content (Nyirenda, 1991). Othieno (1979), working in Kenya, found that a high shoot to root ratio correlated with high yields. By contrast, deep rooting clones had a high advantage in Malawi, where there is severe annual dry season (Nyirenda and Grice, 1990), and in Sri Lanka (Nagarajah and Ratnasurya, 1981).

3.3.1 *Visual selection*

Mother bushes could be selected initially on the basis of visual examination of their exomorphic characters as well as certain biometric parameters (Venkataramani, 1970; Satyanarayana and Sharma, 1981). Leaf characteristics viz. Size, shape, pose, surface, colour etc are used for selection of mother bushes for vegetatively propagated clones or as progenitors for breeding better seed plants with characters for high yield (Banerjee, 1992). Subsequently the number is reduced after recording the crop yield and quality assessment of individual bushes. It was estimated that only about 0.002 percent bushes out of the initial selection come out successful as superior clone (Bezbaruah, 1975).

Since size of leaf is related with weight of shoot and cost of plucking the same has been used as seletion criterion. In USSR, positive correlation between leaf size and yield was observed (Mamedov, 1961). Toyao (1966) also found good correlation between leaf size, leaf number and yield in Japan. These observations were made in areas where the population consists mainly of China and China-hybrid types of tea. No such correlation was observed in other tea growing areas (Visser, 1969; Bezbaruah, 1968). Barua (1989) advocated use of leaf size as a criteria for yield only in sections of small-leaf tea. Visser (1969) opined that total leaf area rather than bush area (Nyirendra and Ridpath, 1984) is a direct determinant of yield. But

Murty and Sharma (1988) observed that leaf area alone may not be a helpful indicator in predicting yield because light interception depending on leaf posture and its conversion at different profiles influence productivity. However, they observed positive correlation between surface area, leaf area, accumulation of dry matter in foliage and wood in the top 20 cm and yield. Although a casual relationship between yield and leaf area index (LAI) has been reported but it is difficult to predict in getting maximum yield when LAI is optimum. However, Banerjee (1992) opined LAI rather than total leaf area possibly a better selection criterion.

Visual selection for vigour can be judged by diameter of the bush, pruning weight, recovery from pruning, wood formation, root weight, branching habit, flushing behaviour of the bush etc. Willson (1999) opined that the potential of individuals may be indicated by the vigour of their growth, which can be assessed by the number of flushing stems in a given surface area multiplied by the area of the top of the bush. However, he further observed that the spread of bush can be affected by local soil variations and the vigour of adjoining bushes. In selecting mother bushes visual selection has been used in different countries like Sri Lanka (Visser, 1969), north-east India (Barua, 1963a), south India (Sharma, 1982), Indonesia (Arifin, 1978) and east Africa (Nyirenda and Ridpath, 1984; Njuguna, 1990).

3.3.2 *Bush size and stem numbers*

The size of bush with well established frames has a definite bearing on selection. Large well filled bushes are advantageous provided that the number of stems per unit area are high and their distribution over the entire area of bush should be as uniform as possible. The individual stem should be as thick as possible with a minimum variation of thickness and stem should not be whippy and thin. Scientists have established significant correlations between yield and bush area, and between number of shoots and bush area in Assam (Barua and Dutta, 1971), Sri Lanka (Visser, 1969), Malawi (Nyirenda and Ridpath, 1984) and east Africa (Magambo and Waithaka, 1985). While considering shoot number, relative proportions of active and *banjhi* (dormant) shoot on the bush surface should be kept in mind. Cannell *et al.* (1977) opined that bush size is considered to be an important parameter in mixed tea stands but it is a poor indicator for yield if the same is not considered along with flushing capacity (Nyirenda and Ridpath, 1984).

3.3.3 *Bud-break on pruned stick*

The number of buds to break on pruned stick determines the number of primaries which throw out new growth. The evenness of bud-break and number of primaries are considered as useful selection critera for the yield. The main branch and the primaries developing from them after a prune should have an orthotropic growth habit (i.e. straight or errect growth) and not plagiotropic (i.e. oblique or horizontally spreading). Bushes with orthotropic growth habit and early secondary lateral branching appear to be high yielders.

3.3.4 *Evenness of flush*

All the buds on a pruned bush should come out at the same time and grow vigorously without entering *banjhi* stage. The occurrence of *banjhi* shoots and time duration for unfolding of successive leaves have also been considered as yield criteria. This apart, if the flush are evenly distributed over the plucking surface, shoots ready for plucking can be easily distinguished from those which are immature.

3.3.5 *Density of plucking point*

It is a useful criterion for quick selection of the yield potential of a mother bush. The density of plucking points on the plucking table is an estimate of the number of shoots actually harvested from a bush. Tea bushes where plucking points at higher density per unit area spread upto the periphery from the centre should be considered as selection criterion.

3.3.6 *Production of dry matter*

Despite of the variation in leaf and bush size, and density of plucking points, productivity in tea is mainly related to the production of dry matter and its partitioning to that part of the plant which makes up the commercial yield (Magambo and Cannell, 1981). Study reveals that the clone with high yield of green leaf do not always necessarily project the high content of dry matter since clonal difference exists in partitioning of dry matter (Kumar *et al.*, 1993). Therefore, this aspect is also one of the effective criteria for selection.

3.3.7 *Resistance to biotic and abiotic factors*

Tea bushes showing varying degrees of tolerance or susceptibilities to blister blight and other diseases, insect pests, drought and other environmental stresses are seen in the field. Resistance against these factors should be taken into consideration in the field at the time of selection. Although little progress in breeding for resistance to pest and diseases has been made but breeding specially for resistance to drought has not been attempted seriously (Banerjee, 1992). However, it is recognised that difference in rooting depth (Nagarajah and Ratnasurya, 1981), accumulation of proline (Raj Kumar *et al.*, 1993) and rate of photosynthesis (Ghosh Hajra and Kumar, 1999) in the tea leaves, the epicuticular wax content and its formations on the leaf surfaces (Kabir *et al.*, 1991), are important factors in drought resistance. Carr (1977a) on the basis of field measurements in Tanzania and Kenya on different cultivars of tea, proposed that xylem water potential measurements alone, or in combination with some measures of stomatal resistance, could be used to select clones for drought tolerance at an appropriate stage in the selection programme. Carr (1977b) using this technique showed, though, that the response of young clones to dry weather in Kenya could not have been predicted from similar measurements made in the heterogenous ortets. Extremely negative values of xylem water potential during drought have also been correlated with visual observations of drought susceptibility in Kenya (Othieno, 1978), Assam (Handique and Manival, 1986) and Burundi (Renard *et al.*, 1979). Drought tolerant clones are characterised by the partitioning of a greater proportion of dry matter to the root system and frame (Magambo and Cannell (1981). To overcome the problem of drought, the technique of grafting high yielding clone on to the root stock of drought hardy clone has also been perfected in south India. So far, seven graft combinations have been released to the industry in south India (Muraleedharan, 1999).

During winter, tea in Darjeeling, Japan and some other countries remains dormant and early flushing after dormancy is considered to be a useful selection criterion as the early flushers are quick growers (Toyao, 1965) and generally have the higher order of volatile flavour index (Hazarika *et al.*, 1984) which is important component of quality.

When selection for frost hardiness is to be made then late flushing characteristic would be ideal (Harada *et al.*, 1956). However, such plants may not be high yielder or of high quality value (Banerjee, 1992).

3.3.8 *Stability of yield*

Tea yield not only varies seasonally (Ghosh Hajra and Kumar, 1999) but also with genotype, altitudes, climatic and edaphic factors (Carr, 1972). The exact contribution of each factor to total variability of yield is not known but the effect of environment seems to be quite profound. Since in selection and breeding genotype-environment interactions assume greater relevance (Wickramaratne, 1981) therefore, a plant selected solely on the basis of yield in a particular environment should invariably be tested under different agroclimatic conditions. The plant selected for high yield and or quality should also have a high adaptation to the range of environments.

3.4 Selection criteria for quality

Quality is an inherent character and important criteria in selecting potential clones. Tea bushes in the field vary widely with respect to this character. Tea quality also varies geographically (Yamanishi *et al.*, 1968) and seasonally (Cloughley *et al.*, 1982). Wight and Barua (1954) sought to relate morphology with quality from the fact that a selection with pubescent (hairiness on the undersurface of young leaves and terminal bud) flush generally produces tea with better quality than one with glabrous (non-hairy) flush irrespective of leaf colour. A significant correlation between the ordered arrangement of pubescence and quality of orthodox tea was reported in Assam (Wight and Gilchrist, 1959), south India (Venkataramani and Padmanabhan, 1964) and China (Wu, 1964). The distribution of pubescence i.e. pubescence on whole leaves, pubescence near midrib and on midrib was studied in central African clones and no clear relationship was found between quality and pubescence characteristics (Anonymous, 1995). The correlation between pubescence and quality has also not been observed using CTC method of manufacture in Assam (Anonymous, 1964). It could be due to the CTC roller which does more mechanical damage to the leaf tissues and might produce undesirable aromas during cut-leaf manufacture (Wight and Gilchrist 1959) than the conventional roller of the orthodox system (Barua, 1989) where tippiness counts. In south India, it has been reported that the clone B/6/29 has rich pubescence and produces good strong, brisk and flavoury teas both by orthodox and CTC methods (Kanthamani, 1969). However, the appearance of the tea made by these two methods markedly differ—'tippy' in the orthodox method and 'wolly' in the CTC method but liquor characters of the teas were good. Kanthamani (1969) further investigated that the leaf hairs contained large number of polyphenols and amino acids. A good number of organic acids present in the leaf portion were also found in the leaf hairs. Caffeine was also found to be present to an appreciable extent. Besides this, a large number of carbonyl compounds (aldehydes and ketones) which are important for the development of aroma was detected in the hairs and very few of these compounds were traced in the non-hairy portion of the leaf (Kanthamani, 1969). In case of Sri Lankan teas, Visser (1969) opined that a combination of slight pigmentation (by anthocyanin) with pubescence influences quality.

Leaf colour was shown to be correlated with quality of processed tea (Wellensiek, 1947; Wight *et al.*, 1963; Venkataramani and Padmanabhan, 1964). The poor quality of tea is usually associated with dark-green and pale-green leaves. Light coloured leaves generally produce better teas than the dark coloured ones (Venkataramani, 1970). However, whether leaf colour is solely responsible to influence quality is still a matter of debate.

The frequency of calcium oxalate crystals in the parenchymatous cells of the leaf petiole designated phloem index (Wight, 1961) is a useful parameter indicating quality. Plants with low phloem index are usually of lower quality value in comparison with higher phloem index. But its use as a selection criterion is restricted since large variations due to sampling and environmental factors are involved in the determination of phloem index (Visser, 1969). Levels of theaflavin, thearubigin, caffeine, aluminium and volatile flavoury compounds (VFC) are the chemical parameters for selecting quality of black tea (Owuor and Obanda, 1998). Aroma is also an important tea quality parameter.

Chloroform test (Sanderson, 1963) for fementation ability of the bushes is also a useful indicator for quality and it is recommended in many countries like Japan, Sri Lanka, India, Central Africa and Kenya.

3.5 Selection for seed varieties

The general approach in the selection process are evaluation of progenies raised from seeds produced by crossing or hybridizing between pairs of morphologically different bushes for vigour and uniformity. The selected parents are then grown to produce hybrid seeds by natural transfer of pollens. The highly heterogeneous F1 seeds produced are evaluated for vigour and more vigorous progenies are reevaluated in different stages of growth and productivity (Wight, 1956). The parameters considered for evaluation are diameter of stem, leaf size, recovery rate, branch thickness, yield, cup quality, resistance to pests and diseases, and environmental stresses

(Visser, 1969). The selection and breeding for seed varieties was a time consuming and laborious process as it often took about 25 years to develop improved varieties (Bezbaruah, 1981). A modified method involving selection of parents with decidedly better characteristics for yield, and their cultivation in isolated seed baries to produce seeds under natural condition reduce the time to about 14 years (Bezbaruah, 1981).

As the seed plants are produced from the crosses between different parents, it introduces vigour in the hybrids and helps to broaden the genetic base as gene from plants with different genetic linkage could be brought together. In comparison with clone the seed plants due to their hybrid vigour and good root system are in general more adoptable to divergent ecological and growing conditions.

3.6 Vegetative propagation and development of clones

In seed-grown tea, existence of wide variation are well known to the tea breeders and this probably made them aware of the advantages of clonal propagation for establishing large and uniform population. Clones are plants genetically identical to the parent originating by vegetative propagation such as by leaf cuttings or grafts taken from tea bush. Possibilities of propagation by single leaf and internode cuttings was explored in India (Tunstall, 1931), Sri Lanka (Tubbs, 1932) and Indonesia (Wellensick, 1938a, 1938b). The technique streamlined subsequently, assisted in the rapid multiplication of selected mother bushes leading to the development of clones (Kehl, 1963; Venkataramani, 1970; Wight, 1960). This encouraged tea breeders to develop clonal varieties. However, it took another 20 years to standardise the technique of vegetative propagation and develop selection procedure for clones. Clonal selection also involves considerable time and is a laborious process but widely adopted method of plant improvement in tea. It usually takes 7-10 years before a clone could be released for commercial planting (Barua, 1963b; Bezbaruah, 1968). The initial selections were done on the basis of colour and size of shoot, pubescence, branching habit, bush area, recovery rate following pruning etc. Visual appraisal of bush and shoot traits was followed by screening the selection for yield and quality. Therefore, clonal selection is basically exploitation of the variability in the existing population of tea by considering the best out of them and afterwards rapid propagation of the selected plants. Plant improvement can also be made by recombination between clones to create new and improved strains, which in turn, may be used either as a seed source, or developed into superior clones by taking advantage of the newly created variability (Bezbaruah, 1968).

In clones, rooting character to a great extent is inherent and three types of root systems viz. shallow or horizontal, vertical and intermediate are commonly known. In the shallow and horizontal root system, all roots develop at the surface and run almost parallel to the ground which is obviously undesirable in a perennial rain-fed crop like tea. In the vertical system, one of the roots tend to go down almost vertically, gradually thickening and functioning as tap root. This type is rarely found. The intermediate root system is most commonly come across and in this type atleast one of the root grows at an angle of 30° to 45° from the horizontal and penetrates deep into the soil (Venkataramani, 1970). Further, the contents of carbohydrates and polyphenols exihibit positive correlation with the rooting performance of clones (Satyanarayana, 1990).

Box 3.1 : Characteristics of tea clone and seed varieties

	Clone	*Seed*
1.	*Needs special technique in the nursery for growing*	*Can grow easily and no special technique is required*
2.	*pH and organic matter content of the soil affect rooting of the cuttings and overall growth of the plant*	*No reflection of soil characteristics in germination of seeds*
3.	*Almost genetically uniform*	*High genetic variability*
4.	*Mostly fibrous root and have more number of primary, subsidiary and feeder roots than seedling plants*	*Presence of strong tap root*
5.	*Yield fluctuations less pronounced*	*Yield fluctuations are marked*
6.	*Mostly specific to an environment*	*More adoptable to divergent ecological and growing conditions.*

3.7 Hybridization

In tea, due to self-sterility, pure lines breeding by conventional techniques is not possible. Tea can be genetically upgraded through hybridization. This can be done by transferring desirable characters from the parents by hybridizing them to develop seed cultivars. The seed obtained from crosses between two clones is known as biclonal seed which show considerable variation in genotypic characters while retaining a certain amount of phenotypic uniformity. Such inherent variation is important for the development of clones, which in turn could be used in evolving seed stocks. Therefore, a clone-seed cycle is important for continued plant improvement and its appropriate commercial exploitation (Banerjee, 1992; Satyanarayana and Sharma, 1993). After evaluation of germinability and growth performance of biclonal progenies in the nursery on the basis of height, stem diameter, root and shoot ratio; the field performance of the progenies are evaluated in the same manner as that described under clonal selection. Hybridization between genetically diverse parents introduces improved hybrid vigour which often reflected higher yield in the progenies. Hybridization also helps in broadening the genetic base as genes from plants with different genetic linkages could be brought together. However, only a limited success has been achieved in developing cultivars through sexual gene transfer because of certain limitations such as large size of the plant and long-duration, generative cycles ranging from 5 to 10 years (Subramaniam, 1995).

Production of synthetic seed varieties was one of the early breeding efforts at Tocklai, Assam. A group of selected clones were planted in isolated seed orchards to produce the seed under natural conditions. Such seed are commonly known as polyclonal seed. These seeds were found to be of unpredictable performance and more unstable than biclonal seeds (Singh, 1982). Moreover, it did not allow much recombination of genes to broaden the genetic base of the cultivars as more or less the same progenitors were used in various combinations (Banerjee, 1992).

3.8 Interspecific hybridization

In tea natural interspecific hybridization between different *Camellia* species are not common and limited attempts have been made for artificial induction of interspecific hybrids. Wight and Barua (1957) attempted interspecific hybridization between Wilson's *Camellia* (*C. irrawadiensis*) and *C. sinensis* var. *assamica*. Experimental F1 hybrids resembled the later in many morphological characters but were inferior in quality which was not accepted commercially. The strong vigour observed in the hybrids indicated the possibility of their utilization in further breeding (Bezbaruah, 1971) but their use as a source of genetic material was not carried out because the produce was not of an acceptable quality (Banerjee, 1993). A

> **Box 3.2 : Self and cross-compatibility variations in tea**
>
> *Tea (C. sinensis L.) is generally a cross pollinating crop. Individuals even within a variety vary in degree of self and cross-compatibility. Irrespective of varieties, some clones were self-incompatible, while as high as 10 percent fruit-set was recorded from selfing in certain clones. In general, self-incompatibility was high in the China variety and low in the Cambod variety, the Assam variety occupying the intermediate position. Some of the inbreed progenies found to have higher percentage of success from selfing than their parents. Cross-compatibility was found to be common both in intra and inter-varietal crosses, although certain combinations were seen completely incompatible. Tea was found to be compatible with number of related species and this would be considered as an advantage for introduction of desirable genomes from other sources for breeding and improvement of the tea plant.*
>
> *Source: Bezbaruah and Saikia (1977)*

successful hybridization between *C. sinensis* and *C. japonica* has been reported by Bezbaruah and Gogoi (1972). F1 hybrids showed close morphological similarity with parents but yield and quality were low. Generally, hybridization with species of *Camellia* from which acceptable quality of tea can be manufactured have been encouraging. But there are exceptions. For example, a high yielding clone TV 24 in Assam produced from the crossing between F1 hybrids from *C. irrawadiensis* x *C. assamica* Shen and TV 2, an Assam-China hybrid (Bezbaruah, 1987).

3.9 Non-conventional breeding techniques

There may be scope for introducing non-conventional approaches to tea improvement. But the potential of non-conventional methods is limited by lack of knowledge of some aspects of tea genetics.

3.9.1 *Mutation*

Attempts at inducing mutation in tea seeds, cuttings, pollen grains with physical (X-rays and G-rays) and chemical mutagens (ethyl methane sulphonate) have not yielded desired success to produce superior mutants. Mutants produced by these treatments have stunted growth, reduced vigour, and a lesser number of foliage and branches (Singh, 1982). The use of irradiated pollen also caused fruit drop (Satyanarayana and Sharma, 1993). The difference of responses to G-rays was also observed in tea clones. Cambod origin are comparatively less tolerant to G-rays than China and Assam origin (Singh, 1980). The apparent genetic variation in the response of mutagens suggests that the technique has potentiality in tea breeding.

3.9.2 *Polyploidy*

Most of the cultivated tea all over the world is diploid (2n = 30). Natural triploids (2n = 45) resulting from open pollination are reported from Japan (Simura and Inabe, 1953), north-east India (Anonymous, 1970), south India (Venkataramani and Sharma, 1975), Georgia (Kapanadze and Eliseev, 1975); and tetraploids (2n = 60), pentaploids (2n = 75), Aneuploids (2n ± 1 to 29) from north-east India (Bezbaruah, 1968). These are commonly found in tea populations but in low frequency (Singh, 1982). Triploids and tetraploids had bigger leaf and havier shoots but poor in cup quality than diploids (Sarmah and Bezbaruah, 1984), and pentaploids and aneuploids were week bushes and poor rooters (Bezbaruah, 1968). Density and morphology of sclereids and stomata proved to be useful indices in determining ploidy level, tetraploids shows highest density and size of sclereids and having the largest stomata. Chromatin length could also be correlated with the level of ploidy (Chaudhuri, 1979). In south India, a natural triploid (clone UPASI - 3) is found to be vigorous, high yielder with overall quality and the clone UPASI - 20 which is another triploid reported to be moderate yielder having drought tolerant capacity (Satyanarayana and Sharma, 1993). Polyploid plants (16 triploids and 2 tetraploids) have also been identified among the east African tea clones (Wachira, 1994). However, polyploids are in general vigorous and hardier than diploids but they do not always contribute high yield (Banerjee, 1992). Polyploids can be produced by treating with colchicine (Visser, 1969) and crosses between triploid and diploid clones. Polyploids with favourable characters could be cloned and used in the improvement of the genetic stock. In Sri Lanka, triploid clone HS-10A did well in very high altitudes compared to the diploid clones. Sundaram and TV29 are triploid clones gaining popularity in India, Sri Lanka and Kenya.

3.9.3 *Tissue culture*

The potential of tissue culture in various aspects of plant improvement have already been recognised and attracted the attention of scientists. Rapid multiplication of propagation materials through tissue culture, particularly in the initial selection, assumes importance in plant improvement programmes. Tissue culture offer opportunity to utilize genetic, physiological and biochemical procedures in developing ideotypes (Banerjee, 1986). It also holds promise in the development of pure lines through haploid technology, production of triploids through endosperm culture, culture of embryos of incompatible crosses and isolation of somaclonal variants (Singh, 1978). Banerjee (1992) suggested further possibilities of genetic manipulation for production of pure line of tea for inbreeding, breeding and propagation of interspecific hybrids derived from the various combination of crosses by cotyledon and embryo culture, fusion of protoplast from different species to produce hybrid synkaryons which then could be regenerated into organised plants by using techniques of cell and tissue culture. Once the plants are regenerated through *in vitro* techniques, it is necessary to establish the genetic variability using Random Amplified Polymorphic DNAs in somaclonal plants. Since they are directly linked with DNA, these markers are more stable and are not influenced by the environmental conditions and are foolproof (Raj Kumar *et al*., 1997). RADP markers linked to any of the economic characters like tea quality and yield may be identified and used in breeding programme to study the inheritance pattern and also screening of genotypes for specific purposes. Possibility of transferring low caffeine genes from other *camellia* sp. should be explored both through conventional hybridization and transgenic approaches to the modification of the biosynthetic pathway for caffeine synthesis, using antisense RNA technology to selectively switch off individual genes and block biochemical pathways. The other approach is through accelerated particle gene delivery (using 'gene gun') which has brought many recalcitrant crops within range for genetic manipulation (Iyer, 1995). Somatic cell hybridization using protoplast fusion has been tried to transfer the Darjeeling tea flavour of 'China' clones to the Assam cultivars possessing strong and brisk liquors (Banerjee, 1986).

Tissue culture studies in tea is being carried out at a number of centres in Taiwan, Japan, Sri Lanka, Thailand and the former USSR (Wu, 1976; Kato, 1986; Arulpragasam and Latiff, 1986; Koretskaya and Zaprometov, 1975). Tissue culture studies on tea varieties cultivated in India is summarised in Table 3.1.

3.10 Plant improvement in certain tea producing countries

Seeds were the only source of propagation in the early years of tea cultivation. In Assam and other parts of north-east India, the need for better planting materials of tea prompted to start unplanned tea breeding by way of collecting superior types of Assam tea from the wild tea patches in the

Table 3.1: Tissue culture studies in tea in India

Explant use	Reference
Callus	Complete plantlets (Das *et al.*, 1990; Jain *et al.*, 1990)
Two and a bud	Shoot formation (Palni *et al.*, 1991; Sood *et al.*, 1991)
Nodal segments	Direct and indirect shoot bud differentiation (Phukan and Mitra 1984; Rajasekaran and Mohon Kumar, 1992; Rajasekaran, 1996; Rajakumar and Ayyappan, 1992). Shoot multiplication and formation of rooted shoot (Banerjee and Agarwal, 1990; Jha and Sen, 1991; Palni *et al.*, 1991)
Leaf segments	Callusing and direct rooting (Palni *et al.*, 1991; Sood *et al.*, 1991)
Stem segments	Callusing (Palni *et al.*, 1991)
Root segments	Callusing, shoot bud and embryo differentiation (Palni *et al.*, 1991)
Anther	Callusing and proembryoid formation (Raina and Iyer, 1983; Palni *et al.*, 1991). Callusing (Sood *et al.*, 1991). Callusing and differentiation of shoot apices (Bhattacharya and Saha, 1992)
Cotyledon segments	Non-organogenetic white friable callus (Bano *et al.*, 1991; Palni *et al* 1991). Organogenetic callus (Palni *et al.*, 1991). Somatic embryogenesis (Bano *et al.*, 1991; Jha and Sen, 1991; Palni *et al.*, 1991; Balasubramanian *et al.*, 2000) and formation of synthetic seeds (Palni *et al.*, 1991). Direct shoot bud formation (Palni *et al.*, 1991; Sood *et al.*, 1991). Rooting (Palni *et al.*, 1991)
Immature zygotic embryo (embryo rescue)	Complete plantlets (Palni *et al.*, 1991; Sood *et al.*, 1991)

hills of Assam, Nagaland, Manipur and Burma. The seed gardens were established using these collections (Wight, 1961). The first attempt to produce a standard source of tea seed as the basis of production of cultivars was made by two German brothers Amos and Julia Stiefalhagen in 1860, soon to be followed by a wide application of mass selection mainly on the basis of leaf shape, size, texture, growth features and flushing behaviour in nursery seedlings for choosing parent plants for developing seed orchards which is known in the industry by the Assamese word 'bari'. This approach in tea breeding resulted in the development of several seed varieties that were superior in yield, quality and morphological uniformity. Such seed varieties having distinct characteristics were popularly known as jats like Dhonjan, Betjan, Tingamira and Rajghar after the location of a particular seed bari. These jats remained the primary source of planting materials for over a century for establishing tea estates in north-east India as well as all over the world (Singh, 1982). The successive mass selection however failed since it was restricted mostly to the same jat or variety, this led to lack of diverse genetic variability. Towards the end of 19th century hybridization efforts of Sir George Watt led to the development of Rajghar jat in which the high quality of light-leaf Assam jat and the vigour of a dark-leaf Manipuri jat were combined (Wight, 1962).

Tea breeding works were only limited to collection of germplasm even after the establishment of the scientific department of Indian Tea Association and its experimental station at Tocklai in 1900. The actual work in fact could not be started till 1946 due to war and other reasons. The main approaches in order to evolve superior planting materials were (a) selection of outstanding bushes for production of improved seed varieties and experimental strains (b) selection of promising bushes with high rooting ability, yield and quality for vegetative production (Singh, 1982).

The first lot of three vegetative clones (TV 1, TV 2 and TV 3) were released to the industry in 1949. Subsequently another twenty seven (TV 4 to TV 30) vegetative clones have been released from time to time for planting in plains garden (Anonymous, 1994). Clones TV1 to TV13, TV18 and TV19 were selected from seed grown populations in Tocklai and Borbhetta and the remaining clones from the progenies of various crosses made in connection with the project for the production of clonal seed stocks. Clones released are grouped under 'Standard', 'Yield' and 'Quality' categories according to their evaluation for yield and cup quality.

A polyclonal variety [TS 203 (Gourishankar)] was released to the industry for north-east India in the mid fifties for the production of seed as an interim source of hardy seed until biclonal seed stocks of superior yield and quality became available to the industry.

From the performance of the progress resulting from artificial crosses, a few pairs of clones were multiplied vegetatively and micro seed baries were planted with each pair of clones. When seed from these baries became available trials were laid out in different areas for final evaluation of progeny. Two biclonal seed stocks (TS 449 [TV1 x 19/31/14] and TS 450 [TV2 x 270/2/13]) were released for commercial cultivation in plains gardens in 1970 followed by release of three more stocks, one (TS 397 [TV1 x 19/35/2]) in 1976 and other two (TS 462 [TV1 x 124/48/8] and TS 463 [TV1 x 107/14]) in 1980 and 1884 respectively. Other released stocks for plains gardens are TS 464 [TV1 x 19/29/2] in 1984 and TS 491 [TV1 x S3A1] in 1989. TS 462 and TS 463 are high yielders whereas TS 464 is of above average yield with good drought tolerance character. TS 520 [TV19 x TV20] and TS 506 [TV1 x 19/22/4] were released in 1992 and 1994 respectively are high yielders and above average quality, suitable for CTC and orthodox manufacture. TS 589 [TV20 x Heeleakah 22/14] also released in 1995 for plains garden has uniform growth habit and above average yield and quality.

Recently, a survey covering 39234 hectares from 170 tea estates located in north-east India has been made by Tea Research Association. It was reported 53 % of the area under clone TV1, TV22, TV23, TV25 and TV26 and seed stocks of TS397, TS463, TS464, TS506 and TS520. The yield of clone, biclonal seed stocks and seed jats are 2500 kg/ha, 1800 kg/ha and 1400 kg/ha respectively.

Since 1967, one hundred and forty one clones and seventeen biclonal seed stocks were evaluated in twentyone trials in Clonal Proving Station (CPS), Ging tea estate, Darjeeling which was opened in 1967 by Tea Research Association. Out of them twentyeight clones and three biclonal seed stocks were released for planting in Darjeeling. The first biclonal seed variety TS 378 [14/5/35 x 14/6/28] (Nanda Devi) bred for Darjeeling was released in 1968 from Tocklai, Assam. Then TS 379 [14/5/35 x 14/12/16] was released in 1989 from CPS. These are Chinary seed stock having the characters of Darjeeling flavour and higher yield potential than the Chinary bushes cultivated in Darjeeling. During 1995 two other seed stocks (TS 557 [AV2 x Teen Ali 17/1/54] and TS 569 [AV2 x Tukdah78]) were also released for Darjeeling hills. All of them have been evaluated at the Ging tea estate belonging to the then Darjeeling Company Limited and presently Darjeeling Plantation Industry Limited.

In the late 1950s initiation of clonal selection marked the beginning of the tea plant improvement in south India. Over 350 selections with desirable attributes were developed and their performance in the field was evaluated. Twentyfour clones were released in between 1969 and 1973 for commercial cultivation (Venkataramani and Sharma, 1975). Simultaneous attempts by the tea estates of south India to develop superior plant material resulted in the development of over 60 clones. Further, clones developed in other regions (eg. TRI-2024 and TRI-2025 from Sri Lanka) were also introduced and evaluated with a view to widening the choice of clones for commercial cultivation. A few of these clones gained wider acceptance in the industry, while the others remained useful for specific situations (Satyanarayana and Sharma, 1993). Wide field evaluation of the available clonal material continued with the simultaneous resumption of clonal selection in the year 1974. The priority in the resumed clonal selection programme was selection of drought tolerant in addition to yield and quality (Sharma and Satyanarayana, 1987). Clonal selection between 1974 and 1982 has led to the identification of 627 promising mother bushes; 377 vegetative progenies belonging 310 of these selections are under different stages of field testing. Based on the data generated, 10 selections were identified as promising and pilot commercial screening has been done. Three clones from these selections ultimately released for commercial planting in 1988 (Sharma and Satyanarayana, 1988). With this, the number of released clones has gone upto twentyeight. Most of them are of above average quality, good rooters and exihibit ortho or semi-orthotropic or plagiotropic growth. A few clones have been selected by some estates.

Development of biclonal seed stock was initiated in the late 1970s and early 1980s. Over the years progenies of biclonal seed stocks of sixtythree clonal combinations have been planted in the field to evaluate their performance. Fourteen selections have given higher yields than those of the six released clones planted as standard check. Among these, five biclonal seed stocks viz. BSS-1 [UPASI-10 x TRI-2025], BSS-2 [UPASI-2 x TRI-2025], BSS-3 [UPASI-9 x TRI-2025], BSS-4 [UPASI-15 x TRI-2025] and BSS-5 [Cr-6017 x UPASI-8] have been released by the UPASI Scientific Department in 1993 (Satyanarayana and Sharma, 1994). These seed stocks are tolerant to drought with moderate to high yield and above average quality.

In Sri Lanka, no concentrated effort till late thirties was made for breeding and improvement of the varieties and most of the seeds for extension and replanting were obtained from haphazardly raised seed bearers (Elliott and Whitehead, 1926). Mass selection was restricted to choosing the outstanding mother bushes which could be propagated vegetatively to produce high yielding uniform stand (Visser and Kehl, 1958). Unlike in Assam, there was not so much emphasis for selection of seed bearers in Sri Lanka (Visser, 1969). In the sixties tea breeding programme were undertaken with two fold objectives: (a) selection of potential combiner clones and their utilization for production of vigorous and uniform hybrid seed varieties and (b) selection of golden bushes for vegetative propagation (Richards, 1966). Mutation breeding also been tried in Sri Lanka.

Bangladesh Tea Research Institute developed eight clonal cultivars. These are BT1 to BT8. Two biclonal seed stocks are also developed by BTRI. These are BTS1 and BTS2. BTS1 is a cross between BT1 and TV1. TV1 is Tocklai released clone.

In Indonesia, effort was made during the early part of the century to establish seed gardens with visually selected bushes on the basis of performance in the nursery and field but the progress was not satisfactory due to lack of knowledge on grafting technique and vegetative mode of reproduction (Visser, 1969). However, in early thirties bud-grafting method was standardised and seed gardens with selected bushes were subsequently established. Later with the discovery of the suitability of vegetative propagation of tea from single-leaf cuttings from the late forties the main attention was diverted towards selection of high yielding clones with pest and disease resistance characters (Bezbaruah, 1975). In Indonesia, the standards set for the selection of mother trees were: average production at least 300 percent above from the average yield of the plantation; average of minimum 50 g processed leaves per pluck; consistent yields from one plucking round to next. In west Java, 0.2 percent plants selected from about one million trees produced three times higher from the average yield of the plantation (Purseglove, 1968). Since 1980s ten clones have been released by Research Institute for Tea and Cinchona (RITC) namely GMB 1 to GMB 10 (GMB stands for Gambung where RITC is located). GMB 1 to GMB 5 have the potential to produce 3500 kg/ha/year of made tea in their third year. GMB 6 to GMB 10 have a productivity potential of 5000 kg/ha/year, resistance to blister blight, and better quality potential (Arifin and Semangun, 1999).

In China and Japan, hybridization and systemic selection were the main breeding methods used for selection of clones for yield, quality, resistance to frost and insects (Chen, 1987; Harada, 1961). Big shoot size was another criterion but *assamica* x *sinensis* hybrids having bigger leaf than China and China hybrids were found to be less tolerant to low temperature (Tomo *et al.*, 1966). In China, there are more than 600 cultivars, of which, 76 are approved nationally, 77 are approved provincially, and 20 are the native and the precious cultivars (Yongming, 1999). However, Chen (1987) reported that there have been 52 good tea varieties approved and popularised for planting in the tea areas. In Japan, production of mutants and breeding of polyploids were also considered in tea breeding programme. At present production vegetative clones which have been selected from hybrid progenies raised by crossing different types of tea appear to be the main emphasis in Japan. In Japan, currently about 80 % of the tea areas are covered by clonal plants. The area devoted to the green tea clone, Yabukita, selected from native Japanese tea plants for making sen-cha, accounts for over 60 % of the total (Takeo, 1992).

In Taiwan, like Japan a number of good clones have been developed from F1 progenies of crosses between different races of Assam and China tea. Separate selections for different kinds of tea such as black, oolong, poachong etc was made (Wu, 1981).

The Vietnam Tea Research Institute had collected over 50 materials initially for breeding. Many good clones and seed jats such as PH1, TH3, 1A have been evolved through selection and hybridization (Tien, 1993).

In the former USSR, selection work was carried out in the beginning of 1925. Selection work was carried out under the guidance of V. N. Pokrovski and later K. E. Bakhtadze in the three steps: (a) selection of bushes and hybridization and study of generations, (b) sort study and sort testing, selective tea introduction into practice, and (C) study of quality of the raised population of sorts. The first selective sorts were raised in 1943 and they represented hybrids of Indian and Chinese and Chinese and Indian tea named as Georgian - 1 and Georgian - 2 (Todua, 1990). A frost resistant strain Georgian - 8 developed by crossing between China and Japan varieties. It can resist upto -22^{0}C even when field covered with snow (Dey, 1972). In the subsequent years a group of frost resistant sorts numbered 6, 7, 8, 10, 11 and 12 were selected. In order to increase the resistance to frost sort 11 and 12 were breeded with local selections by N. Ya Gvazava (Todua, 1990). Simultaneously, scientists Koleley-shivili and Mutovkina, using clonal selection with generative clones, had selected 'Kolkhida' which was comparable to world standards. Physical and chemical mutagens were used and a few mutants have been developed. At the beginning of the 1980s Kerkadze from Georgia had bred mutants which exceeded Kolkhida in yield.

In Africa, a small isolated plot of pure Assam type, called Betjan, had been planted in Mulanje Tea Research Station, Malawi in 1953. By 1960, these trees were well grown and bearing flowers. Dr. R. T. Ellis selected vigorous tea bushes from the original stands of seedling tea planted in Mulanje and plants were tested for fermentation ability. Initially, over 2000 bushes were selected and from these the SFS (Swazi Field Selections) clones, MFS (Mimosa Field Selections) clones, the CL (Chitakali Local) clones and the SL (Swazi Local) clones were developed. The 'local' clone being the China type while other selections tended toward being Assam type or intermediate between the two (Ridpath and Nyirenda, 1982). The progeny derived from these clones grown at Nsuwadzi estate were evaluated for high yield with some quality. From 1968 to 1975 nine clones (SFS number) selected from the field bushes were released and recommended for commercial use. Of these only SFS150 and SFS204 became standard clones and now planted extensively throughout southern Africa, although others are used to a lesser extent. Since the mid 1970s the only other selections used in the breeding programme have been of China hybrid clones (SL numbers) with average flavour (Ellis and Nyirenda, 1995). Further, the cross pollinations are made on parent bushes and allowed to grow into trees for production of seeds. Initially parents used were of the Assam type (e.g. U6 x S17). Although the progenies were of good tea making potential but the establishment was not satisfactory and had tendency to grow few large shoots despite low yield. Crosses were also made between Assam type and China type to achieve the intermediate type of progeny. By 1988 some 160 improved clones (labelled PC clones) have been identified. Early in the breeding programme PC1 and PC81 were released before there was a whole series of successful materials between PC105 and PC119 — all of this series emanating from the same cross (PCI x SFS204) and were all seedlings (Whittle, 1997). Most of the recent selected progeny clones (PC) have been derived from open pollinations of two high quality and reasonably high yielding clones (SFS 204 and PC1). Many of the progenies (eg. PC105) have been found to be higher yielder than their parents as well as excellent cup quality. Recently SFS150 have been crossed with CL12 which is a high quality and flavoury clone (Ridpath and Nyirenda, 1982). In the selection programme, good rooting ability of the cuttings (>80%) and large heavy shoots are considered important in the nursery. Other attributes considered in the field are adaptability to local environment including drought resistance for dry areas, resistance to pests and diseases, off season production etc.

In Kenya, selection of mother bushes was initiated in 1955 by the Tea Research Institute of East Africa (Goodchild, 1956), the fore-runner of Tea Research Foundation of Kenya. There was not much progress on V.P. and clonal selection was made in between 1955 to 1960 but formal trials were launched in 1961(Green, 1962). Many mature seedlings in the fields of the neighbouring estates and young seedlings in the nursery at Timbilil estate were selected and developed into clones (Green, 1964). A clonal selection scheme was initiated from young seedling progenies of clone 6/8 (Njuguna, 1990). The characters used to select mother seedlings were vigour, leaf colour and pose, fermentability as determined by the chloroform test. The selected seedlings were then tested in four stages (rooting trials and clonal field trials etc). After two pruning cycles of the last testing stage (2nd clonal field trial), seven clones with high yield potential and better quality than parent have been released to the tea industry in Kenya (Njuguna, 1990).

Although the prospect of production of high yielding plants during last two decades have improved but it did not help in widening the genetic variability of tea as attempts were mostly restricted to selecting elite mother bushes or progenitors from within the natural hybrid populations of tea (Banerjee, 1992). Use of indigenous tea for developing clonal and seed varieties was sought as a partial solution (Bezbaruah, 1974; Satyanarayana and Sharma, 1986).

3.11 Conservation and utilization of genetic resources

The collection of tea germplasm in India in fact began with the discovery of Assam tea in 1823. The collections were also made from Burma, Cambodia, China, Japan and Vietnam. At present over 1200 genetic stocks of tea including its wild species and weedy relatives are maintained at Tocklai Experimental Station (Bezbaruah and Dutta, 1977). This collection has undoubtedly helped India as well as other tea growing countries in evolving superior plant materials (Singh, 1979). The same have also been extensively used in hybridization (crossing) programmes for the development of clonal varieties (Bezbaruah, 1974) and its exploitation thus helped in broadening the genetic base of certain clones. In south India, a majority of the selections available in the germplasm at present have been selected from some of the commercial tea estates of that region (Satyanarayana and Sharma, 1986). East Africa's tea germplasm is predominantly of the Assam type (*C. sinensis var. assamica*), and although appreciably diverse, the variability is not sufficiently broad since many of the clones are genealogically related (Wachira *et al.*, 1995). The preservation of tea germplasm has great importance primarily because of the seed-grown sections of tea are being massively uprooted and the seed sources of which no longer exist and lost for ever. There is a vast genetic variability of plantations of Darjeeling hills as still over 95% of the tea area is seed originated and old. The wide variability in the commercial tea populations offers scope for the selection of elite mother bushes with desirable attributes in practical plant improvement programme (Richards, 1966; Bezbaruah, 1975). Therefore, it needs to be preserved and utilized judiciously. Tea areas although at a slow rate are replanted with limited vegetative clones and jungles are also being cleared for extension of agricultural activities resulted narrowing the genetic base of Indian tea. Conservation of tea germplasm is done mainly by establishing field gene banks comprising a wide array of indigenous and exotic cultivars will offer the necessary genetic diversity for the improvement of tea plant at a future date.

References

Anonymous, 1964. *Annual Report 1964*. Tea Research Association. Tocklai Experimental Station. Jorhat, Assam, India, pp.36.

Anonymous, 1970. *Annual Report 1969-1970*. Tea Research Association. Tocklai Experimental Station. Jorhat, Assam, India, 40-41.

Anonyrnous, 1994. Tocklai vegetative clones. Tea Research Association. Calcutta, India, 1-60.

Anonymous, 1995. *Annual Report 1994-1995*. Tea Research Foundation (Central Africa). Momosa, 89-90.

Arfin, S. 1978. Preliminary report of mother bush selection at Kayh Aro Estate. Warta BPTKH (Indonesia). *Proc. 2nd Symposium*, Prapat, pp. 265-272.

Arulpragasam, P. V. and Latiff, R. 1986. Studies on the tissue culture of tea (*Camellia sinensis*) (L.) O. Kuntze. *Tea Q.* **55:** 44-47.

Balasubramanian, S., Marimuthu, S., Rajkumar, R. and Haridas, V. 2000. Somatic embryogenesis and multiple shoot induction in *Camellia sinensis* (L). O. Kuntze. *J. Plantation Crops*, **28(1):** 44–49.

Banerjee, B. 1986. Biotechnology - a panacea for tea. *In : Plantation Crops* Vol II (eds. H. C. Srivastava, B. Vatsya and K. K. G. Menon), Oxford & IBH Publishing Co., New Delhi, India, pp. 57-61.

Banerjee, B. 1988. Cup that cheers. *Nature*, **332:** 580.

Banerjee, B. 1992. Selection and breeding of tea. *In : Tea: Cultivation to Consumption* (eds. K. C. Willson and M. N. Clifford), Chapman & Hall, London, 53-86.

Banerjee, B. 1993. *Tea Production and Processing*. Oxford & IBH Publishing Co., New Delhi, India, p. 330.

Banerjee, M. and Agarwal, B. 1990. In vitro rooting of tea, *Camellia sinensis* (L.) O. Kuntze. *Ind. J. Exp. Biol.* **28:** 936-939.

Bano, Z., Rajarathnam, S and Mohanty, B. D. 1991. Somatic embryogenesis in cotyledon cultures of tea. (*Thea sinensis,* L.). *J. Hort. Sci.* **60(1):** 465-470.

Barua, D. N. 1961. Shoot production in cultivated tea (*Camellia sinensis* L.). II. The branch system. *Phytomorphology*, **11:** 98–109.

Barua, D. N. 1963a. Selection of vegetative clone. *Tea Encyclopaedia*, **163:** 32-88.

Barua, D. N. 1963b. Characteristics of the Tocklai released clones. *Two and a Bud*, **10(3):** 26-28.

Barua, D. N. 1989. *Science and practice in tea culture.* Tea Research Association, Calcutta - Jorhat, p. 493.

Barua, D. N. and Dutta, K. N. 1971. Distribution of shoots in plucking surface of tea bush in relation to spacing. *Two and a Bud*, **18:** 8-11.

Bezbaruah, H. P. 1968.Genetic improvement of tea in north-east India — its problems and possibilities. *Indian J. Genet.*, **28A:** 126-134.

Bezbaruah, H. P. 1971. Preserve the valuable tea plants before they become extinct. *Two and a Bud*, **18(2):** 20-21.

Bezbaruah, H. P. 1974. Tea breeding — a review. *Indian J. Genet. (Suppl.)*, **34A:** 90-100.

Bezbaruah, H. P. 1975. Tea breeding — a review. *Two and a Bud*, **22(2):** 123-130.

Bezbaruah, H. P. 1981. Recent advances in tea breeding. *Proc. 29th Annual Conference.* Tocklai, Assam, India, pp. 68-72.

Bezbaruah, H. P. 1987. Use of interspecific hybrids in tea breeding. *Two and a Bud*, **34:** 1-4.

Bezbaruah, H. P. and Dutta, A. C. 1977. Tea germplasm collection at Tocklai Experimental Station. *Two and a Bud*, 24(2): 22 - 30.

Bezbaruah, H. P. and Gogoi, S. C. 1972. An interspecific hybrid between tea (*C. sinensis* L.) and *C. japonica* L. *Proc. Indian Academy of Sciences*, 76, 219-220.

Bezbaruah, H. P. and Saikia, L. R. 1977. Variation in self and cross compatibility in tea (*C. sinensis* L.). Summary of forty years of pollination results. *Two and a Bud*, **24(1):** 21-26.

Bhattacharya, N. M. and Saha, S. K. 1992. Stimulating effects of elevated temperature treatments on production of meristemoids from pollen callus of tea, *Camellia sinensis* (L.) O. Kuntze. *Ind. J. Exp. Biol.*, **30:** 83-86.

Cannell, M. G. R., Nijuguna, C. K., Ford, E. D. and Smith, R. 1977. Variation in yield among competing individual with mixed genotype stands of tea: a selection problem. *J. Appl. Ecol.*, **14:** 969-985.

Carr, M. K. V. 1972. The climatic requirement of the tea plant. A review. *Expl. Agric.*, **8(1):** 1-14.

Carr, M. K. V. 1977a. Changes in the water status of tea clones during dry weather in Kenya. *J. Agric. Sci.*, Cambridge, **89:** 297–307.

Carr, M. K. V. 1977b. Response of seedling tea bushes and their clones to water stress. *Expl. Agric.*, **13:** 317–324.

Chaudhuri, T. C. 1979. Studies on the morphology and cytology of the progenies of triploid tea (*C. sinensis* L.). *Ph.D thesis.* Assam Agricultural University, Jorhat, pp. 176.

Chen, Z. 1987. A perspective review on the tea science. *Proc. Intl. Tea - Quality - Human Health Symp.* 4-9 November, 1987, Hangzhou, China, 3-6

Cloughley, T. B., Ellis, R. T., Pendlington, S. and Humphrey, P. 1982. Volatile constituents of some Central Africa black tea clones. *J. Agric. Food Chem.*, **30:** 842–845.

Das, S. C., Barman, T. S. and Singh, R. 1990. Tissue culture of tea: Plant regeneration of and establishment in the nursery. *Proc. Intl. Conf. Tea Research : Global Perspective*, 11-12 January, 1990, Calcutta, 19-22.

Dey, S. K. 1972. Tea in Russia. *Two and a Bud*, **19(2):** 73-84.

Elliott, E. C. and Whitehead F. J. 1926. *Tea planting in Ceylon*. The Times of Ceylon Co., Ceylon.

Ellis, R and Nyirenda, H. E. 1995. A successful plant improvement programme on tea (*Camellia sinensis*). *Expl. Agric.*, **31:** 307–323.

Ghosh Hajra, N. and Kumar, R. 1999. Seasonal variation in photosynthesis and productivity of young tea. *Expl. Agric.*, **35:** 71-85.

Goodchild, N. A. 1956. Selection of clonal mother bushes. *Annual Report 1955*. Tea Research Institute of East Africa, 24-25.

Green, M. J. 1962.. *Annual Report 1962*. Botany department report. Tea Research Institute of East Africa, 17-23.

Green, M. J. 1964. Investigations connected with vegetative propagation. *Annual Report 1963*. Tea Research Institute of East Africa, 29-40.

Hazarika, M., Mahanta, P. K. and Takeo, T. 1984. Studies on some volatile flavour constituents in orthodox black teas of various clones and flushes in north-east India. *J. Sci. Food and Agric.*, **35:** 1201-1207.

Hadfield, W. 1968. Leaf temperature, leaf pose and productivity of a tea bush. *Nature*, **219:** 282-284.

Handique, A. C. and Manival, L. 1986. Shoot water potential in tea II. Screening Tocklai cultivars for drought tolerance. *Two and a Bud*, **33:** 39–42.

Harada, S. 1961. Tea breeding. *Jap. J. Breed.*, **11:** 181-188.

Harada, S., Watanabe, A. and Sakai, S. 1956. Studies on the frost resistance of tea varieties. *Jap. J. Breed.*, **6:** 39-45.

Iyer, R. D. 1995. Biotech-breeding for plantation crop improvement. *J. Plantation Crops*, **23(1):** 1-18.

Jain, S. M., Das, S. C. and Barman, T. S. 1990. Root induction from regenerated shoots of tea (*Camellia sinensis* L.). *In : Biotek India 90*, New Delhi, pp. 1-12.

Jha, T. and Sen, S. K. 1991. Darjeeling tea : Improvement through *in vitro* techniques. *Proc. Intl. Symp. Tea Science*, Shizuoka, Japan, pp. 390-394.

Kabir, S. E., Ghosh Hajra, N. and Chaudhuri, T. C. 1991. Performance of certain clones under agroclimatic conditions of Darjeeling. *Tea Board India, Tech. Bull.*, **5:** 1-8.

Kanthamani, S. 1969. Tea clones and quality. *UPASI Tea Sci. Dept. Bull.*, **27:** 35-37.

Kapanadze, I. S. and Eliseev, V. A. 1975. Spontaneous and artificial polyploidy in tea (in Russian). *Bull. Acad. Sci.*, Georgian S. S. R. **77:** 173.

Kato, M. 1986. Micro-popagation through cotyledon culture in *Camellia japonica* L. and *C. sinensis* L. *Jap. J. Breed.*, **36:** 31-33.

Kehl, F. H. 1963. *Annual Report 1962. Report on the selection, propagation and testing of clones*. Tea Research Institute, Ceylon, **2:** 107-129.

Koretskaya, T. F. and Zaprometov, M. N. 1975. The tissue culture of tea plant (*Camellia sinensis*) as a model for studying conditions of phenolic compound formation (Russian). *Fiziologia Restenii*, **22:** 282-288.

Kumar, N., Rai, R., Ghosh Hajra, N. and Chaudhuri, T. C. 1993. Dry matter content of certain clones under agroclimatic conditions of Darjeeling. *J. Plantation Crops*, **21(3):** 30-37.

Magambo, M. J. S. and Cannell, M. G. R. 1981. Dry matter production and partition in relation to yield of tea. *Expl. Agric.*, **17:** 33-38.

Magambo, M. J. S. and Waithaka, K. 1985. The effect of pruning at different heights on yield, dry matter production and partitioning clonal tea (*Camellia sinensis*) in Kenya. *Expl. Agric.*, **21(1):** 67-72.

Mamedov, M. A. 1961. Tea selection in Azarbajdzan (Russian). *Agrobiologia*, **1:** 62-67.

Muraleedharan, N. 1999. Tea research in south India. *Planters' Chronicle*, **95(10):** 481–483.

Murty, R. S. R. and Sharma, V. S. 1988. Canopy characters as criteria in determining productivity of tea bush. *J. Plantation Crops*, **18:** 190-196.

Nagarajah, S. and Ratnasurya. 1981. Clonal variability in root growth and drought resistance in tea (*Camellia sinensis*). *Plant and Soil*, **60:** 153-155.

Njuguna, C. K. 1990. Clonal selection from young tea seedlings in the nursery. *Proc. Intl. Conf. R & D in Tea*. 11-12 January, 1990, Calcutta, Tea Research Association, 26–31.

Nyirenda, H. E. 1991. Use of growth measurements and foliar nutrient content as criteria for clonal selection in tea. *Expl. Agric.* **27:** 47–52.

Nyirenda, H. and Grice, W. J. 1990. TRF (CA) recommended clones. *In* : *Tea Planters Handbook*. Section 2.2: 1–5. Tea Research Foundation (Central Africa), Mulanje, Malawi,

Nyirenda, H. E. and Ridpath, V. E. T. 1984. Criteria for clonal selection in tea (*Camellia sinensis*) I. The relation between mother bushes and their clonal derivatives for bush area, shoot number and yield. *Expl. Agric.*, **20(4):** 339-343.

Othieno. C. O. 1978. Supplementary irrigation of young clonal tea in Kenya II. Internal water status. *Expl. Agric.*, **14:** 309–316.

Othieno. C. O. 1979. Shoot/root ratios and yields of clonal tea. *Tea in East Africa*, **19:** 6-7.

Owuor, P. O. and Obanda, M. 1998. The use of chemical parameters as criteria for selecting for quality in clonal black tea in Kenya : achievements, problems and prospects. Review. *Tea*, **19(1):** 49-58.

Palni, L. M. S., Sood, A., Chand, G., Sharma, M., Rao, D. V. and Jain, N. K. 1991. Tissue culture studies in tea (*Camellia sinensis*) (L.) O. Kuntze. *Proc. Intl. Symp. Tea Science*, Shizuoka, Japan, pp. 395-399.

Phukan, M. K. and Mitra, G. C. 1984. Regeneration of tea shoots from nodal explants in tissue culture. *Curr. Sci.*, **53(16):** 874-876.

Pochet, P. and Flemal, J. 1982. Relationship between leaf anatomy and yield in tea. *Café Cacao The'*, **26:** 161–182.

Purseglove, J. W. 1968. *Tropical Crops—Dicotyledons*. Longman group., Singapore, p.719.

Raina, S. K. and Iyer, R. D. 1983. Multicelled pollen embryoids and callus formation in tea anther cultures. *J. Plantation Crops (Suppl.),* **2:** 63-67

Rajasekaran, P. 1996. An *in vitro* and *ex vitro* rooting of micropopagated shoots of tea *(Camellia* spp.). *Sri Lanka J. Tea Sci.*, **64(1-2):** 12-20.

Rajasekaran, P. and Mohon Kumar, P. 1992. Rapid micropropagation of tea (*Camellia* spp.). *J. Plantation Crops (Suppl.),* **20:** 248-251.

Raj Kumar, R. and Ayyappan, P. 1992. Micropropagation of *Camellia sinensis* (L.) O. Kuntze. *J. Plantation Crops (Suppl.),* **20:** 252-256.

Raj Kumar, R., Manivel, L. and Marimuthu, S. 1993. Evaluation of clonal characteristics in mature tea. *J. Plantation Crops (Suppl.),* **21:** 241-247.

Raj Kumar, R., Marimuthu, S., Balasubramanian, S. and Muraleedharan, N. 1997. Evolving new elite tea cultivars - biotechnological approaches. *Planters' Chronicle*, **92(12):** 541-545.

Renard, C., Flemal, J. and Barampama, D. 1979. Evaluation de la resistance a lasecheresse chez le their ai Burubdi. *Café Cacao The'*, **23:** 175–181.

Richards, A. V. 1966. The breeding, propagation and selection of tea. *Tea Q.*, **37:** 154–160.

Ridpath, V. E. T. and Nyirenda, H. E. 1982. Plant improvement — what's it all about? *Quarterly Newsletter*, Tea Research Foundation (Central Africa), Malawi, **67:** 13-14.

Sanderson, G. W. 1963. On the nature of the enzyme catechol oxidase in tea. *Tea Q.*, **36:**103-111.

Sarmah, P. C. and Bezbaruah, H. P. 1984. Triploid breeding in tea. *Two and a Bud*, **31:** 55-59.

Satyanarayana, N. 1990. Certain factors influencing rooting of tea (*Camellia L.* spp.) clonal cuttings. *Ph.D. thesis*, University of Mysore, India, 133 pp.

Satyanarayana, N. and Sharma, V. S. 1981. Biometric basis for yield prediction in tea clonal selection. *Proc. PLACROSYM IV*, 3-5 December, 1981, Mysore, India, 237–243.

Satyanarayana, N. and Sharma, V. S. 1986. Tea (*Camellia* L. spp.) germplasm in south India. *In* : *Plantation crops: opportunities and constraints* (eds H. C. Srivastava, B. Vatsya and K. K. G. Menon), Oxford IBH Publishing Co., New Delhi, India, pp. 173-179.

Satyanarayana, N. and Sharma, V. S. 1993. An overview of tea plant improvement in south India. *In* : *Tea culture processing and marketing* (eds M. J. Mulky and V. S. Sharma), Oxford IBH Publishing, New Delhi, India, pp. 36-44.

Satyanarayana, N. and Sharma, V. S. 1994. UPASI biclonal seed stocks. *Newsletter*, UPASI Tea Research Institute, **3(2):** 2-4.

Sebastiampillai, A. R. 1963. *Annual Report 1962*. Report on plant breeding. *Tea Research Institute of Ceylon*, 87-90.

Sharma, V. S. 1982. Vegetative propagation in tea - a review. *Proc. PLACROSYM V*, 15–18 December, 1982, Kasaragod, India, pp. 73-87.

Sharma, V. S. and Satyanarayana, N. 1987. Tea plant material and its improvement in south India. *Planters' Chronicle*, **81:** 28-33.

Sharma, V. S. and Satyanarayana, N. 1988. Salient features of the three new UPASI tea clones. *UPASI Sci. Dept. Bull.*, **42:** 63-67.

Simura, T. and Inabe, T. 1953. Studies on polyploidy of the tea plant. *Tokaikinki National Agricultural Experiment Station*. Research Progress Report, **1:** 1-14.

Singh, I. D. 1978. Application of plant cell, tissue and organ cultures in tea research and development. *Two and a Bud*, **25:** 65-69.

Singh, I. D. 1979. Indian tea germplasm and its contribution to the world tea industry. *Two and a Bud*, **26:** 23-26.

Singh, I. D. 1980. Non-conventional approaches to the breeding of tea in north-east India. *Two and a Bud*, **27:** 3-6.

Singh, I. D. 1982. Advances in tea breeding in north-east India. *Proc. PLACROSYM V*, 15-18 December, 1982, Kasaragod, India, pp. 88-106.

Sood, A., Dwivedi, S. and Chakrabarty, D. N. 1991. *In vitro* propagation of tea (*Camellia sinensis* (L.) O. Kuntze) — a preliminary study. *J. Plantation Crops (Suppl.)*, **18:** 132–134.

Squire, G. R. 1985. Ten years of tea physiology. *Tea*, **6:** 43–48.

Subramaniam, Bala, 1995. Tea in India. Publications & Information Directorate & Wiley Eastern Ltd., New Delhi, India, p. 123.

Takeo, T. 1992. Green and semi-fermented teas. *In* : *Tea: Cultivation to Consumption* (eds.K. C. Willson and M. N. Clifford), Chapman & Hall, London, 413-457.

Tien, D. M. 1993. Tea industry in Vietnam. Proc. *Intl. Symp. Tea Sci. Human Health.* January 11–14, 1993, Calcutta, Tea Research association, India, pp.103–106.

Todua, G. V. 1990. Efficiency of Georgian selective sorts of tea within northern subtropical regions of Georgia. *Proc. Intl. Conf. R & D in Tea*. Tea Research Association, 11-12 January, 1990, Calcutta, 36-39

Tomo, N., Watanabe, A., Mitsui, M. and Masuda, K. 1966. Studies on the usefulness of Assamese varieties in breeding II. On resistance to low temperature (in Japanese). *Study of tea*, **32:** 9-17

Toyao, T. 1965. Correlation and genetical analysis of the several characteristics at the stage of individual selection of the tea plant II. On the correlation among yield, number of leaves per plant and characters of leaf in the clonal varieties for green tea (in Japanese). *Study of tea*, **31:** 1-4.

Toyao, T. 1966. Studies on Koro tea I. The inheritance of Koro type characters and the estimation of the degree of selfing of the tea plant (in Japanese). *Study of tea*, **32:** 18–22.

Tubbs, F. R. 1932. A note on vegetative propagation of tea. *Tea Q.*, **5:** 154-156.

Tunstall, A. C. 1931. A note on the propagation of tea by green shoot cuttings. *Quarterly Journal Indian Tea Assoc.*, pp. 49-51.

Venkataramani, K. S. 1970. Clonal selection method of improvement of tea planting material. *UPASI Tea Sci. Dept. Bull.*, **28:** 65-74.

Venkataramani, K. S. and Padmanabhan, T. S. 1964. A preliminary assessment of the relationship between certain leaf characteristics and cup quality. *Annual Report 1963–64*. UPASI Tea Sci. Dept., pp. 50-63.

Venkataramani, K. S. and Sharma, V. S. 1975. Notes on the UPASI tea clones approved by Tea Board and released for commercial planting. *Planters' Chronicle*, **70:** 119–121.

Visser, T. 1969. Tea *Camellia sinensis* (L.) O. Kuntze. *In* : *Outline of Perennial Crop Breeding in the Tropics* (eds. F. P. Ferwards and F. Wit), H. Veenman & Zonen, N. V., Wageningen, pp. 459-493.

Visser, T. and Kehl, F. H. 1958. Selection and vegetative propagation of tea. *Tea Q.*, **29:** 76-86.

Wachira, F. N. 1994. Triploidy in tea (*Camellia sinensis*) : Effect on yield attributes. *J. Hort. Sci.*, **69:** 53–60.

Wachira, F. N., Waugh, R., Hacklett, C. A. and Powell, W. 1995. Detection of genetic diversity in tea (*Camellia sinensis*) using RAPD markers. *Genome*, **38:** 201–210.

Wellensick, S. J. 1938a. Research on quantitative tea selection. IV. Mother tree selection (in Dutch with English Summary). *Archief. voor de thee Cultur*, **12:** 2-70.

Wellensick, S. J. 1938b. Observations with budding young tea bushes (in Dutch with English Summary). *Archief. voor de thee Cultur*, **12:** 107-120.

Wellensick, S. J. 1947. Foundation of general plant breeding. *Breeding of tea* (in Dutch). Tjeenk Willink and Zn., Haarlem 2nd edn., 305-347.

Whittle, A. H. 1997. Clonal tea in Southern Africa. *Tea & Coffee Trade Journal*, **169(12):** 62-66.

Wickramaratne, M. R. T. 1981. Genotype-environment interactions in tea (*Camellia sinensis* L.) and their implications in tea breeding and selection. *J. Agric. Sci. (Cambridge)*, **96:** 471-478.

Wight, W. 1939. *Annual Report 1939*, Indian Tea Assoc. Sci. Dept., Tocklai, Assam, India, pp. 32-35.

Wight, W. 1956. Commercial selection and breeding of tea in India. *World Crops*, **8:** 263–268.

Wight, W. 1960. Green and brown cuttings. *Two and a Bud*, **1(2):** 10.

Wight, W. 1961. Combiners for tea breeding. *Two and a Bud*, **8:** 19-21.

Wight, W. 1962. Tea classification revised. *Curr. Sci.*, **31:** 298-299.

Wight, W. and Barua, P. K. 1954. Morphological basis of quality in tea. *Nature*, **173:** 630-631.

Wight, W. and Barua, P. K. 1957. What is tea. *Nature*, **179:** 506-507.

Wight, W. and Barua, D. N. 1939. The tea plant industry: some general principles. *Tropical Agriculture (Ceylon)*, **93:** 4-13.

Wight, W. and Gilchrist, R. C. H. H. 1959. Concerning the quality and morphology of tea. *Annual Report 1959*. Tea Research Association. Tocklai Experimental Station. Jorhat, Assam, India, pp. 69-86.

Wight, W. and Gilchrist, R. C. H. H. and Wight, J. 1963. Note on the colour and quantity of tea leaf. *Empire J. Expl. Agric.*, **31:** 124-126.

Willson, K. C. 1999. *Coffee, Cocoa and Tea*. CABI Publishing, Wallingford, UK, p. 300

Wu, C. T. 1964. Studies on the heredity variation and morphology of pubescence on young tea shoots. *J. Agric. Assoc. China*, **47(1):** 22.

Wu, C. T. 1976. Studies on the tissue culture of tea plant (in Chinese). *Bull. Taiwan Tea Exp. Stn.*, **72:** 30-42.

Wu, C. T. 1981. Tea industry in Taiwan. *Industries Monthly, Taiwan*, **13(5):** 18-19.

Yamanishi, T., Kobayashi, A., Nakamura, H., Uchida, A., Mori, S., Ohsawa, K. and Sasakura, S. 1968. Flavour of black tea. Part V. Comparison of aroma of various types of black tea. *Agric. Bio. Chem.*, **32:** 379-386.

Yongming, Yu, 1999. Agrotechnology of tea in China. *In* : *Global Advances in Tea Science* (ed. N. K. Jain). Aravali Books International (P) Ltd., 481–500.

Chapter 4

Climatic requirements

4.1 Introduction

The original habitat of tea plant in south-east Asia experiences a warm, wet summer and a cool, dry (or less wet) winter. From the main centres of cultivation in south-east Asia tea has been introduced into many tropical and subtropical countries and now grown under a wide range of climates. The tea areas of today range from the most northernly, as Georgia, situated in the Trans-Caucasian province of commonwealth of independent states (former USSR) with a latitude of 43^0 N where the winter is long, severe and snow is common, the most southernly as Corrientes in Argentina with a latitude of 27^0 S. Tea is cultivated from almost sea-level to altitude of 2460 m. The Assam-type of tea plant, with relatively large, horizontally held leaves is thought to have developed in the understorey of a dense tropical rain forest south of the Himalayan mountain, whilst the China-type of tea plant, with small semi-erect leaves is thought to have indigenous in a less heavily forested environment north of the Himalayas. These give an indication of the conditions natural to the species.

Spread over such a vast land of diverse topography, the soil–climatic environments of the tea areas differ considerably. Some indication on the climatic conditions in different tea growing countries has been given in Chapter 1. Laycock (1964) described a 'good growing season' for tea as one having 'warm days, long sunshine hours, high humidity and adequate (*sic*) rainfall, preferably in overnight showers'. Carr (1972), and Carr and

Stephens (1992) reviewed the effects of climate (and weather) on the yield of tea. Now, an attempt is made to quantify the range climatic conditions under which tea is cultivated.

4.2 Air temperature

Tea is grown under a regime of air temperature that varies between –8° and 35°C. Temperature strongly influences the growth, development and yield of tea plants though the relationship between temperature and shoot growth varies according to growing conditions. It is difficult to obtain a clear-cut relationship between yield and mean daily temperature because of the synchronized flushing nature of the tea bush (Tanton, 1992). In Georgia, very low temperature is experienced during the winter and can often be below 40°C, whereas Krasnodar and the Azerbaijan regions minimum winter temperature can be as low as –8 to –12°C. The maximum summer temperature in July / August can be around 21°C. Labedev (1961) reported stoppage of growth when the mean air temperature fell below 13°C. Nakayama and Harada (1962) did not observe any growth of potted tea plants in Japan when kept at a constant temperature of 12°C. Carr (1972) reported virtual cessation of growth of tea in southern Tanzania below 14°C mean air temperature. In Darjeeling, the extension growth stops at monthly mean maximum and minimum temperatures of 19.4°C and 12.4°C respectively in November and it starts flushing during end of March when mean maximum and minimum temperatures exceed 21°C and 14°C respectively (Ghosh Hajra and Kumar, 1999). In Malawi, poor yield is often associated with the number of days with temperature below 21° or above 30°C, in some situations leaf expansion is reduced at temperatures below 14.5°C (Fordham, 1977). Monthly mean maximum and minimum temperatures in Kenya rarely exceed 24°C and 11°C respectively, at any time of the year but the tea plants flush throughout the year and produce yearly yields of the same order as plants in many warmer regions (Barua, 1989). Thus, there is agreement between observations made in different parts of the world that extension growth of the tea plant ceases below a minimum temperature of 13°C, but there is also some evidence of considerable clonal variation (7-15°C) (Carr and Stephens, 1992). In tea, there are two distinct phases to the temperature response, a base temperature effect and the linear response. A cultivar may not grow below a certain base temperature to which it is sensitive and it has been postulated that growth stops due to observed temperature induced phase changes in membrane lipids (Raison, 1974). Tanton (1982a) using controlled environment facilities in Malawi, confirmed that the base temperature for shoot extension (of clone SFS 150) was in the range 12-13°C. It is assumed that about 475 day degrees above the base temperature of 12.5°C are needed for a bud released from apical dominance to reach a harvestable three leaves and a bud size (Carr and Stephens, 1992). During the cold season only the old, slow growing shoots would have remained on the bush but once temperature started to rise a new generation of shoots would have started to develop. It has been observed that certain clones yield relatively well in hot weather than in cool, while others yield relatively well in all weather. These observations could be explained that temperature response of tea is not only location specific but it may even vary genetically between the cultivars under comparable condition of growing (Banerjee, 1993).

In Darjeeling, highest yield was achieved in June when mean maximum and minimum temperatures were 23.5°C and 18.3°C respectively (Ghosh Hajra and Kumar, 1999). The rapid decline in yield in Darjeeling during October and then stops during November until end of March indicates that low temperature is one of the major climatic variable, limiting yield. It has been reported that the higher yield in Darjeeling was achieved during the period from June to September when the differences between maximum and minimum temperatures were least in comparison with the rest of the year (Ghosh Hajra and Kumar, 1999). In the tea areas of Sylhet and Chittagong of Bangladesh, temperature reaches up to a maximum of 40°C. The extreme minimum in the Sylhet region drops in January to 1.7°C rising to 2.8°C in February. On the other hand, the extreme minimum in Chittagong zone drops to the extent of 7.8°C to 8.9°C. Hasan *et. al.* (1965) also achieved best growth of tea in Pakistan (now in Bangladesh) when the differences between maximum and minimum temperatures were least. Low temperature is also the vital factor causing low yields in the cold season in Malawi (Tanton, 1992). In China, it has been observed at Huangyan country in Zhejiang province, if the mean air temperature in May is lower than about 20°C, spring tea yield decreases, if the mean air temperature in June is lower than about 24°C, yield of summer tea is increased (Xu, 1983). Huang (1989) reported that the optimum temperature for most varieties grown in China is in the range of 20 to 30°C and the maximum temperature is 35°C (or 30°C of mean daily air temperature).

The air temperature remains above 30°C for a greater part of the harvesting period in the plains of north-east India and south India, and hence planting shade trees were found beneficial to keep the temperatures below a level where it cannot affect photosynthesis adversely.

4.3 Soil temperature

In many instances soil temperature is of greater importance to plant life than air temperature and soil temperature influences growth and yields of tea. Fordham (1970) in Malawi, confirmed how growth of shoot was positively correlated with soil temperature (at 0.3 m beneath a grass surface) up to 25°C, but the rates of shoot growth decreased at soil temperature above 25°C. It perhaps linked to large saturation deficits of the air (Carr and Stephens, 1992). In Malawi during winter months, Tanton (1982b) was unable to maintain high rates of shoot growth when soil temperatures raised from 18°C to 25°C (at 0.2 m). In the two years after planting in Kericho, Kenya, the results of Othieno (1982) showed that application of polythene mulches increased soil temperature, or reducing it with organic mulches, stone chippings, had direct effects on stem diameter, yield (higher soil temperatures resulted in more axillary buds developing into shoots) and production of total dry matter. The range of soil temperature covered was 14 to 21°C at 75 mm depth. In young tea, grass mulches although reducing run off and soil erosion, conserving moisture and improving nutrient availability, but it reflects more (0.20) incoming radiation than bare (moist) soil (0.11) leading to reductions of soil temperatures of about 2°C (Othieno, 1982). This delays the rate of development of young tea in high altitude areas. Carr and Stephens (1992) concluded that soil temperature is an important variable, with a lower limit of about 20°C, at 0.3 m under short grass surface (or 16°C beneath a canopy tea), below which shoot extension rates will be reduced. The corresponding upper levels are 29°C and 25°C respectively.

4.4 Rainfall

Tea is basically a rain-fed crop. It is grown well in areas where annual rainfall varies from 1150 to 6000 mm. Tea should not normally be grown in areas where the rainfall is below 1150 mm, unless irrigation is available. The effect of precipitation is perhaps more manifested by its influence on moisture status of the soil and in inducing vegetative growth. Therefore, distribution of rainfall over the year is as vital as the total annual rainfall.

The distribution of rain depends on latitude, and is modified by altitude and topographic factor associated with proximity to escarpments, mountains or inland water (Carr and Stephens, 1992). There are two rainy seasons which sometimes merge and two dry seasons of different severity found in the areas close to the equator. The seasons merge into one dry and one wet season at higher altitude. The north-east India receives rain from 1500 mm annually in some parts of Central Assam to more 5000 mm in the few pockets of Dooars. However, the distribution of rain is most uneven. For example, 75 % of the annual rainfall occurs during June to October. November to April is a dry period and parts of central Assam, Cachar, Dooars and Terai are more susceptible to drought than upper Assam. Annual rainfall varies from 2274 (Kalimpong sub-division) to 4082 mm (Kurseong sub-division) in Darjeeling where in May, the southerly winds reach the hills and cause increased precipitation which is at times very high. The plants were able to recover fully from the water stress with the onset of the monsoon rain in June. There is some residual effect of monsoon in November but November and December are almost rainless and the light showers which fall in January and February occur when shallow depressions are passing eastward over the plains. In October, northerly winds begin, cloud is much less than in previous months and rainfall occurs, mainly owing to cyclonic storms which generally recurve towards north Bengal at the end of the season. Tea areas in Himachal Pradesh receive rainfall of 2000 to 5000 mm annually with a distinct dry period persist in May–June. The annual rainfall in the tea growing region of south India varies between 900 and 8000 mm. The tea areas of south India are located on the western face of the Western Ghat Mountains (eg. Anamallais, Travancore etc) receives 80 % of the total rain from the south-west monsoon while the Nilgiris facing mainly east gets more than half the amount of rain from the north-east monsoon.

In Sri Lanka, the tea district of Uva receives more rain from the north-east monsoon but areas like St. Coombs, Ratnapura, Nuwara Eliya etc located on the west are fed mostly by the south-west monsoon. However, rain in Kandy area is exposed to both the monsoon current and more evenly distributed. The annual rainfall of Bangladesh ranges from 1400 to 5000 mm. The districts of Chittagong and Hill districts receive rainfall not exceeding 3000 mm. The tea growing areas of Sylhet and Chittagong fall in the wet area. In Indonesia, rainfall is evenly distributed throughout the season and varies between 2500 to 5000 mm. Malaysia being in the humid tropics, enjoys rain throughout the year with higher intensities during April/ May and September/October; the annual rainfall ranges from 2300 to 2550 mm in the lowlands and 2800 to 3300 mm in the highlands. The major tea belt of Japan receives 1800 to 2500 mm of rain annually and distribute fairly evenly. The mean annual precipitation of the tea producing region of China varies from 750 mm to 2000 mm. Annual rainfall in the tea areas of former USSR varies between 1000 to 2500 mm. Western Georgia experiences highest rainfall of about 2500 mm, whereas in the Krasnodar and Azerbaijan regions the annual rainfall could be at or below 1000 mm. Even within the Georgia, easternly or northernly situated plantations experience less rainfall than those situated in the west (Dey, 1972).

The total rainfall in most of the African tea areas is not adequate. This apart its uneven distribution throughout the year often limits annual tea yield. For example, Mufindi of Tanzania, Mulanje of Malawi and tea areas of Mozambique is much in excess of the total evaporative demand but most of it comes during the months of December to May which severely restricts tea shoot growth over this period. The mean annual rainfall in Kenya is about 2100 mm of which about 90 % falls between mid March and mid November. The months of December to March in Kenya is a relatively dry period. In Tanzania, mean annual rainfall is about 1700 mm, rainy season lasting from November to May, a cool dry season from June to August and warm dry season from September to October. In Malawi, rainfall declines with distance from mountain averaging about 2300 mm at points close to the mountains and declining to 1500 mm 3 to 5 km away (Carr and Stephens, 1992).

The pattern of distribution of crop does not always correlates with main period of rainfall and it is difficult to calculate the exact requirement of water in tea at different phases of its growth and development. However, the amount of rainfall necessary to ensure that water is available to meet the evaporative demand depends on its distribution and on temperature and other parameters in the dry season (Willson, 1999).

Squire and Callander (1981) idealized annual water balance for mature tea plantation based largely on data obtained from Africa. For total precipitation of 1700 mm, transpiration accounted for 1050 mm (E_T =

0.75 E_0, E_0 = 1400 mm) and interception for 200 mm (200 rainy days, canopy storage 1 mm); stream flow totalled 450 mm (430 mm resulted from deep drainage and 20 mm (1 % of precipitation) from runoff. In this example the annual change in soil water content was only ± 30 mm. Since annual totals of E_0 in many current tea areas lie in the range of 1300 to 1600 mm, rainfall should on average be at least 85 % of this, or 1150 to 1400 mm if irrigation is not to be essential. These values are similar to the minimum rainfall totals suggested previously by Carr (1972).

Sri Lanka and south India where south-west monsoon is heavy, excessive rain water logs the soil and slows down the growth of tea. When soil is saturated particularly at the peak of the rainy season, some rainfall is lost by runoff (for example Darjeeling) until the ground cover has reached at least 60 % (Othieno, 1975). The adverse effect of either excess or lack of adequate rainfall can be compensated by drainage or irrigation respectively.

4.5 Humidity

It has influence in determining the loss of moisture by evapotranspiration. The invisible water content of the air is expressed as relative humidity (RH), saturation vapour pressure deficit (SVPD) or dew point. Relative humidity is the moisture content of a volume of air expressed as percentage of the maximum quantity that the air can hold at the prevailing temperature. RH of 80–90 % is favourable during the growth period of tea plants, below 50 %, shoot growth is inhibited and below 40 % growth is adversely affected (Huang, 1989). High humidity reduces water loss and low humidity increases loss of water, induces leaf water stress and causes reduction of growth rate of tea during dry period. Low humidity decreases photosynthesis probably because of rapid evaporation from guard cells and causing stomata to close (Rosenberg, 1974). But it may be advantageous in relation to temperature because it reduces leaf temperature by increasing transpiration rate. It was found that cell sap concentration (CSC) of tea shoot increased by 0.6 % for each 10 % fall in RH. Tea grew rapidly at a CSC of 8 or 9 % but at a CSC above 10 % growth rate were reduced (Filippov, 1968). Due to heavy and fairly well distributed rainfall in south India, Indonesia, Sri Lanka and other countries of south-east Asia, mean RH rarely drops below 60 % at any time of the year. Despite the unbalanced distribution of rainfall in north-east India and Darjeeling (71 % of the annual rainfall occurs during June to August), even during the driest part of the year in March and April, mean RH never drops below 60 % in this region (Ghosh Hajra and Kumar, 1999). Hadfield (1968) reported that high temperatures associated with high VPD inhibit tea growth in north India. In Darjeeling, higher temperatures in June and August were associated with lower values of VPD and, since the temperature was within the optimum range, highest yield was recorded in June (Ghosh Hajra and Kumar, 1999). In the northern latitudes of the former USSR and Japan atmospheric humidity remains fairly high (Barua, 1989). An overall relative humidity in Sylhet and Chittagong tea zone of Bangladesh varies from 56 to 80% and 68 to 80%, respectively. The situation is, however, different in many parts of Africa. In Malawi, during the hottest part of the year SVPD may rise up to 45 mbar (equivalent to 4.5 kPa) (Ellis, 1971) and irrigation or misting would have to be more frequent in order to keep the leaf temperature low and humidity high during the hot and dry part of the year. Tanton (1982b) reported that VPD was the major factor decreasing the crop yields in the hot dry months in Malawi and excessive leaf temperatures under irrigation do not contribute appreciably to a decline in yield. He showed that the apparent critical saturation deficit value (the difference between the saturated vapour pressure at the given temperature and the actual vapour pressure) which when exceeded caused a reduction in relative shoot growth rates was 2.3 kPa (relative humidities of 28 % at air temperatures of 25°C or 45 % at 30°C). With a high saturation deficit, the normal relationship between shoot growth and air temperature does not hold. Although there may be sufficient water in the soil (through rain or irrigation) growth and yield will fall under such conditions (Willson, 1999).

4.6 Solar radiation

It is the source of energy which sustains organic life on the earth surface. Crop production is in fact an exploitation of solar radiation. The intensity and duration of sunshine has also an important influence on the growth of tea plant. The amount of radiation available at the surface of a tea canopy at high altitude near the equator can be up to 600 W m^{-2} (Squire and Callander, 1981). The vast majority of this short-wave radiation is intercepted by leaves in the top 0.3 m of the canopy, almost regardless of foliage geometry below 0.1 m (Hadfield, 1974), with only about 5 % reaching the ground (Carr and Stephens, 1992). Healthy plant with a flat canopy completely shading the ground absorb most of the incident radiation.

Photosynthesis for the whole plant will continue up to high levels of radiation depending upon canopy structure, leaf area and its orientation. Only a very small proportion of the net available energy is used in the process of photosynthesis; the vast remaining energy is dissipated as heat through the process of evaporation and heating to surrounding air (Carr and Stephens, 1992). Therefore, evaporative cooling can be of critical importance, but the stomatal conductance reduces at high saturation deficits, as the stomata narrow, which limits evaporative cooling (Willson, 1999). Leaf temperature will rise significantly, up to 12°C for every 100 W m^{-2} (equivalent to 0.14 cal cm^{-2} min^{-1}) of radiation (Squire and Callander, 1981).

The highest radiation intensity on a bright clear day in summer recorded at Tocklai, Assam was 1.0 cal cm^{-2} min^{-1} (equivalent to 698 W m^{2}) (Barua, 1989). In Mimosa, Malawi, Squire (1977) reported 900 W m^{-2} (equivalent to 1.29 cal cm^{-2} min^{-1}) irradiance at noon on clear days. The mean radiation intensity at the high altitudes of Southern Tanzania remains around 0.5 cal cm^{-2} min^{-1}. The hours of bright sunshine are fewer in the rainy season of Darjeeling hills, Bangladesh and other places due to the overcast clouds.

i. Shade

This has been discussed in Chapter 8 (section 8.8).

ii. Day length

There are remarkable variations in day length between 9.4 and 15 h prevailing in the tea growing areas. In the tea growing belt of north India, day length during different seasons varies from 10 to 13 h. There is evidence that winter dormancy sets in tea shoots when the day length falls below about 11 h 10 m duration and last for at least six weeks (Barua, 1969). The longer the period of short days, longer is the dormancy (Laycock, 1969). Growth of tea in different latitudes show a weak but positive relationship between day length and yield which suggested that growth ceases when days are less than 11 h (Laycock, 1969). In Malaysia, tea areas are within 6^{0} north of the equator, the difference in day length between the longest and the shortest day is approximately 35 minutes and there is no complete dormancy period in the growth of tea. In Darjeeling, highest yield was recorded during June when average day length was 13 h 49 m. In fact, 50% annual crop is produced in Darjeeling during June to August when day length obtained in between 13 h 49 m to 12 h is a factor contributing to high productivity during this period. Day length in Bangladesh is shortest in December, the average lowest being 10.6 h to average longest in June, being 13.7 h. Vegetative growth ceases during winter periods from November until the end of March when day length was falling from 11 to 10 h 28 m. Considerable increase in winter crop under experimental conditions was noticed when a low level of illumination used in Assam (Barua, 1969) or night break treatments used in Malawi (Herd and Squire, 1976) to give a day length of 13 h. These studies although suggest the importance of day length in tea yield but it is not confirmed whether light or temperature was crucial. In Malawi, Tanton (1982b) reported that day length does not affect shoot extension when the nights are cool (10°C), but growth rate is depressed by short days (11h) when nights are warm (20°C). It appears that decreasing day length and/or short days caused a small drop in growth rate but the effect was insignificant when compared with the effect of declining temperature at the end of the season (Tanton, 1981 and 1982a). The small response to day length when the nights were cool suggests that in the regions of the world where yields drop at the onset of winter, the yield depression is more likely to be the result of cold night temperatures than of day length. Day length appears to be more important where nights are warm, but there can be few tea growing regions of the world where short days and warm nights are experienced and thus from a practical point of view it is likely to be unimportant (Tanton, 1992).

4.7 Wind

Wind is an omnipresent part of the environment and has both harmful and beneficial effects on plants. It causes injury by toppling and uprooting plants, breaking stems and branches, causing stem malformation, and injuring leaves. In addition, wind erodes soil, inhibit growth by adversely affecting physiological processes, and increases risk of insect attack, disease and fire. The complex effects of wind on tea plantation are exerted by its physical force as well as by its role in turbulent transfer of heat, water vapour, CO_2, spores etc. Wind velocity could be useful in reducing high leaf temperature and it affects the rate of transpiration which would otherwise affect photosynthesis. However, transpiration is not a passive process and the plants exercises control over transpiration by closing the stomata of the leaf when loss of water becomes very rapid. The velocity of wind increases with elevation. In Darjeeling hills, the warmer air starts penetrating through the valleys from the plains during March and register high velocity in June (Ghosh Hajra and Kumar, 1999). However, very little work on the effect of wind turbulence on physiology of growth and productivity of tea has been reported. It was found in Mufindi of southern Tanzania that 3–6 years old clonal tea plants sheltered from wind by 4 m high *Hakea saligna* hedges experienced greater water stress than plants which were more exposed to the wind during extended periods of dry weather, as indicated by shoot water potential measurements (Carr, 1971), estimates of stomatal opening, measured soil water deficits, visual appearance and yield (Carr, 1985).

4.8 Altitude

The China variety in particular and the hybrids in general produce very fine, flavoury tea when they are grown above an altitude of about 1100 m as in the case of Darjeeling and in the high-grown areas of south India and Sri Lanka. Surprisingly, such flavour does not develop at high altitudes in Java of Indonesia and east Africa. This is probably because of the uniform weather conditions that prevail resulting in harvesting of crop throughout the year (Arunachalam, 1995). China hybrid bushes from Darjeeling fail to reproduce the full hill muscatel flavour if grown in the plains of Assam and Dooars. It may thus be stated that the Darjeeling teas owe their unique flavour partly to the type of bush and partly to the climate. Altitudes affect shoot growth rate and influence quality of processed tea (Owuor *et. al.*, 1990). They compared a number of quality variables for clonal tea grown within a 10 km radius on four estates located at different altitudes from 1860 to 2180 m in the Kericho district of Kenya and found that increases in altitude improved the quality of flavour (as measured by the relative changes in the concentration of Group I and II volatile flavour components (VFC)), the concentration of the theaflavins and thearubigins (with the exception of one clone), and caffeine. Tasters' evaluations for all clones also improved with increase in altitude (Carr and Stephens, 1992). An increase in the VFC of Darjeeling tea with altitudes has also been reported by Mahanta *et. al.* (1988).

4.9 Hail

It is unpredictable and of a localized nature. Hail in tea growing districts viz. Cachar, Dooars, Darjeeling, Terai of north eastern India are almost a regular phenomenon every year during spring and cause severe losses. Hailstroms occur in Darjeeling hills in one valley or other during March to

Fig. 4.1: Tea plant injured by hail stones.

May, but end of March to early April is the period when it occurs most. Other tea areas, especially in Kericho and Nandi Hills of Kenya, Bangladesh and China are prone to hail storms. Mwakha (1983) reported occurence of an average of 100 hailstorms in a year in Kericho and Nandi Hills areas and hail is potentially very damaging with annual yield losses of up to 30 % (average 10 %) (Carr and Stephens, 1992). Attempts have been made to suppress hail using explosive rockets, seeding of potential hail clouds with silver iodide in Africa but success has not been achieved.

The severity of damage ranges from shattering of leaves to peeling of bark in trunks, young stems break and bushes are defoliated (Fig. 4.1 and 4.2). It can be particularly harmful when the bushes are recovering from the effect of a severe drought and are refoliating after pruning. Despite destruction of current crop it damages stems which lead to disease and pest problems. Damaged bushes become liable to be infected by fungal diseases, wood parasites like *Nectria* sp., *Poria hypobrunnea*, *Macrophoma theicola* etc. In Darjeeling hills, loss of higher priced quality crop during first and second flush due to hail damage is actually the real damage in terms of income. The plucking table of the severely affected bushes should be raised

Fig. 4.2: Soil of the tea plantation in Darjeeling hill covered with hail stones occurred during April.

by one full leaf/fish leaf. Light prune, deep and medium skiff teas after recovery from hail damage should be tipped at the recommended measure, i.e. no additional height is to be provided.

Establishment of a pure stand of *Indigofera teysmani* at a closer spacing (3 m x 3 m) in young tea and growing wind breaks of trees at right angles to the path normally followed by hailstorms are suggested (Anonymous, 1988). Henderson (1966) reported an investigation into hail suppression.

Hail damaged areas are susceptible to mites, thrips and blister blight. All possible care should be taken to keep them free from pests and blight. All hail damaged areas should as quickly as possible be sprayed with copper fungicide mixed with spreading agent. Young tea and pruned section must have priority over other sections. Plucking should be done very liberally on these areas.

References

Anonymous, 1988. *Tea Encyclopaedia. Serial No. I4, Tocklai Experimental Station*, Jorhat, Assam, India: Tea Research Association.

Arunachalam, K. 1995. *A hand book on Indian tea*. Arunachalam Associates (Plantation consultants), Wellington Bazar, The Nilgiris, Tamil Nadu, India, p. 271.

Banerjee, B. (1993). *Tea production and processing*. Oxford & IBH Publishing Co., New Delhi, India, p. 336.

Barua, P. K. 1969. Seasonal dormancy in tea. *Nature*, **224:** 514.

Barua, D. N. (1989). *Science and practice in Tea culture*. Tea Research Association, Calcutta - Jorhat, p. 493.

Carr, M. K. V. 1971. The internal water status of the tea plant (*Camellia sinensis*): Some results illustrating the use of the pressure chamber technique. *Agricultural Meteorol.*, **9:** 447–460.

Carr, M. K. V. 1972. The climatic requirements of tea plant: A review. *Expl. Agric.*, **8:** 1–14

Carr, M. K. V. 1985. Some effects of shelter on the yield and water use of tea. *In : Effects of Shelter on the Physiology of Plants and Animals. Progress in Biometeorology* (ed. J. Grace), vol. 2, Swets and Zeitlinger B. V, Lisse, pp. 127–144.

Carr, M. K. V. and Stephens, W. 1992. Climate, weather and the yield of tea. *In : Tea: Cultivation to consumption* (eds. K. C. Willson and M. N. Clifford). Chapman & Hall, London, 87–135.

Dey, S. K. 1972. Tea in Russia. *Two and a Bud*, **19(2):** 73–84.

Ellis, R. T. 1971. The general background to plant / water relation investigations on tea in Central Africa and some findings to date. *In: Water and the tea plant. Proc. of Symp. of the Tea Research Institute of East Africa* (eds. M. K. V. Carr and Susan Carr), Kericho, Kenya, 19–57.

Filippov, L. A. 1968. A method for assessing the influence of air temperature and humidity on the water regime of the tea plant. (Engl. Transl.), *Sov. Plant Physiol.*, **15:** 668–672.

Fordham, R. 1970. Factors affecting tea yields in Malawi. Ph.D. thesis, Bristol University.

Fordham, R. 1977. Tea. *In : Ecophysiology of Tropical Crops* (eds. Paulo de T. Alvim and T. T. Kozlowsky). Academic Press, London and New York, 333–349.

Ghosh Hajra, N. and Kumar, R. 1999. Seasonal variation in photosynthesis and productivity of young tea. *Expl. Agric.*, **35:** 71–85.

Hadfield, W. 1968. Leaf temperature, leaf pose and productivity of the tea bush. *Nature*, **219:** 282–284.

Hadfield, W. 1974. Shade in north-east Indiantea plantations II. Foliar illumination and canopy characteristics. *J. Appl. Ecol.*, **8:** 179–199.

Hasan, K. A., Chaudhury, S. H. and Halim, M. A. 1965. Effect of climatic factors on the yield of tea. *Tea Journal of Pakistan*, **3:** 4–13.

Henderson, T. J. 1966. Hail suppression project utilizing anti-hail rockets. *Tea*, **7(1):** 11, 13, 15, 17, 19, 21.

Herd, E. M. and Squire, G. R. 1976. Observations on the winter dormancy of tea in Malawi. *J. Hort. Sci.*, **51:** 267–269.

Huang, Shoubo. 1989. Meteorology of the tea plant in China: A review. *Agric. Forest Meteorol.*, **47(1):** 19–30.

Labedev, G. V. 1961. The bush under irrigation. (in Russian). *Izvestia Akadamia Nauk*, SSSR, Moscow.

Laycock, D. H. 1964. An empirical correlation between weather and yearly tea yields in Malawi. *Tropical Agric.* (Trinidad), **41:** 277–291.

Laycock, D. H. 1969. Latitude, day length and crop distribution. *Proc. 24th Tocklai Conf.*, Tocklai Experimental Station, Jorhat, Assam, India, 1–9.

Mahanta, P. K., Barua, S., Owuor, P. O. and Murai, T. 1988. Flavour volatile of Assam black teas manufactured from different plucking standards and orthodox teas manufactured from different altitudes of Darjeeling. *J. Sci. Food Agric.*, **45:** 317–324.

Mwakha, E. 1983. Hail reports from Kericho and Nanditea areas (1977–1982). *Tea*, **4:** 8–10.

Nakayama, A. and Harada, S. 1962. Studies on the effect of temperature on the growth of the tea plant. Part 4. The effect of temperature on the growth of the young tea plants in summer. (in Japanese). *Bulletin of the Tea Research Station*, **1:** 28–40.

Othieno, C. O. 1975. Surface run-off and soil erosion in fields of young tea. *Tropical Agric.* (Trinidad), **52:** 299–308

Othieno, C. O. 1982. Diurnal variations in soil temperatures under tea plants. *Expl. Agric.*, **18:** 195–202.

Owuor, P. O., Obaga, Samuel O. and Othieno, Caleb O. 1990. The effects of altitude on the chemical composition of black tea. *J. Sci. Food Agric.*, **50:** 9–17.

Raison, J. K. 1974. A biochemical explanation of low-temperature stress in tropical and sub-tropical plants. In *Mechanisms of Regulation of Plant Growth* (eds. R. L. Bioleski *et al.*), *Royal Society of Newzealand Bulletin*, **12**: 487–497.

Rosenberg, N. J. 1974. *Microclimate: The biological environment*. John Wiley & Sons, N. W., London, pp. 206–22.

Squire, G. R. 1977. Seasonal changes in photosynthesis of tea (*Camellia sinensis* L.). *J. Appl. Ecol.*, **14:** 303–316.

Squire, G. R. and Callander, B. A. 1981. Tea plantations. *In: Water Deficits and Plant Growth* (ed. T. T. Kozlowski), vol. 6, Academic Press, New York, pp. 471–510.

Tanton, T. W. 1981. Growth and yield of the tea bush. *Expl. Agric.*, **17:** 323–331.

Tanton, T. W. 1982a. Environmental factors affecting the yields of tea (*Camellia sinensis*). I. Effects of air temperature. *Expl. Agric*., **18:** 47–52.

Tanton, T. W. 1982b. Environmental factors affecting the yields of tea (*Camellia sinensis*). II. Effects of soil temperature, day length, and dry air. *Expl. Agric*., **18:** 53–63.

Tanton, T. W. 1992. Tea crop physiology. *In Tea: Cultivation to consumption* (eds. K. C. Willson and M. N. Clifford). Chapman & Hall, London, 173–199.

Willson, K. C. 1999. *Coffee, Cocoa and* Tea. CABI Publishing, UK. p. 300.

Xu, Changjin, 1983. The climate and tea production in Huangyun country.(in Chinese). *Zhejiang Meteorological Science Technology*, **2:** 10–14.

Chapter 5

Tea soils

5.1 Introduction

Soils are the natural bodies on which plants grow. To the planters the soil is more than useful – it is indispensable. To ensure optimum growth of plants, the soil has to be maintained in proper physical, chemical and biological conditions. These are strongly influenced by climate, plant and animal life, and human activities. Planters need to be aware of how these processes are influenced and how they can be manipulated in order to grow healthy, productive tea plants. Soil is often described as consisting of solid particles, water, gaseous elements, humus and raw organic matter. Soil is also the dwelling place for a large variety of living creatures (see Box 5.1).

Box 5.1 : The importance of soil life

A rule of thumb is that, under favourable conditions, one tenth of the organic matter in a soil is made up of soil animals. Thus, a layer of 10 cm of a hectare of soil with 1 % of organic matter contains roughly 1500 kg of soil fauna. This equals the weight of 3–4 cows.

Source: Dalzell et. al. (1987).

5.2 Formation, classification and characteristics of tea soils

Tea is grown in a variety of soil types that are widely different in origin and geographical formation. These soil types have developed from diverse parent rock material and under different climatic conditions. Tea in China, Georgia, Indonesia, Sri Lanka and Turkey is mostly grown on sedimentary soils derived from gneiss or granite. In Taiwan, tea occupies a tract of tertiary rocks derived from a residual formation. Most of the soil types developed from volcanic ash in Kenya, Uganda, Burundi, Rwanda, Tanzania and Congo (Othieno, 1992). However, tea in these countries is also grown on soils derived from gneiss and granite. Malawi's tea is grown on soil types that have been washed from a granite massif and on sedimentary gneissic soil types (Othieno, 1992). The soils of Bangladesh, China, Sri Lanka and Taiwan are red-yellow podzolic and reddish-brown lateritic. Bangladesh soils have abundance of Kaolinite, quartz, hematite, goethite and gibbsite minerals. These have low cation exchange capacity as well as low nutrient releasing capacity and show strong phosphate fixing capacity (Alam, 1999). Besides, tea soils of Bangladesh contain over 85% aluminium in the exchangeable sites (Ahsan, 1984), and an overall content of nitrogen and organic matter varies respectively from 0.07 to 0.09 percent and 1.0 to 1.2 percent. The younger soils of the flats consist of sands, silts, and clays brought down by the river systems draining part of the Himalayas and the hills in eastern India. The Japanese tea soils are classified as red-yellow podzolic and volcanic (Othieno, 1992). In Kenya, and parts of Tanzania and Uganda, the soils are nitisols in the FAO-UNESCO classification system (Sombroek *et al.*, 1982). But in east African countries there are pockets of acrisols and ferrasols on which tea is grown. In Indonesia, tea is grown in three different soil types – andosol, red-yellow podzolic and regosol. Andosol is generally preferred and most of the plantations are found in this soil type. About 18 and 15 percent of the tea areas are under red-yellow podzolic and regosol type respectively. The soils of tea areas in Malaysia are mostly serdang, malacea and rongan series which are derivations from shade and granite (Dutta, 1972). However, all these soils are acidic (Table 5.1). The physical proportions of some soils on which tea is grown are presented in Table 5.2.

In north-east India, except in Darjeeling, the alluvial deposits laid down by rivers like the Brahmaputra (Assam valley), Barak (Cachar valley), Toorsa and Teesta (Dooars) during the recent geological times form the tea soils. These alluvial deposits in different parts of this region vary in their physical and chemical characteristics which are depicted in Table 5.3. The soils of the Brahmaputra valley can be classified into three broad groups (Barua, 1989) viz. (1) the soils which owe their origin mainly to the Brahmaputra river and they are usually very sandy with poor nutrient. The

Table 5.1 : General chemical properties of some soils in tea-growing areas.

Country (area)	Soil depth (cm)	pH (soil water 1:2.5	Organic matter (%)	Total N (%)	C/N (ratio)	Available P (p.p.m.)	Exchangeable bases (mequiv. $100g^{-1}$) K	Mg	Ca	Exchangeable H (mequiv. $100g^{-1}$)	CEC (mequiv. $100g^{-1}$)
Kenya											
Kericho	0 to 5	5.1	8.46	0.6	8.2	14	2.08	2.5	0.9	1	-
	5-10	4.5	8	0.58	8.02	8	1.9	1.4	0.6	3.9	-
	10-15	4.6	7.09	0.5	8.24	7	1.74	0.7	0.2	3.8	-
	0-23	4.75	3.68	0.1	0	4	3.08	2.01	0.46	-	20.78
Malawi,											
Newly planted tea	0-23	5.3	3	0.17	10.5	14	1.16	1.31	2.46	-	2.7
Replanted tea	0-23	4.15	2.66	0.14	11.46	14	0.69	-	-	-	-
Sri Lanka		4	5.4	0.33	9.63	24	0.36	-	1.125	-	-
Taiwan											
Horizon	I	4.4	1.53	0.09	9.9	1.5	0.15	0.08	1.71	18.5	-
	ii	4.4	1.38	0.08	10	1	0.15	0.07	0.41	18.5	-
	iii	4.6	0.53	0.4	7.8	0	0.16	0.16	0.1	0.53	18.5
Papua New Guinea*	0-15	5	-	-	-	13	1.05	1.16	4.8	-	21
	0-15	4.3	-	-	-	36	0.74	0.48	2	-	24

N, Nitrogen; C, carbon; P, phosphorus; K, potassium; Mg, magnesium; Ca, calcium; H, hydrogen; CEC, cation exchange capacity; - Data not available.

Source : Othieno (1992); *K. C. Willson (unpublished)

Table 5.2 : Physical properties of some tea soils of the world (0 - 30 cm depth on average).

Parameters	A	B	C	D	E	F	G	H
Texture – Fractions (%)								
Coarse sand/gravel	20	34	24	21	2	2	-	-
Fine sand	19	33	39	32	4	6	14	49
Silt	36	22	29	10	11	9	34	30
Clay	25	10	6	37	82	83	52	1.7
Organic matter	-	-	-	4.2	8.5	6.8	2.3	1.7
Texture grade	SiL	SaL	SaL	SaL	C	C	CL	CL
Particle density	-	-	-	2.61	2.60	2.55	2.71	2.66
Apparent density	-	-	-	1.23	0.85	0.77	1.39	1.42
Porosity	-	-	-	53	73	71.3	48.7	46.8
Water-holding capacity (%)	-	-	-	22.5	59.0	60.9	28.2	20.8

A Central Province, Sri Lanka; B Uva, Sri Lanka; C Pengalengan, Java, Indonesia; D Usambaras, Tanzania; E Kiambu, Kenya; F Kericho, Kenya; G Hsing-hua series, Taiwan; H Ho-kang series, Taiwan

- Data not available

Textural grades: SaL Sandy loam; SiL Silty loam; CL Clay Loam; C Clay

Source: Othieno (1992).

soils are rarely used for tea cultivation, (2) the soils deposited from the adjacent hills by the tributaries of the Brahmaputra are generally rich loam suitable for tea, and (3) the older red soils occurring largely in the north bank of Brahmaputra are rich in nutrients and good for tea.

The Dooars and Terai comprise a narrow piece of land immediately to the south of the foothills of Himalaya. The oldest alluvia on the basis of nature of sediment and age is the Dooars red bank soil on a high plateau on west-central Dooars. These soils are highly acidic, well-withered and rich loamy of great depth. The youngest in the series is the grey sandy-loam soils of eastern Dooars, which are generally rich in lime, magnesium and phosphoric acid, but low in organic matter and nitrogen content in general (Banerjee, 1993). The rest of the alluvium is the intermediate between these two extremes. The plateau series in Hantapara is moderately old deposit with clear signs of withering. The *Mal* sands are washed down from the Himalaya and are of sedimentary origin. They are rich in organic matter and characterised by a high percentage of coarse sand. But in parts of Terai, the sands are heavily oxidised and the colour changes to red from normal reddish dark brown (Banerjee, 1993). The physical and chemical

Table 5.3 : Physical and chemical properties of some north-east Indian soils*
(Expressed as percent on oven-dry basis)

Constituent	Upper Assam loam	Mid-Assam sandy	Lower Assam sand	Tezpur	Teela Cachar	Clay flat Cachar	Plateau Cachar	Peat (Bheel) Cachar
Coarse sand	23	1	52	23	23	Nil	13	5
Fine sand	26	57	25	24	38	9	30	5
Silt	22	21	13	13	14	15	27	9
Fine silt	8	13	4	16	8	35	8	22
Clay	15	4	3	18	10	28	16	10
Loss on irrigation	5.8	3.1	3	5.2	5.5	11.8	5.1	46
Organic matter (Grandeau)	1.55	1.35	1.2	2.2	2.2	4	1.6	25
Nitrogen	0.106	0.085	0.071	0.096	0.12	0.22	0.13	0.6
Total phosphoric acid	0.075	0.051	0.06	0.046	0.3	0.072	0.1	0.2
Available phosphoric acid	0.011	0.009	0.012	0.003	0.14	0.005	0.025	0.05
Available potash	0.009	0.008	0.004	0.007	0.12	0.015	0.005	0.04
Available lime	0.02	0.013	0.033	0.03	0.05	0.08	0.02	0.3
Acidity (Hopkins)	490	450	320	720	800	1600	1000	3000
pH of water extract	5.7	5.6	5.6	5.2	5.6	5.2	-	4.8
pH of KNO_3 extract	4.6	0.5	4.6	4.4	4.6	4.5	-	4
Insoluble siliceous matter	84	92	90	80	92	78	84	53

* The data quoted are very old and the system for expressing the fractions of mechanical composition used then was the old British one : coarse sand, 1.0- 0.2 mm; Fine sand, 0.2- 0.04 mm; Silt, 0.04- 0.01 mm; Fine silt, 0.01- 0.002 mm; clay, <0.002 mm.
Source : Mann and Gokhale (1960).

properties of the typical tea soils of Dooars are presented in Table 5.4.

In south India, most of the tea growing areas are located along the mountainous range bordering the west coast of the Peninsula. These soils are mainly derived from gneissic rocks containing large amount of mica. The soils have undergone very considerable weathering resulting in low lime content. Though the soils of different parts of south India vary considerably, all are sedimentary soils (Arunachalam, 1995), except in the Wynaad where the detritus from Nilgiris and other surrounding hills has got mixed with the soils formed *in situ*.

The soils of High Ranges, Anamallais, central Travancore and south

Table 5.4 : Physical and chemical properties of typical tea soils of Dooars and Terai* (Expressed as percent on oven dry basis)

	Dooars	Mal	Grey	Hantapara
Coarse sand	10	55	8	25
Fine sand	9	19	28	30
Silt	20	8	29	13
Fine silt	20	10	26	17
Clay	32	4	7	11
Loss on ignition	9.8	4.0	3.5	4.3
Organic matter (Grandeau)	3.10	2.40	1.50	2.0
Nitrogen	0.135	0.130	0.085	0.100
Total phosphoric acid	0.130	0.090	0.160	0.120
Available phosphoric acid	0.012	0.021	0.055	0.044
Available potash	0.018	0.008	0.015	0.015
Available lime	0.025	0.021	0.150	0.035
pH of water extract	5.4	5.3	5.8	5.6
pH of KNO_3 extract	4.2	4.5	4.8	4.6
Insoluble siliceous matter	68	86	85	85

*The data quoted are very old and the system for expressing the fractions of mechanical composition used then was the old British one: Coarse sand 1.0–0.2 mm; Fine sand 0.2–0.04 mm; Silt 0.04–0.01 mm; Fine silt 0.01–0.002 mm; Clay <0.002 mm

Source: Mann and Gokhale (1960).

Travancore are open in texture, coarse and fine sand fractions together comprising 62 to 67 percent of the soil. Such open soils readily permit percolation of water and thus drainage takes place easily. The aeration is free in these soils and do not form a pan. The soils of the Nilgiris and Wynaad are characterized by clay loam type with the clay fraction preponderating. To some extent, the presence of higher amounts of organic matter in the soil of the Nilgiris offsets the disadvantages of clay. These soils are heavy in character, puddle easily with water, the particles adhere more closely and hence water does not drain through well. They form hard clods in dry weather and since percolation of water is slow they require good drainage. For improvement of drainage and lightening the soil, trenching and filling the trenches with organic matter will be useful. Since there is a tendency to form pan, planting of deep rooted shade trees is advisable. A good deep forking will be useful after pruning (Anonymous, 1961). The soils of south Travancore and Wynaad are chemically poorer than the soils of other districts. In general, all soils are distinctly acidic. Tea soils of south India in general, are rich in nitrogen and are classified as latosols (Ranganathan and Natesan, 1985). The physico-chemical characteristics of south Indian tea soils are presented in Table 5.5.

Table 5.5 : Physico-chemical properties of tea soils of south India (Expressed as percent on oven dry basis)

Parameters	High Range	Anama-llais	Nilgiris	Central Travancore	South Travancore	Wynaad
Gravel and stone	30	35	20	30	15	15
Coarse sand	38	35	25	28	43	26
Fine sand	27	32	23	36	19	18
Silt	20	15	18	13	9	20
Clay	15	18	34	23	25	35
Loss on ignition	17.0	15.2	14.8	15.0	11.7	12.0
Organic matter (% Carbon x 1.72)	6.0	5.4	5.5	4.4	3.6	3.4
Nitrogen	0.350	0.280	0.240	0.220	0.150	0.170
Total phosphoric acid	0.130	0.140	0.110	0.130	0.090	0.064
Available phosphoric acid	0.010	0.009	0.004	0.008	0.005	0.003
Total potash	0.580	0.430	0.290	0.370	0.300	0.280
Av. Potash	0.040	0.030	0.023	0.024	0.025	0.022
pH	5.7	5.1	5.0	5.6	5.5	5.5
Bulk density	1.070	1.060	1.080	1.070	1.000	0.900
Particle density	2.080	2.060	2.280	2.100	2.100	2.240
Pore space ml/100 cc of soils	53.00	54.00	55.80	58.00	51.20	64.40
Water holding Capacity %	58.00	56.00	57.00	61.00	52.00	80.00
Permeability	Good	Good	Gool	Good	Good	Good
Soil texture	Loams	Loams	Clay/ Loam to Clayey	Loams	Loams/ sandy Loam	Heavy/ clay loam to clayey

Source: Mann and Gokhale (1960) and UPASI, Tea Research Institute.

The tea soils of Darjeeling belongs to the Daling series which rests under a variety of foliated and banded metamorphic rocks, partly sedimentary and partly igneous in origin (Wadia, 1989). These rocks are generally known as Darjeeling gneiss and composed of mica-schists and gneisses. Some of the gneisses have been formed by injections of granite fluid along the micaceous layers of the schists. Where soaking has been through, the gneisses approach granites in composition and are made of biotite muscovite, quartz and foldspars. The sedimentary varieties of the Darjeeling gneiss contain such minerals as garnet, sillimanite. The presence of kainite and staurolite indicates that the rocks were subjected to higher temperature and pressure than the Daling rocks. The Darjeeling gneiss also carries subordinate bands of quartzite. Saha *et al.* (1995) identified the following four soil series during the survey of the soils of Darjeeling which is under tea for more than hundred years. The study area comprise of Kurseong, and parts of Darjeeling and Kalimpong sub-divisions. The series are:

(1) Monteviot series – Umbric Dystochrept; (2) Thurbo series – Typic Udorthent; (3) Chamong series – Umbric Dystochrept; (4) Teesta valley series – Entic Haplumbrept

The physico-chemical properties of Darjeeling soils enumerated in Table 5.6 reveal narrow to wide ranging variations among above four series. The soils of Monteviot, Chamong and Teesta valley series are more coarse in nature as compared to the Thurbo series. The clay content is also higher in the Thurbo series. Saha *et al.* (1995) observed that the organic carbon content in all the four series shows a general trend of decline down the profile. PH values shows the reverse trend of increasing down the profile which indicates the leaching of bases downward. The carbon exchange capacity values are higher in Thurbo series as compared to others. Calcium forms the preponderant cation in all the series followed by magnesium and potassium. Thurbo series shows relative abundance of calcium ions but the free iron oxide contents are higher in the Monteviot and Chamong series. The mobile iron and aluminium recorded highest in Monteviot and Chamong series respectively. The bulk density values of Teesta valley and Monteviot series are comparatively higher than the other two series.

The dominant mineral found in Monteviot and Chamong series is chlorite and a considerable amount of vermiculite mineral has also been found although the abundance of illite group precedes the former. However, rutile and anatase are also found but in lesser amount. Kaolinite and illite are the dominant minerals found in Teesta valley series. Smectite has also been detected in Teesta valley series but no trace of rutile and anatase are observed. The chlorite, rutile and anatase minerals have been observed in most of the tea garden soil-clays of the quality sections belonging to Monteviot, Chamong and Thurbo series. The non-quality sections of most

Table 5.6. Physico-chemical properties of tea soils of Darjeeling (as per USDA soil taxonomy)

Parameter	Monteviot* series	Thurbo* series	Chamong* series	Teesta valley* series
Texture fractions %				
Coarse sand	52.48	34.42	49.82	40.58
Fine sand	21.5	21.64	21.74	23.98
Coarse silt	8.22	16.6	10.2	11.7
Fine silt	6.14	7.9	7.48	10.38
Clay	11.86	19.36	10.76	13.36
Texture grade	Sandy loam	Loam	Sandy loam/ Loamy sand	Sandy loam/ Loam
Bulk density (g cm^{-3})	1.36	1.19	1.30	1.33
Water holding capacity (%)	37.26	40.53	39.15	40.28
Porosity (%)	38.29	41.51	40.23	41.07
Organic matter (%)	1.83	1.96	2.40	1.87
pH (soil water 1: 2.5)	5.09	5.24	4.71	5.35
Total nitrogen (%)	0.09	0.08	0.12	0.08
Available phosphate (%)	0.02	0.04	0.02	0.00
C /N (ratio)	11.35	14.21	11.17	12.37
CEC (cmol/kg)	10.86	17.19	9.86	12.25
Exchangeable cations (cmol/kg)				
Ca^{++}	1.25	3.24	1.49	2.17
Mg^{++}	0.30	0.31	0.26	0.25
K^{+}	0.19	0.20	0.16	0.17
Free iron oxide (%)	7.24	5.72	6.58	3.38
Mobile iron (%)	0.74	0.46	0.54	0.43
Mobile aluminium (%)	0.47	0.96	2.05	0.92

*Values are average of all horizons/depth; Source: Saha *et al*. (1995).

of these tea garden soils showed the preponderance of illite mineral with traces or no amounts of rutile and anatase minerals (Saha *et al*., 1995).

The tea soils of Palampur, Himachal Pradesh and Dehra Dun, Uttar Pradesh are alluvial origin and are not always acidic but rich in organic matter. The soils of Kangra valley, Himachal Pradesh are podzolic sandy grey on clay loam with a pH ranging from 5.4 to 5.8. These soils are poor in organic matter content.

5.3 Identification through indicator plants

Tea soils can be identified by the presence of certain plants among the indigenous vegetation. Albizzias have been found to be good indicators for suitability of a soil for growing tea in India and Sri Lanka. In north India, i.e. Assam, Dooars and Darjeeling, *Albizzia stipulata*, *A. procera* and *A. lebbeck* respectively are the best indicator plants and *Albizzia moluccana* is a good indicator plant in Sri Lanka (Mann, 1935). The fern *Gleichenia* is also regarded as an indicator plant in Sri Lanka and south India. The most reliable indicator for tea are the aluminium accumulating species and often have blue flowers. There are number of indigenous plants in Kenya which are associated with good tea soils (Brown, 1966). These are: trees – *Newtonia buchanani*, *Albizzia* sp.; shrubs – *Vernonia auriculifera*, *Triumfetta macrophylla*; herbs – *Boreria princei*; ferns – *Pteridium aquilinum*. In Uganda, *Dissotis violacea*, *D. brazzaei*, *D. irringiana* and *Craterispermum laurinum* have been used as indicators for suitability of tea soils (Chenery, 1955).

5.4 Physical properties

The physical properties of soil, despite from controlling soil moisture and conservation of fertility (Dey, 1966; 1972) it affects root growth and hence productivity (Ghosh *et al.*, 1984). It is governed by number of factors. Some of the important factors are listed below.

5.4.1 *Depth*

It is of vital importance for plant growth. A good depth is desirable to give a high water storage capacity. An ideal tea soil should be well drained and 1.52 m in depth (Mann, 1935). Although there are situations in different parts of the world where tea is cultivated on soils of lesser depth. It is however known that tea grown in high water-table soils and other types of shallow soils which occasionally include drained swamps (Othieno, 1992). Since these soils are problematic, require special attention if economic yield is to be obtained.

5.4.2 *Texture*

The relative proportions of the various soil separate like sand, silt, clay, gravel and coarser fragments in a soil determines the soil's textural class. Cohesion of these individual particles to form aggregates is called as soil structure. The overall textural designation, or class, is determined on the basis of mass ratios of sand, silt and clay (Table 5.7). The proportion of each size group in a given soil (the texture) can not be altered and thus is considered a basic property of a soil.

Table 5.8 : Dry matter production and nutrient content of plant parts formed above pruning cuts of bushes (clone B/6/60) under regular plucking.

Plant part above pruning cut	Cumulative dry matter (tonnes ha^{-1})	Nutrient removal (kg ha^{-1})		
		N	P	K
Up to first year				
Crop	1.20	48	6	24
Foliage	3.30	122	15	60
Twig	0.82	13	2	10
Wood	1.79	41	8	15
Total	7.11	224	31	109
Up to second year				
Crop	4.80	192	18	96
Foliage	4.74	164	12	46
Twig	3.50	64	7	24
Wood	7.02	68	10	76
Leaf litter	0.60	13	2	11
Total	20.66	601	49	202
Up to third year				
Crop	10.80	414	46	201
Foliage	7.33	249	30	81
Twig	6.74	142	25	81
Wood	17.27	162	45	64
Leaf litter	1.80	61	10	7
Total	43.94	1028	166	412

Source: UPASI, Tea Research Association

green crops (rehabilitation crops), decay of weeds, mulch materials etc. are the sources for building up organic matter. The estimated values of some materials are presented in Table 5.9.

5.5 Chemical properties

5.5.1 *Soil colloids*

The most active portions of the soil are those in the colloidal state. Soil colloids can be roughly grouped into organic and inorganic phases. The inorganic phase may consist either of residues of the original minerals present in the parent material or of new formation produced during the weathering of these minerals. To some extent, the nature of the inorganic phase is dependent upon the forces in the biosphere in any specific area. The inorganic phase is present almost exclusively as clay of various kind. The organic consists of fresh or decomposed residues of plant, animal, and

Table 5.9 : Nutrient status of some organic waste materials.

Category	Source	Nutrient content (%)		
		Nitrogen (N)	Phosphoric acid (P_2O_5)	Potash (K_2O)
Animal waste	Cattle dung	0.3 – 0.4	0.10 – 0.15	0.15 – 0.20
	Cattle urine	0.80	0.01- 0.02	0.5 – 0.70
	Sheep and goat dung (mixed)	0.65	5.5	0.03
	Night soil	1.2 – 1.5	0.8	0.5
	Human urine	1.0 – 1.2	0.1 – 0.2	0.2 – 0.3
	Leather waste	7.0	0.1	0.2
Compost	Farmyard manure	0.5 – 1.0	0.15 – 0.20	0.5 – 0.6
	Poultry manure	2.87	2.90	2.35
	Town compost	1.5 – 2.0	1.0	1.5
	Rural compost	0.5 – 1.0	0.2	0.5
	Water hyacinth compost	2.0	1.0	2.3
	Tea waste compost**	3.60	0.60	2.30
Oil cakes	Castor	5.5 – 5.8	1.8	1.0
	Coconut	3.0 – 3.2	1.8	1.7
	Cotton seed	3.9 (6.5)*	1.8 (2.8)*	1.6 (2.1)*
	Ground nut	4.5 (7.8)*	1.7 (1.7)*	1.5 (1.4)*
	Karanj (*Pongamia pinnata*)	3.9 – 4.0	0.9 – 1.0	1.3
	Neem	5.2	1.0	1.4
	Niger	4.8	1.8	1.3
	Mahua butter tree (*Bassia latifolia*)	2.5 – 2.6	0.8	1.8
	Rapeseed	5.1	1.8	1.0
	Linseed	5.5	1.4	1.2
	Safflower	4.8 (7.8)*	1.4 (2.2)*	1.2 (2.0)*
	Sesame	6.2	2.0	1.2
Animal meals	Blood	10 – 12	1.2	1.0
	Meat	10.5	2.5	0.5
	Horn and hoof	13.0	0.3 – 1.5	-
	Raw bone	3 – 4	20 – 25	-
	Steamed bone	1 – 2	25 – 30	-
	Fish	4 – 10	3 – 9	1.8
Miscellaneous Materials	Paddy husk**	0.15	0.13	0.22
	Paddy straw**	0.50	0.07	0.60
	Maize straw***	0.40	1.50	1.60
	Thatch grass**	0.50	0.10	0.50
	Guatemala grass***	1.6	0.60	1.70
	Napier grass***	0.8	0.40	1.30
	Mimosa***	2.6	0.60	1.00
	Tea pruning litter**	2.0	0.50	1.50
	Shade tree droppings**	2.5	0.70	0.80
Wood ash	Ash, coal	0.73	0.45	0.53
	Ash, household	0.5-1.9	1.6-4.2	2.3-12.0

Figures in ( ) are for decorticated material,
Source: Gaur (1994), Chakravartee and Sinha (1994)**, Banerjee (1993)***.

microbial remains that may or may not be combined with the inorganic phase. Nutrient elements in the soil are freely exchanged among the surfaces of plant roots. clay and humus. The nutrient exchange capacity varies directly with the amount and type of clay or humus content in soil.

5.5.2 *Cation exchange capacity*

The interchange between a cation in solution and another cation on the surface of any surface—active material such as clay and organic matter is known as cation exchange. The sum total of exchangeable cation that a soil can absorb is expressed in centimoles per kilogram (cmol/kg) of soil or clay is called as cation exchange capacity (CEC) (Brady, 1984). The CEC varies according to type of clay minerals. The uptake of nutrients is controlled by exchange capacity of soil clay as well as the roots of the plants. CEC of tea soils is generally lower than that of the tea roots. CEC values are of importance in tea in formulating the basis for assessing differential nutritional needs of different genotypes growing under a wide range of soil conditions (Chamuah, and Dey, 1982).

5.5.3 *Acidity*

One of the markable physiological characteristics of the soil solution is its reaction – that is, whether it is acidic, alkaline, or neutral. Acidity or alkalinity of a solution is measured by a scale of pH value. In the scale, 7 denotes neutrality, whilst figures from 7 downward indicate increasing acidity and above 7 denote increasing alkalinity. The term pH is defined as the negetive logarithm of hydrogen ion concentration in moles per litre. Soil pH has significant influences on availability of plant nutrients as well as microbial activity. Microbial activity in the soil is at its optimum above pH 5.5. As the pH of some acid soils is raised from 5 to 7 or 8, nutrients such as iron, manganese, and zinc become unavailable whereas molybdenum availability increases. Phosphorus is never readily available, but its availability is generally highest in a range centering around pH 6.5. Availability of sodium, calcium, potash and sulphur decreases when pH goes down below 6. At pH value below 5, aluminium, iron, and manganese are often soluble in sufficient quantities to be toxic to plant growth. The bicarbonate ion can produce similar negative effects on certain plants at very high pH values (Brady, 1984). The tea plant grows well only on a soil which is acidic in reaction and pH ranges between 4.5 and 5.5. Vigorous growth of tea has been reported even at a pH of 3.2 (Barua, 1989). But this is not only

> **Box 5.4 : Why does tea require an acid soil?**
>
> *Though it is not possible to give a satisfactory answer but views offered by different workers are as follows :*
>
> *Tea is an aluminium accumulator and its content of tea may be 1600 ppm (Chenery, 1955). In the acidic soil where tea is grown, huge reserve of aluminium occur in the exchangeable form but what the plant does with such a large amount of aluminium is still not properly known. However, the availability of aluminium decreases as the pH of the soil rises. The failure of tea on soils of high pH is perhaps connected with the non-availability of aluminium in such soil. Murthy (1994) has suggested that the bacteria which converts the nutrient material in the soil to form assimilable by the tea plant flourish best in solutions of a pH value of 5.5; or tea may be unable to absorb phosphates readily from nearby neutral solutions and as phosphates are more soluble in acid solutions, these may be more suitable; again, it would seem that tea can tolerate a certain amount of manganese and may even require an abnormal amount of this element for satisfactory growth, and it only appears in significant quantities in soil solutions in acidic range.*

undesirable but also sometimes harmful.

a. Source of acidity

Hydrogen and aluminium ions are the main causes of soil acidity. In an extremely leached soil in which tea grows best, most of the ions especially the cations (calcium, magnesium, potassium) are replaced by hydrogen from the water molecule and can be washed off the mineral (Othieno, 1992). In acid soil, aluminium ions on their own or when react with water also produce acidity. The nitrogenous fertilizers used in tea tend to develop an acid residue in soils. This is mainly due to the influence of certain nitrogen carriers that supply ammonia or produce ammonia when added to the soil. The major acidifying effect of ammonia ions exerted when they are nitrified. The ammonium compounds, upon oxidation increase acidity as follows :

$$NH_4^+ + 2O_2 \longrightarrow 2H^+ + NO_3^- + H_2O$$

The urea, upon hydrolysis yield ammonium ions are potential source of acidity. Further, in the process of decomposition of organic matter, a number of organic acids which supply hydrogen ions are released into the soil. The ions exchange with and release bases on the clay minerals. The released bases are subsequently leached out leaving the soil acid (Othieno, 1992).

b. *Correction of pH*

It is generally agreed that tea needs maintenance of optimum range of soil pH for desired growth and productivity. Therefore, sequential soil samples analysis at regular intervals would indicate acidity status and

Table 5.10 : Rate of application of ameliorants to correct soil acidity at various levels.

Type of soil	Dolomite or lime stone (80 – 100 mesh) tons/ha	Purpose	Sulphur (fine mesh) kg/ha	Purpose
Sandy loam	1.25	To raise pH value by one point	500	To lower pH value by one point
Loam	2.0		700	
Silty loam	3.0			

Source: Banerjee (1993).

necessary corrections can be done. If pH values are to be raised, application of lime stone or dolomite is advised and if the values are to be lowered then iron pyrites, sulphur or aluminium sulphate can be used (Table 5.10). In south India, 26 % of tea soils have pH between 4.0 and 4.5 and 5 % of the soils with pH less than 4.0. To raise the pH to about 5, use of lime at the following quantity is recommended (Ranganathan and Natesan, 1987) :

pH	Lime ($CaCO_3$) tonnes ha^{-1}
Below 4.0	3
4.0 – 4.5	2
4.6 – 4.9	1

In Darjeeling hills, 28 % of tea soil have pH below 4.5. The pH of the different soil samples analysed is presented in below :

Box 5.5 : Results of analysis of soil samples of different valleys of Darjeeling for the period from 1992-1996.

pH	*Below 4.5*	*4.5 – 5.5*	*Above 5.5*
Top soil	*28.80*	*59.30*	*11.90*
Sub soil	*26.11*	*66.44*	*7.45*
Carbon	*Less than 1 %*	*1 – 2 %*	*Above 2 %*
	2.84	*37.51*	*59.65*
Available K	*Below 60 ppm*	*60 – 100 ppm*	*Above 100 ppm*
	42.41	*28.64*	*28.95*

(Figure represent the percent of soil sample analysed)

Source : De (1997).

When an acid soil is limed it improves fine textured soils to less sticky and more crumby. It also provides favourable conditions for root growth and movement of water in the soil. Further, the availability of phosphates and molybdates will increase. The exchangeable calcium and magnesium will also increase. Lime stimulates the microbial activity and thereby hastens the decomposition of organic matter. Lime application is recommended only in the mature tea plantations that are in regular plucking. The material for lime is to be broadcasted just prior to pruning. It can also be applied in the second year after pruning without interfering with the manuring schedule. Since the availability of sufficient soil moisture is precondition to get optimum effect of liming therefore the best time for application would be in pre-monsoon months or in the receding monsoon months, so that continuous wet months following the application help to dissolve lime and thereby aid in the rapid neutralization of soils (Verma, 1993). It is better to use dolomite lime stone materials instead of agricultural lime in soil that are deficient in magnesium. Dolomite should be broadcasted uniformly over the soil.

Application of iron pyrites has gained wide popularity due to its low cost and easy availability. The pyrites should be broadcasted evenly over the area. It should be allowed for oxidation for a minimum period of 15 days and then mixed with the soil through forking to a depth of 7 to 15 cm. A light irrigation after forking will be beneficial. Sulphur is not soluble in water and it acidifies soil relatively quickly. But aluminium sulphate is very soluble in water and acidify soil without any adverse effects on tea. However, in excess, it can damage tea plants but its effect is mostly not permanent.

5.6 Nutrient requirements

See Chapter 10.

5.7 Biological properties

A large number of organisms live in soil. Compared to virgin lands, tea fields are generally lower in numbers and weight of soil organisms especially the soil fauna. The activities of specific groups of soil organisms are commonly identified by their number, weight per unit volume or area of soil and their metabolic activity (Brady, 1984). The beneficial effects of the soil organisms is transformation of soil organic matter into plant nutrients especially in acid tea soil. But soil microbial activities are also associated with the harmful effects viz. pest and diseases of the tea plant. The active biological agents in soils are:

5.7.1 *Macroanimals*

Mole rat (Talpa sp.) is making tunnels 8–12 cm below the soil surface

> **Box 5.6 : Natural amendment of soil pH affected by dolomite mining in some tea estates of Jalpaiguri, West Bengal, India**
>
> *It has been observed that the pH values in some tea estates of Jalpaiguri district has recently gone up ranging from 7–8 or even more. This is mainly due to inflowing run off of water carrying dolomite from dolomite hills located at the adjoining country Bhutan. The soils of these tea estates should be flooded with irrigation water (irrigation water must be free from salts of sodium and calcium) and then drained out after 2–3 days. Proper draining should be made for leaching out the dissolved salts. The pH range found in these soils does not warrant application of any soil amendment at this stage. If suitable irrigation water is not available then rain water will suffice the purpose provided the drainage is at its best. However, before the rains/application of irrigation water, the area should be ploughed down to about 25 to 30 cm depth with the help of a tractor. After the soil is leached out with irrigation/rain water pH will definitely decline considerably. Once the pH reaches around 6 then next year before rains aluminium sulphate can be applied and forked into the soil @ 3 tons ha^{-1}. Urea application in this soil is strictly forbidden. The best NPK source is SOA, SSP and MOP. SOA should be applied at least for the coming 8–10 years of time which will bring down pH further and thus application of any soil amendment can be avoided in future. SOA will take care of the pH. Checking of pH of the soil at regular interval should be on.*

and earthworms are important in improving soil productivity. The holes left in the soil serve to increase aeration and drainage. The insects, millipedes, slugs, snails, spiders, mites etc often use more or less undecomposed plant tissues as food and thus initiate the process of decomposition that are continued by soil microorganisms (bacteria and fungi).

5.7.2 *Microanimals*

The microscopic animal life in soils are nematodes and protozoans. Nematodes are classified on the basis of their demands of food into (a) most common and live on decaying organic matter, (b) predatory on soil fauna including other nematodes and (c) attacking plant root causing crop loss.

Protozoa (amoeba, flagellates) are the microscopic and simplest form of animal life that feed mainly on bacteria. They influence the balance of microbial population in the soil.

5.7.3 *Roots of higher plants*

Their roots grow and die in the soil. The dry roots left in the soil are source of dead tissue for nutrition of soil fauna and microflora. The exudates from the roots and various compounds formed at the root surface play an important role in nutrient availability. The living roots also physically modify the soil.

5.7.4 *Algae*

Soil algae in vegetative form is present in the surface layers. It contributes some to the organic matter content of the soil and blue green algae can fix appreciable amount of atmospheric nitrogen.

5.7.5 *Fungi*

It includes beneficial in terms of decomposition of organic matter and produce harmful diseases of tea. Fungi also play a vital role in the transformation of the soil constituents. They also breakdown the resistant cellulose, lignin and gum as well as they are versatile in their ability to decompose organic residues in the soil.

5.7.6 *Micorrhiza*

Micorrhiza meaning 'fungus root'; an association of a fungal mycelium with root of a higher plant. This symbiotic association increases the availability of several plant nutrients especially more phosphate.

5.7.7 *Actinomycetes*

These organisms are numerous in terms of population per unit of soil and have great importance in the decomposition of soil organic matter and the liberation of nutrients therefrom.

5.7.8 *Bacteria*

The activities of bacteria are of great importance. In the soil they are concerned in the decay of plants and animal tissues, making available food materials for higher plants, and certain bacteria are of particular significance in this respect in making available nitrogenous compounds. They are the agents of enzymatic transformations of (a) nitrification, (b) sulphur oxidation, and (c) nitrogen fixation.

5.8 Managing tea soil health

Good soil health is the precondition for good plant health. The maintenance of good soil structure and fertility through proper husbandry practices are vital management considerations for tea soils. Further, soil fertility management is the basis for sustainability in every agricultural production system. Creating a favourable condition for soil life and plant growth, nutrient application and soil conservation are important aspects of soil fertility management. Soil fertility arises from complex interactions between the biological, chemical and physical properties of soil. Moreover, little is known about the complexity of some of these interactions or their rates of change, especially in relation to our ability to predict variations in soil fertility.

Unplanned land use, bad soil structure, inadequate drainage, soil compaction etc reflects decline of productive potential of the tea soils. These adverse conditions can be checked by adopting proper soil management practices. These are:

5.8.1 *Water conservation*

Water may enter into a plantation as rain, flood or irrigation of water, groundwater or air humidity. In general, 93–95 % of the total annual rainfall received during monsoon which sometimes cause waterlogging and drainage problem in many tea estates of Assam and other parts of India. The rest (5–7 %) falls during the period from November to April which sometimes leads high soil moisture deficit. Thus, tea may suffer from excess and deficit of the soil moisture in the same year. In marginal rainfall areas, every drop of rain is important for growth of tea plant. Planters can influence the availability of water in the soil by improving soil structure and storage capacity (eg. managing organic matter, use of catchments), increasing the infiltration capacity and decreasing evaporation (eg. mulching) or draining excess water.

Box 5.7 : Reason for shortage of soil moisture

- *Rainfall is insufficient or irregular with long, dry spells during the rains.*
- *The soil might be sandy or low in organic matter (humus) so that it can not store much water from one rainy period to the next.*
- *The soil might be of a type or in a condition that has a low infiltration rate so that when it rains much is lost as runoff.*
- *The soil might dry out quickly because of high air temperatures and strong winds that increase evaporation of water from the soil and transpiration from the vegetation.*
- *The soil might be on a slope so that most rainfall is lost as runoff when the surface is bare.*
- *The soil might be shallow.*

5.8.2 *Drainage*

Drainage system can aid the removal of surplus

water. Drainage is one of the major inputs for increasing tea productivity. In India, it is estimated that over 200000 hectares of tea land suffers from water logging and requires scientific drainage system (Chakravartee, 1994) to drain excess water below the root zone. The necessity of disposal of surface and sub-surface water in the tea areas is a vital factor for growth. Removal of excess water from the soil prevents serious damage to the rooting system and fertile top soil. Provision of efficient drainage system to maintain water table at a minimum depth of 100 cm is considered adequate for growing tea.

5.8.3 *Uprooting*

Top soil, being near the surface, is the major zone of root development. It carries much of the nutrient available to plants and it supplies a large share of the water used by crops (Brady, 1984). Therefore, efforts should be made to minimize loss of rich top soil during uprooting of old tea bushes and at the time of clearing land for tea planting. It is also important to see that all roots are removed at the time of uprooting. Deep hoeing to a depth of 30 to 40 cm can help in removing left out roots of teas, shade trees and stones. Levelling of land of uprooted area before planting tea ensures that new stand does not suffer localized waterlogged conditions.

5.8.4 *Soil rehabilitation*

It has been observed in north-east India (George and Singh, 1990) and Kenya (Othieno, 1992) that the old tea land could have developed some toxic (allelopathic) substances since many years it remained under tea prior to uprooting. In Kenya, food crops viz. Maize, beans, despite use of fertilizers, failed to grow when planted in such land immediately following uprooting of tea (Othieno, 1992). Therefore, uprooted land should not be kept bare in any case and must be rehabilitated by planting a mixture of pure stands deep rooting grass like Guatemala (*Tripsecum laxum*) (Fig. 5.1) and *Mimosa invisa* or Norcot (Napier) (*Pennisetum purpureum*) (Fig. 5.2). Planting of weeping love grass (*Eragrostis curvula*) has given good result at high elevation and slopes. Soil rehabilitation will help in improving the soil structure and also increase fertility of soil (Templer and Machaga, 1978).

Fig. 5.1: The Guatemala grass (*Tripsecum laxum*) has been planted in the uprooted tea areas for rehabilitation of soil.

Fig. 5.2: Norcot (Napier) (*Pennisetum purpureum*) can be used for aforestation to stabilize landslide and for soil rehabilitation.

5.8.5 *Mulching*

This is one of the age old proven cultural practices. Mulch can be defined as a shallow layer at the soil/air interface with properties that differ from the original soil surface layer. Mulching is an important technique for improving soil microclimate; enhancing soil type, structure and fertility; conserving soil moisture; reducing weed growth preventing damage by impact from solar radiation and rainfall (erosion control), and reduces the respiratory rate of roots thereby effecting economy in the oxidation of carbohydrates. In Sri Lanka, Sandanam and Rajasingham (1982) observed that soil loss with clean weeding during the first and second years amounted to nearly 89 and 51 t/ha respectively compared with 7 and 10 t/ha with Guatemala grass (*Tripsacum laxum*) mulching, and 5 and 2 t/ha with Mana grass (*Cymbopogon confertiflorus*) mulching. Soil losses with intercropping treatments amounted to 32 and 5 t/ha with *Crotalaria striata* and 11 and 2 t/ha with *Eragrostis curvula*, compared to 28 and 2 t/ha respectively under selective manual weeding with minimum soil disturbance. Soil losses in the third and fourth years were recorded to be more than 2 t/ha with all treatments. Mulching can ideally be done towards the end of the monsoon with Guatemala grass (Fig. 5.3), weeping love grass, bracken fern, tea

Fig. 5.3: Young tea mulched with Guatemala loppings.

Fig. 5.4: Tea plantation workers in Darjeeling hill transporting mulch materials.

prunings, shade loppings and other plant materials (Fig. 5.4). Effective mulching of young tea fields could be achieved with 15 to 25 tons of plant materials per hectare. It would be worthwhile planting Guatemala grass in large vacant patches to facilitate rehabilitation of soil and raising mulch material. Possible constraints to mulching are insufficient availability of mulch materials and pest problems (rodents, insects, fungi, persistent weeds).

5.8.6 *Herbicide weeding*

In addition to leaving prunings *in situ* herbicide weeding of tea facilitates minimum disturbance of the top soil where feeder roots are concentrated. This practice helps feeders to grow without frequent hindrance also built up organic matter thus soil fertility.

References

Ahsan, Q. 1984. Evaluation of old tea soils of Bangladesh with particular reference to soil fertility and soil management . *M.Sc. Thesis*, University of Aberdeen, Scotland, UK.

Alam, A. F. M. B. 1999. Profile of tea industry in Bangladesh. *In* : *Global Advances in Tea Science* (ed. N. K. Jain). Aravali Books International (P) Ltd., New Delhi, 1–22.

Anonymous. 1961. Tea soils of south India. *Planters' Chronicle*, **56:** 381–387.

Arunachalam, K. 1995. *A hand book on Indian tea*. All India Printing Press, Pondicherry, India, p. 271.

Banerjee, B. 1993. *Tea production and processing*. Oxford & IBH Publishing Pvt. Ltd., New Delhi, p. 336.

Barua, D. N. 1989. *Science and practice in tea culture*. Tea Research Association, Jorhat - Calcutta, India, p. 493.

Brady, N. C. 1984. *The nature and properties of soil*. Macmillan Publishing Co., Inc., USA, p. 750.

Brown, L. H. 1966. *A report on tea growing potential in Kenya*. Ministry of Agriculture, Nairobi, Kenya.

Canada Development of Agriculture, 1972. *Glossary of terms in soil science. Publication No. 1459*. p. 66.

Chakravartee, J. 1994. Importance of drainage in tea. *In : Drainage in tea* (ed. J. Chakravartee). Tea Research Association, Jorhat, Assam, India, 1–3.

Chakravartee, J. and Sinha, M. P. 1994. Soil management in tea for north-east India. *In: Field Management in Tea* (ed. J. Chakravartee). Tea Research Association, Jorhat, Assam, India, 41–49.

Chamuah, G. S. and Dey, S. K. 1982. Cation exchange capacity of clonal tea roots and its implication on fertilizer response. *J. Sci. Food Agric.*, **33:** 309–317.

Chenery, E. M. 1955. A preliminary study of aluminium and the tea bush. *Plant and Soil*, **6:** 174–200.

Dalzell, H. W., Biddlestone, A. J., Gray, K. R. and Thurairajan, K. 1987. Soil management: compost production and use in tropical and subtropical environments. *Soil Bull.*, **56:** Rome, FAO.

De, S. K. 1997. Limiting factors for productivity in Darjeeling. *Proc. Seminar Darjeeling Tea Industry*. September 5, 1997, DPA & Tea Board, Calcutta, India, 6–18.

Dey, S. K. 1966. Soil fertility: physical and chemical aspects. *Two and a Bud*, **13:** 4–6.

Dey, S. K. 1972. Role of soil depth and texture classification in evolution of productivity index. *Two and a Bud*, **19:** 31–37.

Dutta, S. K. 1972. Tea in Malaysia and Indonesia. *Two and a Bud*, **19(1):** 1–9.

Gaur, A. C. 1994. Bulky organic manures and crop residues. *In: Fertilizers, organic manures recyclable wastes and biofertilizers* (ed. H. L. S. Tandon). Fertilizer Development and consultation Organization, New Delhi, 36–51.

George, U. and Singh, R. 1990. Biological and chemical factors affecting replantation in tea. *Proc. Intl. Conf. R & D in Tea*, Tea Research Global Perspective, 11–12 January, 1990, CalcuttaTea Research Association, Jorhat, Assam, India, 77–83.

Ghosh, P., Goswami, H. and Dey, S. K. 1984. Studies on certain physical properties of tea soils. *Two and a Bud*, **31:** 4–8.

Hipolito, M. C. 1999. Soil acidification: myths or reality? *ILEIA Newsletter*, **15 (1/2):** 25–26.

Mann, H. H. 1935. Tea Soils. *Commonwealth Bureau of Soil Science Communication No. 32*, Harpenden, England.

Mann, H. H. and Gokhale, N. G. 1960. Soils of tea growing tracts of India. *J. Indian Soil Sci. Soc.*, **8:** 191–200.

Murthy, S. B. K. 1994. Soil acidity and tea culture. *Planters' Chronicle*, **89 (3):** 118–119.

Othieno, C. O. 1992. Soils. *In : Tea: Cultivation to consumption* (eds. K. C. Willson and M. N. Clifford). Chapman & Hall, London, 139–172.

Ranganathan, V. 1973. Certain aspects of manuring. *UPASI Tea Sci. Bull.*, **30:** 42–51.

Ranganathan, V. and Natesan, S. 1985. Potassium nutrition of tea. *In : Potassium in Agriculture* (ed. R. D. Munson). *Agronomy Society of America*, 981–1022.

Ranganathan, V. and Natesan, S. 1987. Manuring of tea – revised recommendations. *In: Hand Book of Tea Culture*, Sec. 11 UPASI Tea Research Institute, 1–27.

Saha, R., Mondal, D. and Bisen, J. S. 1995. Tea soils of Darjeeling. *Tea Board Scientific Monograph series no.1*, Darjeeling Tea Research Centre, Kurseong, 1–29.

Sandanam, S. and Rajasingham, C. C. 1982. Effects of mulching and cover crops on soil erosion and yield of young tea. *Tea Q.*, **51(1):** 21–26.

Sombroek, W. G., Braun, H. M. H. and Van der Pouw, B. J. A. 1982. *Exploratory soil map and agroclimatic zone map of Kenya 1980*, Ministry of Agriculture, Nairobi, Kenya.

Surendra Mohon, M. 1995. Fertility status of the tea soils. *Planters' Chronicle*, **90(6):** 239–249.

Templer, J. C. and Machega, J. S. E. 1978. Replanting tea land to Guatemala grass. *Tea*, **18(1):** 16–17.

Verma, D. P. 1993. Nutrient management of tea in south India. *In : Tea culture processing and marketing* (eds. M. J. Mulky and V. S. Sharma). Hindustan Lever Research Foundation, Bombay, 55–69.

Wadia, D. N. 1989. *Geology of India*. Tata McGraw Hill Publication Co. Ltd., (Tenth Reprint). p. 508.

Willson, K. C. 1969. The mineral nutrition of tea. *Potash Review, Subject 27*, 47th Suite, 1–17.

Chapter 6

Physiology of tea productivity

6.1 Introduction

Tea is an evergreen small tree by habit, but has been trained as a shrub/bush through the periodic management practices like pruning and plucking. It is a C_3 plant, that is, the first stable product of CO_2 fixation contains three carbon atom, following Rubisco metabolism (Roberts and Keys, 1978). Tea plant adapted well to the different terrain and climatic conditions. The crop needs acid soil, a humid tropical environment and will not tolerate drought. Tea is traditionally interplanted with shade trees (in India, Bangladesh) which favourably alters the gross and net productivity of the tea bushes through increase in photosynthetic efficiency and favourable soil and microclimatical conditions. But in several areas (in Malawi, Kenya etc) it was established that tea produces its highest yields in full sunlight provided that nutrient supply is nonlimiting. Tea bushes in commercial production are planted closely and pruned to a low flat surface for ease in pluckability and to stimulate development of more new shoots as tender growing shoots form the economic yield. Of the various management practices followed, pruning, tipping, plucking, shade management etc. are unique to tea only and most of these are interrelated and the leniency of one compensates for the severity of the others (Jain and Tamang, 1988). The field management practices of this crop are based on some important physiological capabilities of the plant. Understanding the physiology and biochemistry of the tea plant holds the key for sustained productivity, quality produce and cost effective management. In this chapter the present state of our knowledge of the tea crop's physiology and areas where thorough understanding is needed are discussed.

6.2 Crop distribution

Tea is cultivated in 31 countries, scattered from 45^0 north to 33^0 south of the equator. Certain variations in these areas are caused in physiological behaviour of tea plant due to interaction with environment. One such gross change in its physiological behaviour is the crop distribution depending upon the location (Fig. 6.1 and 6.2).

In north India including Darjeeling, the crop peak is reached during June–August. There are two distinct peak cropping periods in south India,

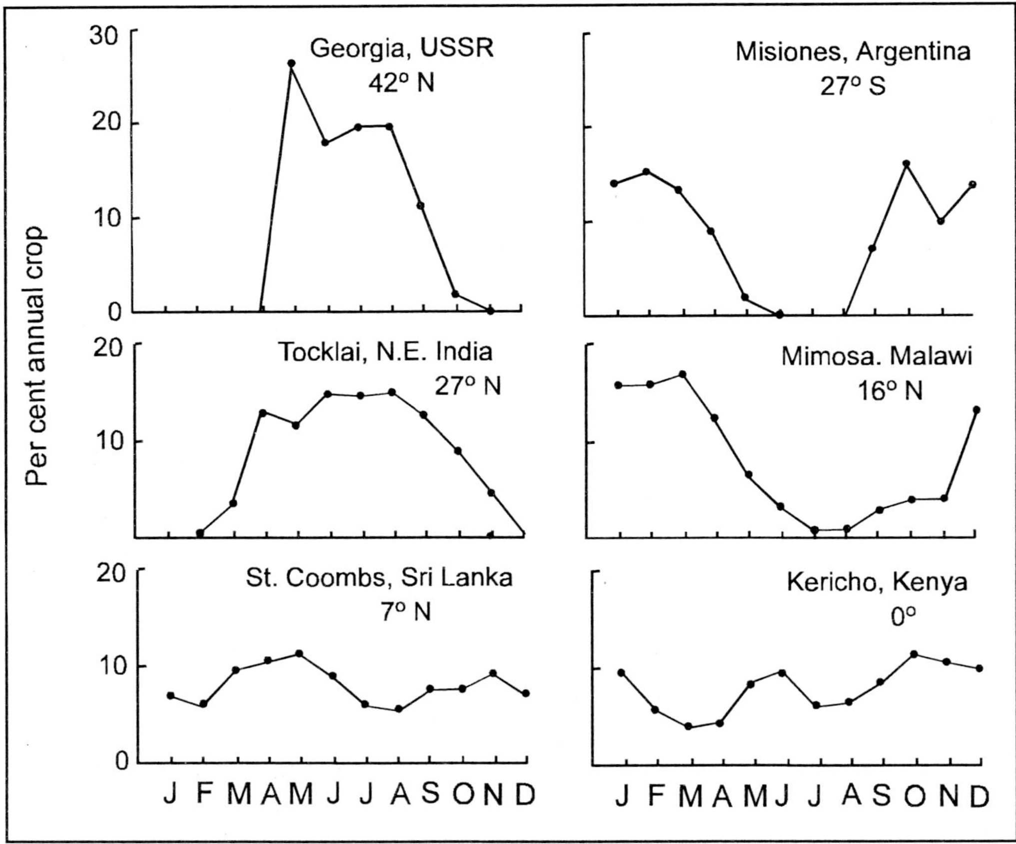

Fig. 6.1 : Monthly average crop distribution at different latitudes.

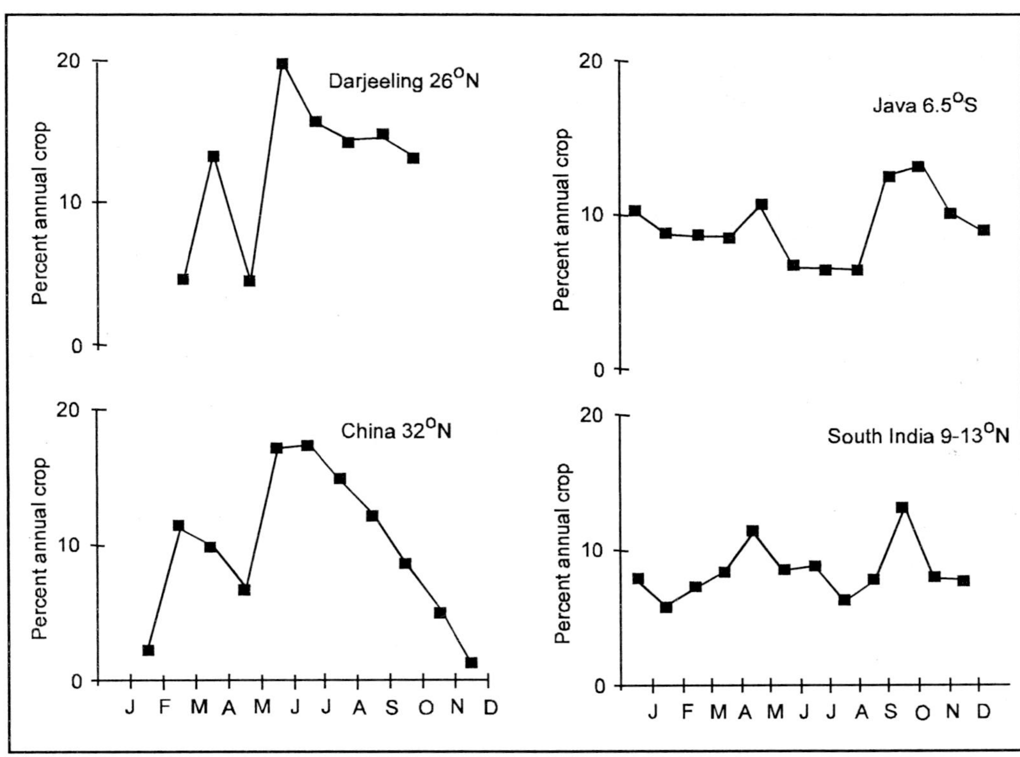

Fig. 6.2 : Monthly average crop distribution at different latitudes.

one pre-monsoon and the other post, south-west monsoon. The variation in crop ranges from minimum of 7.8 percent to maximum of 13.1 percent in different months (Murthy, 1992). However, tea crops throughout the year in most tea growing countries, although weekly and seasonal yield varies considerably. Carr and Stephens (1992) analysed the long term yield records form commercial tea estates in Kenya, Tanzania and Malawi. They observed considerable variations of the seasonal yield distribution in these three sites mainly due to seasonal changes in temperature and the development of large soil water deficits during dry season. Most even yield distribution except small drop of production during the dry season was found in Kericho, Kenya. In contrast, Mufindi, Tanzania and Mulanje, Malawi show marked seasonal variation in yield distribution. In Malawi, more than 70 percent of the annual crop is produced in the rainy season (November to March) in a series of three peaks and troughs, with a cycle period of six weeks, after which the yield drops drastically from April until August and rise briefly again in September (Fordham and Palmer-Jones, 1977).

Crop distribution in a year depends on the pruning time which in turn is decided by the prevailing seasonal conditions. Uneven crop distribution exerts undue pressure on the planters in mobilising the resources and maintaining the quality of the product.

Box 6.1 : Differences between C_3 and C_4 plants

C_3 plants	C_4 plants
1. CO_2 acceptor is RuBP.	CO_2 acceptor is phosphoenol pyruvic acid (PEP).
2. First stable product is phosphoglyceric acid.	First stable product is oxaloacetic acid.
3. Only one type of chloroplasts are present in each cells participating in photosynthesis.	Chloroplasts of bundle sheath are different from that of mesophyll cells.
4. Two pigment systems are present in each chloroplast.	Photosystem II is absent in the chloroplasts of bundle sheath. These are dependent on mesophyll chloroplasts for the supply of NADPH + H^+.
5. Calvin cycle occurs in the chloroplasts of mesophyll cells.	Occurs in the chloroplcsts of bundle sheath cells as the enzymes for Calvin cycle are absent in the chloroplasts of mesophyll cells.
6. Photorespiration occurs.	Absent.
7. Less efficient in utilizing CO_2.	Much more efficient.
8. Requires 18 ATP for synthesis of the glucose molecule.	Requires 30 ATP for the synthesis of one glucose Molecule.

6.3 Components of yield

This has already been discussed in Chapter 3, para 3.3. Examination of the relative importance of these components of yield may give an idea on the effect of cropping practices have on yield.

6.3.1 *Size of harvested shoot*

The size of shoots harvested is an important factor in determining yield (Ellis and Grice, 1976; Tanton, 1979). Tanton (1979) reported that in commercial production of tea in Malawi, shoots are harvested every 7 to 14 days, each harvest consisting of a population of shoots ranging from 5 to 15 cm long. Hence, most shoots are harvested before they reach their maximum rate of absolute growth, and the detrimental effect of harvesting small shoots is more pronounced when considered in terms of dry weight.

If the shoot grew for one more week, its total length would increase and its weight would double. When the number of days between harvests was increased from 7 to 14, yield increased by 38 %. Tanton (1981a) further showed that only 11 % for shoot size and remaining 89 % being accounted for by the number of shoots harvested of the total variation in weekly yield throughout the season, although it is important in accounting for difference in the yield between certain clones. The price realized for processed tea depends on its quality, subjectively determined by taste and appearance, and tends to be inversely related to the size of the harvested shoots (Stephens and Carr, 1990). See details in Chapter 9 and para 9.7.3.

6.3.2 *Size of the developing shoot population*

This is expressed as the number of shoots developing per unit area at any one time. Although throughout the cropping year the size of the shoot population in a carefully harvested close canopy of young but mature tea has been shown to remain fairly constant but the size of the shoot population increases with the onset of favourable conditions after periods of dormancy or semi-dormancy which are caused by periods of stress. A possible explanation of this distribution of shoots is that apical dominance or similar phenomena, which are known to be important in affecting patterns of shoot growth in plucked tea (Tanton, 1981b), maintain a constant shoot density while there is active shoot growth. At the onset of favourable growth conditions after a period of stress when shoots were dormant or semi-dormant, many shoots are stimulated to grow and new level of basal shoot population is formed and thereby causing an increase in shoot density throughout a pruning cycle (Tanton, 1981a). The uniform size of the shoot populations suggests a one-to-one replacement of shoots, i.e. when one shoot is harvested another shoot grows from the topmost shoot of the remaining shoot butt (Tanton, 1992). Tanton (1992) further stated that some shoots do arise from unplucked origins and some stumps have more than one shoot, but these must be balanced by an equal number of shoots released from apical dominance which fail to develop.

6.3.3 *Pattern of shoot development*

The tea plant exhibits alternate phases of growth and dormancy. These growth phases involve peaks when production of shoots are maximal and are obscured by repeated harvesting of the young shoots. This aspect has been discussed in Chapter 2 but physiological significance of the pattern of shoot development will be discussed here.

Barua and Das (1979) explained the periodicity of shoot growth in terms of anatomical changes and Banerjee (1993) stated that it is an interaction between physiological and anatomical changes. Thus, Bond (1945) observed reduction of the vascular tissue below the growing apex of a rapidly elongating periodic tea shoot and further concluded that the restricted supply of water and nutrients to the apex was the cause of periodicity of shoot growth. Due to reduction of vascular tissues, the bud enters into a state of apparent rest causing the shoot to become dormant after unfurling the last leaf of the current flush. The vascular area below it continues to increase in size and after sometimes when the same has enlarged sufficiently, the bud starts getting the required quantum of water and nutrients causing the newly formed initials inside the bud to swell up and finally break for new flush of growth. The process continues. In fact, the spell of a *banjhi* therefore can not be regarded as a period of inactivity, because during this stage, the primordia for an ensuing cycle of flush are laid. Wight and Barua (1955) felt that apart from the 'bottleneck' in the supply of water and nutrients, a remote cause must exist in the proximal parts of the plant. Tanton (1979) analyse shoot growth rate data in terms of an exponential rate of shoot extension up to harvestable size, while Herd and Squire (1976) and Squire (1979) analyse shoot growth as having an exponential phase until shoots grow to 2–3 cm and then a linear phase to a harvestable size of 10–15 cm.

6.3.4 *Growth of shoot and yield*

It has been examined in several studies that shoot growth is a major determinant of yield. Like many tropical crops, tea exhibits periods of intensive leaf growth alternating with periods of inactivity. Fordham (1970) and Fordham and Palmer-Jones (1977) have postulated that these alternating leaf growth in the level of crop production were caused by the synchronous development of a basal population of shoots. A vegetative shoot that is allowed to grow uninterrupted unfolds leaves in a normal manner until leaf production ceases with the appearance of *banjhi* bud. Following a period of apparent dormancy, described as the *banjhi* period, the onset of favourable conditions growth is resumed with further leaf unfolding, known as the flush period, after which the cycle is repeated (Fordham, 1977).

These favourable conditions are created by the onset of rains and rapid increase of temperature in Malawi (Tanton, 1992) and Darjeeling respectively. At the onset of the favourable conditions, a rapid, synchronized growth phase of apical shoots results in a peak in tea production when the majority reach harvestable size. Subsequently there is a lull until replacement shoots, released from apical dominance by plucking during the first production peak, are ready for harvest. Once initiated these fluctuations in yield continue throughout the growing seasons though they tend to become less marked with time. Tanton (1992) observed a population of shoots reached harvestable size over a few weeks when shoots are growing fast in the most favourable conditions for growth, but shoots growing slowly were harvested over many weeks. This effect has a considerable influence on the distribution of crop.

6.4 Photosynthesis

Growth of all green plants is dependent on photosynthesis, absorption of moisture and nutrients from the soil. Tea leaves when they are fully expanded attain self sufficiency in photosynthetic capacity. Photosynthetic efficiency of a leaf is maximized when it is about four weeks old and optimal efficiency lasts for about six months and thereafter the efficiency gradually declines. The small fish leaf is also photosynthetically very active and found that four times more efficient per unit leaf area. Photosynthetic efficiency of tea leaf in fixing the atmospheric carbon dioxide and partitioning of assimilates are governed principally by the environmental conditions, the management practices and the genetic potential of the plant (Barua, 1953; Marimuthu *et al.*, 1994; Sanderson and Sivapalan, 1966; Zelitch, 1971). Among the weather variables, both intensity and photoperiod incident light affect the photosynthetic efficiency of tea leaf and this has direct influence on the assimilation capacity, partition of assimilates, growth rate and quality constituents (Barua, 1964; Manivel and Hussain, 1986).

6.4.1 *Effect of light intensity*

Intensity of light affects the rate of photosynthesis. In most of the plants, photosynthesis is maximum in bright diffused sunlight. It decreases in strong light and again slows down in the light of very low intensity. If the supply of CO_2 is maintained constant, the rate of photosynthesis increases with the intensity of light till maximum is reached. At this stage plant becomes light saturated. After this, the rate of photosynthesis is controlled by carbon dioxide. In Tocklai, it was observed that enrichment with CO_2 and greater light intensity can increase the rate of photosynthesis by 20 to 45 percent (Barua, 1976). Smith *et al.* (1993) initially did not find any significant reduction in photosynthesis at high illuminance but photo-inhibition was only revealed when nutritional status of tea was taken into account. In Japan, Sakai (1987) estimated photosynthetic light saturation to be between 500 and 1000 µ mol m^{-2} s^{-1} PPFD (Photosynthetic photon flux density). Light saturation was estimated at above 350 W m^{-2} (equivalent to 735 µ mol m^{-2} s^{-1}) in Malawi (Squire, 1977), and, 1340µ mol m^{-2} s^{-1} in Darjeeling (Ghosh Hajra and Kumar, 1999). This relatively high light saturation point suggests that in the absence of limiting factors such as excessive leaf temperature or poor plant nutrition, tea grown without shade should fix more carbon dioxide than shaded plants. This could be confirmed from the data obtained from the shade experiment conducted in Malawi. Tea under *Grevillea robusta* shade trees yielded 1034 and 1017 kg ha^{-1} N (±113) with 45 and 112 kg ha^{-1} N respectively, while unshaded tea yielded 1003 and 1759 kg ha^{-1} N (±113) at the same low and high levels of nitrogen (Tea Research Foundation, 1969/70). In south India, about 12 percent increase in net photosynthetic rate was observed under regulated shade of *Grevillea robusta* than the plants grown under unshaded environment. Hence, growing shade trees in tea plantations in south India as per the recommendation would be advantageous and proper management will allow adequate quantum of light at bush level besides keeping the bush under a favourable microclimate.

The interception and distribution of photosynthetic radiation within the tea canopy also affect net photosynthesis. It revealed that the top leaf is most efficient in photosynthesis due to its age, adequate incident light and proximity to the growing shoot. The rate of photosynthesis declines from the top of the canopy downwards due to paucity of light and physiological maturity.

6.4.2 *Effect of temperature*

Like all other physiological processes, photosynthesis also needs a suitable temperature. In the presence of plenty of light and carbon dioxide, photosynthesis increases with the rise of temperature till it becomes maximum. The effect of temperature on photosynthesis for different areas can not be generalised as the optimum range would possibly vary depending on the overall environmental conditions of an area. Although the relation between temperature and photosynthesis can be defined in the laboratory, it is difficult to do so in the field where high temperature probably occur with low leaf water potential. In Japan, Sakai (1987) found a much lower temperature optimum for photosynthesis at 25°C for laboratory-grown tea, although Hadfield (1975), also using laboratory-grown plants in north- east India, found a temperature optimum at around 35°C, with a rapid fall above 37°C and no net uptake of CO_2 above 42°C. Barbora (1994) reported temperature optimum for net photosynthesis under controlled condition in a growth chamber rested within 25 to 30°C in high and low yielding clones of Assam, China and Cambod types. On the other hand, working with leaf discs of tea suspended in bicarbonate solution, Aoki (1979) found in Japan that 30°C to be the optimum temperature for net photosynthesis. In Darjeeling, since air temperature is never lethally high, it was not a limiting factor for photosynthesis. Thus, the available information is too limited to generalize on the temperature requirement of the whole range of tea plants. Tea cultivars with horizontal, semi-erect and erect leaf pose show varying responses to natural light. However, the optimal temperature of 25°C and 30°C for most tea plants refer to that of the leaf not to ambient temperature (Barua, 1987). The temperature of the tea leaf tends to be 10 to 15°C higher than that of the ambient under strong light and high tempearure (Hadfield, 1968).

Unlike extension growth, photosynthesis does not stop at 13°C. In north-east India and Darjeeling, photosynthesis was observed to continue during winter when minimum temperature fell below 13°C and the tea bushes were dormant. In Japan, photosynthetic activities during winter were depressed by 40 % in leaves of all ages (Okano and Matsuo, 1994). Squire (1977) reported that in Malawi, the shoot growth of tea dropped to about 10 percent during the winter months of May to August (of the peak values obtained in summer months).

6.4.3 *Effect of water deficits*

Drought stress appears to affect rates of photosynthesis in tea significantly less than in other crops, with the data from Smith *et. al.* (1993) showing a 13 % drop in net assimilation per unit leaf area when ô (water stress factor) fell to as low as 0.05. The data of Squire (1977) for photosynthetic rates show a larger drop of about 30 % in the dry season compared with the wet season in Malawi, where values of ô are probably similar to, if not lower than those encountered in Tanzania. In Darjeeling, December to April could be considered as a moisture stress period. During pre-monsoon (April) when atmosphere was dry and plants were suffering from moisture stress, minimum photosynthetic rate was recorded in comparison with other seasons. With the onset of the monsoon rain in June the plants were able to recover fully from the water stress and an increase in the water content of the leaves resulted in a corresponding increase in photosynthetic rate in general. The increased water content helps in maintaining the turgidity of the assimilatory cells and the proper hydration of their protoplasm. During a long, severe drought, bushes might exhaust their reserves of carbohydrate and the growth of the shoots would then be limited by shortage of reserves when drought ends. Barua (1987), working in Assam, observed in the case of water-stressed plants, normalcy in terms of photosynthetic rate was restored after 4-5 days of watering.

6.4.4 *Photosynthesis and yield*

Since it is the vegetative young shoot that is harvested from the tea bush, it has been presumed that productivity of tea is more directly linked to photosynthetic rate than in other perennial crops (Roberts and Keys, 1978). Further, the importance of carbon input in tea productivity can be appreciated from the fact that 95 percent of dry weight of plant is derived from photosynthesis (Banerjee, 1993). Therefore, increase in net photosynthesis and more efficient partitioning are logical steps for increasing productivity.

Several attempts have been made to correlate photosynthesis or factors related to photosynthesis with crop yield (Devanathan, 1975; Hadfield, 1975) but interpretation of results in terms of photosynthesis and yield is not possible since climatic conditions favourable to photosynthesis are also favourable for growth of shoot. Hadfield (1975) showed that China-type tea, with a photosynthetically favourable semi-erect leaf pose, produced four times the yield of Assam tea, which has a horizontal leaf pose, concluding that this difference was related to photosynthetic efficiency. However, the semi-erect leaf types produced up to six times the number of shoots per unit area as the horizontal type, and hence had a much larger and more active sinks (Tanton, 1979). Squire and Callender (1981) suggested that tea yield is independent of current photosynthesis. Their observation was based on the low harvest index of tea, the fact that harvesting method can determine yield, and the lack of correlation between shoot growth and photosynthetic rate in the short term. This uncoupling of photosynthetic rate and shoot growth could be due to large soil water deficits (Stephens and Carr, 1991a), high saturation deficit or low temperature (Squire, 1979). When such constraints are removed at the onset of favourable growing conditions, carbohydrate reserves built up during the period of growth limitation can be mobilised to promote the growth of the emerging shoots (Hakamata and Sakai, 1980). In this way yield would be uncoupled from carbon assimilation in time, but not be completely independent of it (Smith *et al.*, 1993).

A proportion of the carbon fixed by photosynthesis is used in maintenance respiration, and the remainder is then available for partitioning into shoot or root growth or into storage reserves. If photosynthesis and yield are closely coupled, the largest yield in Darjeeling would be expected in October when photosynthetic rate was observed to be maximal. But the highest yield was obtained in June. Similarly in mature tea plantation of the Southern Highlands of Tanzania, Smith *et al.* (1993) recorded highest rate of photosynthesis from irrigated plots when potential soil water deficit reached 50 mm and nitrogen applied at the rate of 225 kg ha^{-1}. However, largest yield was recorded in the plots where nitrogen applied at the dose of 375 kg ha^{-1}. Squire and Callender (1981) based on the low harvest index in tea, concluded the fact that harvesting method can determine yield, and the lack of correlation between shoot growth and photosynthetic rate in the short term.

6.5 Respiration and photorespiration

The size of the tea bush makes it difficult to get a direct estimate of respiration rates, but since direct measurement of plant mass indicates that tea accumulates 17.50 t ha^{-1} $year^{-1}$ of dry matter, of which approximately 16.00 t ha^{-1} $year^{-1}$ is estimated as CH_2O, respiratory loss in Malawi was estimated as 67 percent of total photosynthetic production (Tanton, 1979). In Assam, Barua (1987) reported 34 to 37 percent of the dry matter produced by a plucked tea bush in a year remained in the plant, the rest 66–63 percent was lost in respiration. Close agreement between these two estimates might be due to similarity of ambient temperatures of Malawi and Assam. Respiratory loss could be less with corresponding gain in dry matter in colder climates of Kenya or high elevation Sri Lanka (Barua, 1987).

Decreasing respiration would result in increases in net photosynthesis because respiration utilizes a high proportion of the fixed sugars. As no energy-rich compound is produced during this process, photorespiration acts to undo the work of photosynthesis. However, dry matter accumulation would benefit from a reduction in respiration rate.

The operation of photorespiration in the tea plant has been demonstrated by Roberts and Keys (1978). They observed rise of oxygen tension in the photosynthetic medium stimulated photorespiration, reduced photosynthesis and raised the CO_2 compensation point. At the atmospheric levels of O_2 and CO_2, photorespiration was 19 percent of photosynthesis. Banerjee (1993) opined that if the ratio between photorespiration and net photosynthesis in different tea genotypes can be worked out, it might even be possible to develop biochemical inhibitors which will decrease photorespiration and increase net photosynthesis.

6.6 Winter dormancy

The young shoots of a tea plant which form the major part of the harvest is called as flush. The rest period when a whole plant or a plant tissue shows no growth between two successive flushes is known as the dormant or *banjhi* period. Tea enters into phase of dormancy during winter in Assam and Dooars (December to January), Darjeeling (middle of November to end of March) and other countries (Table 6.1). Tea cultivated near the equator produces almost the same yield every month, but further from the equator winter harvest gradually declines and at latitude beyond about 16°, there is almost complete winter dormancy. Barua (1969) showed that tea bushes go dormant when the average day length remains approximately below 11 h 15 min for a period of about six weeks and this happens beyond about 16^0 north or south latitudes. The dormant tea plant resumes growth when the day length rises about this critical level in the spring. Short days alter the balance of the endogenous growth regulators in favour of dormancy and long days in favour of growth. When dormant tea plants were injected with gibberellic acid at 10 and 40 ppm in December, the treated plants started growth within a fortnight but became dormant after unfolding two to four leaves. These plants resumed growth for second flush in early March which coincided with bud break and growth of the untreated plants for the first flush. However, research findings have not

Table 6.1 : Period of winter dormancy in some tea growing regions.

Region	Mean latitude	Months of dormancy
Argentina	30⁰ S	3 - 4
Darjeeling	27⁰ N	4
Georgia	42⁰ N	6
Indonesia	5⁰ S	0
Iran	37⁰ N	5 - 6
Kenya	0⁰	0
Kangra (HP)	32⁰ N	4
Malawi	16⁰ S	0
Mauritius	20⁰ S	2 - 3
N. E. India	26⁰ N	3 - 4
South India	11⁰ N	0
Sri Lanka	6⁰ N	0
Turkey	41⁰ N	6

conclusively proved that growth promoters can be recommended for getting over the phenomenon of winter dormancy.

In certain tea growing areas at higher latitudes as in Georgia (42⁰ N), Darjeeling (27⁰ N), Iran (37⁰ N) etc, the minimum temperature in winter may drop even below freezing point. Thus at higher latitudes, short day length and low ambient temperature are confounded during the winter season when visible growth of the tea plant completely stops. But temperatures prevailing in certain tea growing areas suggest that low temperature can not be its sole reason. For example, mean range of maximum and minimum temperature at Tocklai (Assam) during winter is 22⁰C-26⁰C and 10⁰C-16⁰C respectively where winter dormancy is experienced for about 3-4 months, whereas at Kericho in Kenya, the temperature varies from 19⁰C-25⁰C and 8⁰C-9⁰C respectively there is no dormancy.

Exposure to 13 hr photoperiod from November to March under ambient temperature condition at Tocklai also failed to prevent dormancy but bud break was advanced by about a month (Barua, 1969). Lack of water or nutrition was also ruled out as major cause of dormancy. Kinetin and indole acetic acid has no effect but tricontanol has been reported to reduce the dormancy level (Ranganathan *et al*., 1983). Nandi and Palni (1993) reported the involvement of cytokinins in the control of winter dormancy and subsequent bud break. However, the problem of winter dormancy in tea not yet resolved. Low temperature and day length apparently interact in inducing dormancy but their respective roles are not clear.

6.7 Partition of photosynthates and productivity

A large fraction of the dry matter synthesized by plants is lost (64 percent) in respiration and the balance (36 percent) is distributed among various plant organs. Prunings alone accounted for above 57 percent of the loss. The percentage of dry matter in plucked shoots in relation to the gross weight of the entire plant (total weight of dry matter produced) is the harvest index of tea. The harvest index of tea is reported to be 9 to 12 percent in plains of north-east India (Rustagi and Barman, 1994), 16 percent for clone UPASI 3 in south India (Murty and Sharma, 1986), 24 percent observed for clone S15/10 for the period two to four years after field planting in the Mufindi district of southern Tanzania (Burgess and Carr, 1996a) and 8 percent reported for clone 6/8 at high altitude in Kenya (Magambo and Cannell, 1981). A high harvest index does not necessarily mean high yield. But increasing the harvest index helps in reducing the inputs without reducing the yield. The increment in photosynthesis, reduction in respiratory loss, mobilization of food materials to new shoots, and reduction in the formation of non-productive plant parts are important for increasing the harvest index. Improved nutrition and, on some tea estates of eastern Africa, the introduction of irrigation will have increased the efficiency of conversion of intercepted radiation to dry matter. The effect of low (and high) temperatures and large saturation deficits of the air could reduce photosynthesis and dry matter production. Values of the conversion coefficient for most existing tea cultivars appear to be low by comparison with other crop (Carr and Stephens, 1992). In southern Tanzania, Burgess and Carr (1996a) reported that net dry matter production in young tea appears to be less sensitive to low temperatures than shoot extension, the reduced 'demand' for photosynthates by the shoots at low temperatures results in increased partitioning of dry matter to the roots. This translocation of photosynthates to the roots during cool weather has also been demonstrated for tea using radio-carbon tracing techniques in Assam (Manivel and Hussain, 1982), whilst active root growth during the cool season in Tanzania has been observed in underground root chambers by Carr (1971). The processes are reversed when air temperatures increase.

High temperature leads to fast shoot extension rates and the resultant 'demand' of the shoots for photosynthates reduce the diversion of assimilates to the roots. In tea, preferential translocation of dry matter to the canopy after a cool season has been reported in Malawi (Fordham 1972; Herd and Squire, 1976), which explains, in part, the large peaks in yield observed from irrigated tea during October in southern Tanzania (Stephens and Carr, 1991b; Burgess and Carr, 1996b).

Dry matter production and partitioning also vary in tea cultivars. Kabir *et al*. (1992) while studying growth parameters of six China-type high quality clones found a significant correlation between leaf area and shoot dry weight which showed the growth of the plant will give an indication of its efficiency for dry matter production. Amongst the clones stem diameter, plant height, dry weight of root and shoot were observed to be maximum in B157 and minimum in AV2. In southern Tanzania, large initial yields from clone S15/10 resulted from a high harvest index, which was associated with greater dry matter partitioning to leaves and harvested shoots in comparison with other three clones (Burgess and Carr, 1996a). A similar positive relationship has also been reported in Kenya (Othieno, 1982).

Magambo and Cannell (1981) observed that plucking significantly decreases the proportion of dry matter partitioned to stems but increases its proportions to leaves and roots. However, it was reverse in unplucked bushes.

The process of dry matter production and partitioning in tea are still not well understood. Some uncertainty exists in estimates of the radiation conversion efficiency, which ranges from 0.25g MJ^{-1} to 0.49g MJ^{-1} (Burgess, 1992). Much of this variability could be due to differences in respiratory loss, differences in the size of the bushes and the ambient temperature. There is even more uncertainty in the way in which newly produced assimilates are partitioned to various parts of the plant (Matthews and Stephens, 1998).

References

Aoki, S. 1979. $^{14}CO_2$ fixation in leaf discs of *Camellia sinensis* (L) O. Kuntze (Japanese). *Study of Tea*, **56:** 1–5.

Banerjee, B. (1993). *Tea production and processing*. Oxford & IBH Publishing Co., New Delhi, India, p. 336.

Barbora, A. C. 1994. Effect of temperature on photosynthesis and respiration in tea

(*Camellia sinensis* L.). *Two and a Bud*, **41(1):** 12–18.

Barua, D. N. 1953. Effect of light intensity on the growth and assimilation of tea (*camellia sinensis*) seedling. *Ph.D. Thesis*, Cambridge University, England.

Barua, D. N. 1964. Effect of light intensity on the assimilation characteristics of detached tea leaves. *J. Agric. Sci.*, **63:** 265–271.

Barua, D. N. 1976. Reporting on new developments in tea physiology at Tocklai Conference, 1976.

Barua, D. N. 1987. Tea. *In : Tree Crop Physiology* (eds. M. R. Sethuraj and A. S. Raghavendra). Elsevier Science Publishers, The Netherlands, 225–246.

Barua, D. N. 1969. Seasonal dormancy in tea (*Camellia sinensis* L.). *Nature*, **224:** 514.

Barua, D. N. and Das, S. C. 1979. Mechanism of growth periodicity in tea. *Two and a Bud*, **26:** 36–40.

Bond, T. E. T. 1945. Studies in the vegetative growth and anatomy of the tea plant (*Camellia thea* Link) with reference to phloem. II. Further analysis of flushing behaviour. *Ann. Bot.*, **9:** 183–216.

Burgess, P. J. 1992. Response of tea clones to drought in southern Tanzania. *Ph.D. Thesis*, Cranfield Institute of Technology, Silsoe, UK.

Burgess, P. J. and Carr, M. K. V. 1996a. Response of young tea (*Camellia sinensis*) clones to drought and temperature. II. Dry matter production and partitioning. *Expl. Agric.*, **32:** 377–394.

Burgess, P. J. and Carr, M. K. V. 1996b. Response of young tea (*Camellia sinensis*) clones to drought and temperature. I Yield and yield distribution. *Expl. Agric.*, **32:** 357–372.

Carr, M. K. V. 1971. An assessment of some results of tea-soil-water studies in southern Tanzania. *In: Water and the Tea Plant* (eds. M. K. V. Carr and Susan Carr). Kericho, Kenya: Tea Research Institute of east Africa, 21–48.

Carr, M. K. V. and Stephens,W. 1992. Climate, weather and yield of tea. *In : Tea: Cultivation to Consumption* (eds. K. C. Willson and M. N. Clifford). Chapman & Hall, London, 97–135.

Devanathan, M. A. V. 1975. Weather and the yield of crop. *Expl. Agric.*, **11:** 183–186.

Ellis, R. T. and Grice, W. T. 1976. Plucking policy and techniques. *Tea Research Foundation (Central Africa), Quarterly Newsletter*, **41:** 3-14.

Fordham, R. 1970. Factors affecting tea yields in Malawi. *Ph.D. Thesis*, Bristol University.

Fordham, R. 1972. Observations on the growth of roots and shoots of tea (*Camellia sinensis* L.) in southern Malawi. *J. Hort. Sci.*, **47:** 221–229.

Fordham, R. 1977. Tea. *In: Ecophysiology of Tropical Crops* (eds. Paulo de T. Alvim and T. T. Kozlowsky).Academic Press, London and New York, 333–349.

Fordham, R. and Palmer-Jones, R. W. 1977. Simulation of intra seasonal yield fluctuations of tea in Malawi. *Expl. Agric.*, **13:** 33–42.

Ghosh Hajra, N. and Kumar, R. 1999. Seasonal variation in photosynthesis and productivity of young tea. *Expl. Agric.*, **35:** 71–85.

Hadfield, W. 1968. Leaf temperature, leaf pose and productivity of the tea bush. *Nature*, **219:** 282–284.

Hadfield, W. 1975. The effect of high temperatures on some aspects of the physiology and cultivation of the tea bush (*Camellia sinensis*) in north-east India. *In: Light as an Ecological Factor* II (eds. G. C. Evans, R. Bainbridge and O. Rackham). London: British Ecological Society, 477-495

Hakamata, K. & Sakai, S. 1980. Translocation and redistribution of $^{14}CO_2$- photosynthates assimilated in winter leaves in the young tea plant. *Studies of Tea,* **58:** 11-20.

Herd, E. M. and Squire, G. R. 1976. Observations on the winter dormancy of tea (*Camellia sinensis* L.) in Malawi. *J. Hort. Sci.*, **51:** 267–279.

Jain, N. K. and Tamang, M. B. 1988. Physiological basis of tea management practices. *Proc.Intl. Cong. Plant Physiol.* February 15–20, 1988, New Delhi, India, 469–479.

Kabir, S. E. and Ghosh Hajra, N. 1992. Analysis of growth components of Darjeeling tea clones. *Phytobreedon*, **8(1):** 54–62.

Magambo, M. J. S. and Cannell, M. G. R. 1981. Dry matter production and partitioning in relation to yield of tea. *Expl. Agric.*, **17:** 33–38.

Manivel, L. and Hussain, S. 1982. Photosynthesis in tea II: Direction of movement of photosynthates. *Two and a Bud*, **29:** 49–52.

Manivel, L. and Hussain, S. 1986. Relative sink capacity of developing tea shoots. *Two and a Bud*, **33:** 30–33.

Marimuthu, S., Raj Kumar, R. and Manivel, L. 1994. Factors affecting partitioning of assimilates in tea. *J. Nuclear Agric. Biol.*, **23:** 219–223.

Matthews, R. B. and Stephens, W. 1998. Cuppa-tea: A simulation model describing seasonal yield variation and potential production of tea. 2. Biomass production and water use. *Expl. Agric.*, **34:** 369–389.

Murthy, S. B. K. 1992. Harvesting in tea – precepts and practices. *Proc. 30 th UPASI Sci. Conf.*, September 14, 1992, 43–62.

Murty, R. S. R. and Sharma, V. S. 1986. Canopy architecture in tea (*Camellia* L. spp.). *J. Plantation Crops*, **14:** 119–125.

Nandi, S. K. and Palni, L. S. 1993. Shoot growth and winter dormancy in tea. *J. Plantation Crops*, 21(supplement): 328–332.

Othieno, C. O. 1982. Supplementary irrigation of young clonal tea in Kenya. III. Comparative dry matter production and partition. *Tea*, **3(1):** 15–25.

Okano, K. and Matsuo, K. 1994. Winter-depression of photosynthesis in tea plants (*Camellia sinensis*) with special reference to cultivar and leaf age. *Japanese J. Crop Sci.*, **63(2):** 240–246.

Ranganathan, V., Raman, K. and Nateson, S. 1983. Nutritional and physiological interactions with lengths of plucking rounds and banjhiness. *UPASI Tea Sci. Dept. Bull.*, **38:** 67–69.

Roberts, G. E. and Keys, A. J. 1978. Mechanism of photosynthesis in tea plant (*Camellia sinensis L.*). *J. Expt. Bot.*, **29:** 1403–407.

Rustagi, P. N. and Barman, T. S. 1994. Physiological processes influencing growth and productivity of tea. *In : Field Management in Tea* (ed. J. Chakravartee). Tocklai Experimental Station, Jorhat, Assam, India, 10–16.

Sakai, S. 1987. Studies on photosynthesis and dry matter production of tea plants. *Bulletin of the National Research Institute of Tea*. Kanaga Shizuoka, Japan: National Research Institute of Tea.

Sanderson, G. W. and Sivapalan, P. 1966. Translocation of photosynthetically assimilated carbon in tea plants. *Tea Q.*, **37:** 140–153.

Smith, B. G., Stephens, W., Burgess, P. J. and Carr, M. K. V. 1993. Effect of light, temperature, irrigation and fertilizer on photosynthetic rate in tea (*Camellia sinensis*). *Expl. Agric.*, **29:** 291-306.

Squire, G. R. 1977. Seasonal changes in the photosynthesis of tea (*Camellia sinensis* L.). *J. Appl. Ecol.*, **14:** 303-316.

Squire, G. R. 1979. Weather, physiology and seasonality of tea (*Camellia sinensis*) yields of Malawi. *Expl. Agric.*, **15:** 321-330.

Squire, G. R. and Callender, B. A. 1981. Tea plantations. *In : Water Deficits and Plant Growth* 6 (ed. T. T. Kozlowski). New York: Academic Press, 471-510.

Stephens, W. and Carr, M. K. V. 1990. Seasonal and clonal differences in shoot extension rates numbers in tea (*Camellia sinensis*). *Expl. Agric.*, **26:** 83–98.

Stephens, W. and Carr, M. K. V. 1991a. Responses of tea (*Camellia sinensis*) to irrigation and fertilizer. II. Water use. *Expl. Agric.*, **27:** 193-210.

Stephens, W. and Carr, M. K. V. 1991b. Responses of tea (*Camellia sinensis*) to irrigation and fertilizer. I. Yield. *Expl. Agric.*, **27:** 177-191.

Tanton, T. W. 1979. Some factors limiting yields of tea (*Camellia sinensis*). *Expl. Agric.*, **15:** 187–191.

Tanton, T. W. 1981a. Growth and yield of the tea bush. *Expl. Agric.*, **17:** 323–331.

Tanton, T. W. 1981b. The *banjhi* or dormancy cycle in tea. *Expl. Agric.*, **17:** 149–156.

Tanton, T. W. 1992. Tea Crop physiology. *In: Tea: Cultivation to Consumption* (eds. K. C. Willson and M. N. Clifford). Chapman & Hall, London, 173–199.

Wight, W. and Barua, D. N. 1955. The nature of dormancy in the tea plant. *J. Exp. Bot.*, **6:** 1–5.

Zelitch, I. 1971. *Photosynthesis, photorespiration and plant productivity*. Academic Press, London, p.347.

Chapter 7

Techniques of propagation

7.1 Introduction

The traditional source of planting material is seed. Though it is convenient to use but a seed population is composed of a large number of genetically heterozygous genotypes. However, it can be fitted into a wide range of environmental and field conditions without much change in its overall performance. In order to produce a uniform crop with predetermined characters, vegetative propagation of superior clones are practiced. Vegetative propagation through single-node leaf cuttings is a reliable and economic method. The superiority of binodal cuttings in terms of better growth and success in Darjeeling over single-node cuttings has been reported (Anonymous, 1997). In Malawi, Kayange (1987) found success rates of binodal cuttings over 80% under mist. However, binodal and trinodal cuttings initially give more branches in the field but transpire more in the nursery (Willson, 1992).

Multiplication of clones or seed stocks is a multistep process. A mistake at any stage of propagation may lead to poor success and unsatisfactory results. Therefore, a sound planning for choice of planting materials and adoption of appropriate techniques in raising and management of nurseries is important.

7.2 Location

The nursery should preferably be located between low and mid elevations and also at a convenient place where constant supervision is possible. It would be helpful if the site is located near a perennial water source and is desirable if it is naturally protected from wind. In fact the area should have a shelter belt planted with evergreen trees like *Acacia auriculiformis*, *Cassia siamea*, *Melia azedarach* (Ghora neem) etc to create a favourable microclimate to protect the nursery from dry wind. The slope of the land is also important. Flat land certainly need drains. Since waterlogging leads to several problems in the nursery therefore a proper drainage at the site is essential. It is convenient to situate the nursery not too distant from the area to be planted.

7.3 Seed production and propagation

Selected plants are planted in 5-6 m apart in the seed orchards (known as 'baries' in the industry). The land of the seed bari should be flat or with a gentle slope with good drainage. The plant should be allowed to grow freely, apart from hygienic removal of dead branches with occasional thinning of branches to ensure entry of adequate light through canopy; trees must not interlock. A ground cover should be grown for the first few years and must be kept clear and free from weeds when production of seed starts. Seeds are borne mostly on the surface of canopy of the tree and collected on the ground. Tea seed bearers require sufficient fertilizer (Anandappa, 1986; Anonymous, 1986a). The design of polyclonal orchards has been discussed by Anandappa (1986).

Tea seeds (Fig. 7.1) if kept open lose their viability rapidly with time. The viability can be extended if the seeds are packed in boxes in moist charcoal. Seeds stored safely up to one year in cold storage at 5 ±1°C. Cold stored seeds could be used to sow seed at stake in spring/early summer which would require lesser care than winter sown seed at stake especially in Dooars/Terai. Generally, a kilogram of seeds will have 275 to 450 seeds depending upon their size. The seed should be sorted by floatation immediately before sowing them in the nursery. All those which have sunk within 24 h are the best. Seed which sink during the ensuing 72 h may be used but raised separately; they may produce significant proportion of week plants. The remaining seeds which are floaters must be discarded.

Seeds are spread out in the sun, preferably on black polythene sheet as it retains the heat and kept moist by frequent watering. As soon as a seed cracks it is removed and sown in the nursery. Seeds can also be

Fig. 7.1 : Close up of seeds.

germinated in sand in beds or trays covered with damp hessian. Seeds are usually sown in raised beds prepared after mixing good top soil and sand in the ratio of 1: 3. The beds will be of convenient sizes of 1 metre wide and elevate them about 15 cm above the surrounding paths after digging the whole area to 75 cm and of desired length. On the emergence of the radicle they are planted in the beds with the 'eye' (micropyle) pointing downwards or parallel to the ground surface in order to reduce the percentage of bent tap roots and stems. The soil should be scraped to one side with the hand to a depth of about 2.5 cm and seeds are sown at 1 inch interval in such a way that 3/4 part of the seed is inside the soil and rest 1/4 part is above the soil. After watering the beds, the moist gunny bags or hessian cloth spread on the beds in order to retain the soil moisture, prevent bird damage and warm up the beds to facilitate early cracking of seeds. When roots are formed they should be transplanted either in polythene sleeves (10 cm in diameter and 30 cm long) or in the transplanting bed. If the seedlings are retained in the nursery for a longer period larger sleeve of heavier gauge will be required.

Soil of the transplanting bed should be of good structure, depth and pH should be around 5.0, and free from parasitic eelworms. After sowing the seeds, the beds or sleeves should be covered with a thin layer of mulch (cut grasses, leaves of green crops etc). Planting seeds 12.5 cm apart in a triangular arrangements enable them to grow to stumps. The distance between seeds should be about 20 cm if the seedlings are to lifted with 'bheti' (a cylinder of earth about 15 cm in diameter and 20-25 cm in depth around the roots). The soil of the beds should remain moist. Planting high shade is necessary in the nursery, but this is gradually thinned as the plants grow. Regular application of foliar fertilizer (NPK) help the formation of strong and vigorous plants and recommendations are made for different regions (Anonymous, 1986b; Kathiravetpillai and Kulasegaram, 1986). Seedlings when taken as stumps are pulled or dug out and cut 10 cm and 45 cm above and below ground level respectively. Seedlings are ready to be stumped for planting out after 2-3 years from sowing the seeds depending on growth rate. Further, they should attain at least 1 cm in diameter at the collar and have been atleast 3 months in full sunlight for field planting (Willson, 1999).

7.4 Vegetative propagation

Raising tea plants from single node cuttings has now been widely and successfully adopted economic method of vegetative propagation. The success of vegetative propagation depends upon the selection of plants that are above average yield and quality of processed tea. Selection criteria and breeding have been discussed in Chapter 3.

7.4.1 *Mother bushes*

The nucleus plot of mother bushes of desired clones from which cuttings are to be taken (Fig. 7.2) should be established at a convenient place as separate multiplication plots so that they can be given special treatment. The special treatment helps to increase efficiency in the nursery. A spacing of 120 cm triangular may be adopted for planting Assam and Cambod types of clones whereas for Chinery clones 100 cm spacing could be given. In the nucleus area, planting of number of mother bushes depends on the total requirement of cuttings. A healthy mother bush supplied with balanced nutrients is conducive to good growth of cuttings (Sharma, 1982). Cuttings taken from the centre of the bush usually grow better than those from the periphery. However, this difference in growth rate lasts only for about 16 weeks after which the rooting pattern becomes comparable (Banerjee,

Fig. 7.2 : Cuttings are almost ready in the nucleus plot of mother bushes.

1992). Cuttings can be taken from producing plants in the field but results are likely to be inferior (Willson, 1992).

In north-east India, mother bushes should be manured with 2: 1: 2 or 2: 1: 3 (if soil potash status below 100 ppm) NPK mixture @ 140 to 150 kg N/ha when propagation is done twice in a year. Half the quantity of mixture may be applied to the moist soil in early spring (March-April) when some new foliage appears on the bushes and the balance half after regrowth following pruning/skiffing in autumn (June/July). The nitrogen level may be reduced by 20-30 kg/ha when cuttings are taken only once in a season (Chakravartee and Barbora, 1994). In south India, about 60 g of 4: 3: 3 NPK mixture per bush of young tea (upto 5 years) and about 60 g of 1: 1: 2 NPK mixture per bush of mature tea should be given after each harvest of cuttings in addition to the routine manuring of the field (Sharma, 1976). Zinc and magnesium may need to apply under certain conditions. In Kenya, twice the normal application level is recommended (Anonymous, 1986c). The plants should be protected from pests and diseases in order to obtain healthy cuttings for quicker growth in the nursery.

The mother bush should be clean pruned in mid-October to early February. It may be again deep skiffed in June/July if required. But in Darjeeling, mother bushes should be clean pruned between end of September to mid October every year. Annual pruning in mid or high elevations is not advisable. A cycle of prune and deep skiff is recommended for such situation. Care should be taken in the pruned year to see that branches cut off for making cuttings are cut at 15 cm above the pruning cut. In September/ October in the year after a prune, as skiff is given at about 12 cm above the level of pruning. The removal of *banjhi* shoots after prunning is important. To ensure supply of good cuttings for extended propagation, the time of pruning/skiffing should be staggered by 10-15 days in batches.

Cuttings with active axillary buds root quicker. 15 to 21 days prior to taking the cuttings, the terminal bud should be pinched off to initiate the axillary buds to sprout. To hastened better root development and dry matter accumulation in cuttings, the bush should be sprayed with 1% aqueous solution (1 kg in 100 litres of water) of zinc sulphate @ 50 to 100 ml per bush 21 days prior to taking cuttings. Application of 1% urea following pinching off terminal bud further aids rooting (Sharma, 1976). The source and age of the plant have influence on the rooting of the cuttings. Sharma (1982) observed in south India that all cuttings taken from 2-year-old clonal plants rooted in 14-16 weeks, but only 28-59 % cuttings taken from 30-year-old mother bushes rooted at the end of 16 weeks. Cuttings taken from the old bushes after appropriate pruning, rooting and growth of cuttings can be brought to the level of those from the younger plants (Venkataramani, 1963). In Darjeeling, binodal cuttings (comprising of two nodes) have shown better growth in respect of root and shoot development (Fig. 7.3) as well as frame formation. It has also shown early sprout of roots and shoots. The plants are also ready for planting in the field much earlier than the plants developed from single node cuttings (Anonymous, 1997).

7.4.2 *Type and preparation of cuttings*

Cuttings should preferably be taken from the primaries with dormant apical buds after discarding two or three topmost leaves. In some clones, quite early the lateral shoots are thrown out from the lower portion of the primaries. In such cases, the unbranched portion of the primaries should be taken first and then the laterals when they mature after a few weeks (Borthakur and Banerjee, 1994). In certain tea growing countries, multinode cuttings are exclusively recommended for propagation and were shown to

Fig. 7.3 : Six months old single nodal (right) and binodal (left) plants showing more developed root and shoot systems.

be superior to single node cuttings in rooting and subsequent growth. It was observed that the multinodal cuttings on an average, produced two or three equally developed shoots, although there is tendency to dominate by one of the growing shoots may be noticed in some clonal plants. The presence of two or more well developed shoots from the basal bud should be an advantage to induce ideal frame formation in the plants in the field (Bezbaruah, 1977; Anonymous, 1997). Multinodal cuttings may be advantageously utilized in certain clones in Assam and Darjeeling hills provided sufficient propagation material is available (Bezbaruah, 1977; Anonymous, 1997). In view of the large number of plants required in tea

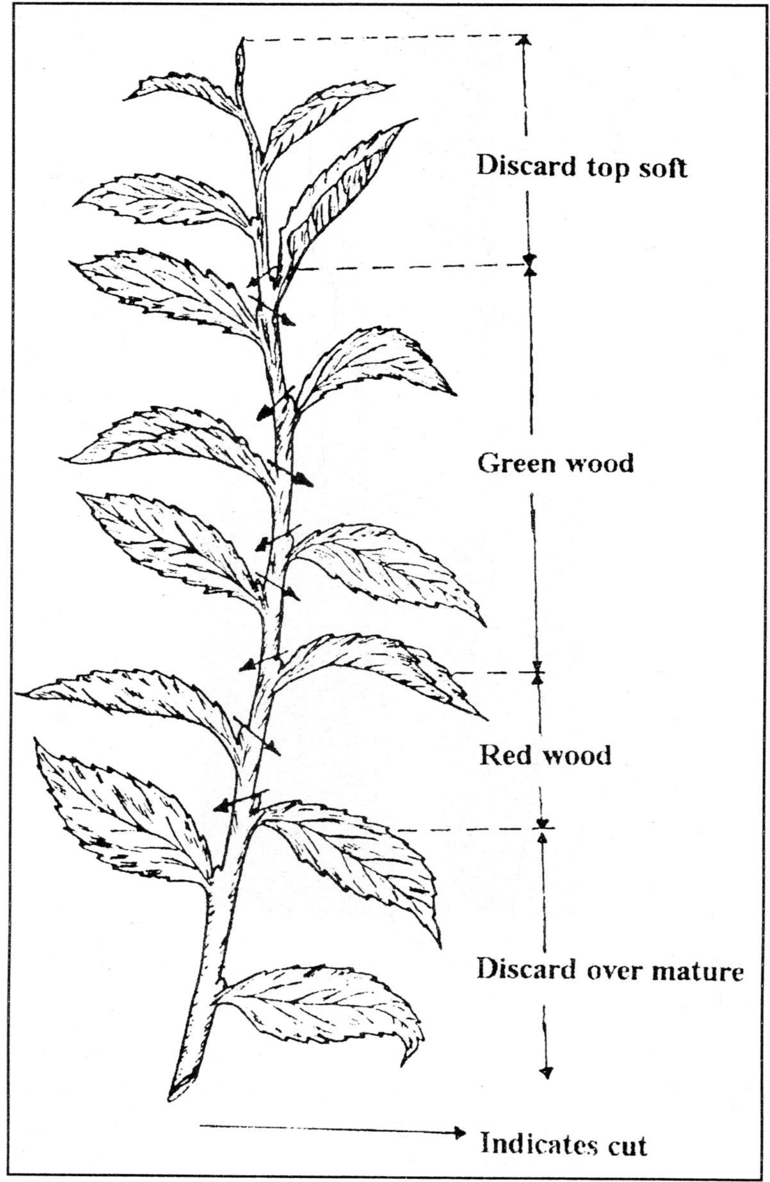

Fig. 7.4 : Method of preparation of single node cutting [Source: Eden (1976)].

plantations, the single-node cuttings were found to be the most economic proposition. Green, semi-hard wood with one leaf and an internode is taken from the stem with a sharp knife without injuring the bark or causing jagged cut as shown in Figure 7.4. The basal cut is made obliquely more or less parallel to the leaf blade and the top cut. Long internodes are cut back to 3 to 4 cm by a second cut. If internodes are less than 3 cm, take binodal cuttings but remove the lower leaf (Willson, 1992). The bottom brown wood portion of the stem may be discarded as they take very long time to root. The most suitable cutting for propagation consisting of a mature leaf with around 2.5 cm of stem below and around 0.5 cm stem above it. In large leaf clones, reduction of leaf area into half of its size results in better rooting (Bezbaruah, 1977). The practice will also help in accommodating more number of cuttings in the nursery. In case of multinode cuttings, irrespective of leaf size, showed progressive reduction in percentage rooting with reduction of leaf area. The increase in amount and length of stem tissue in the multinodal cuttings possibly requires the support of entire leaf area for uptake of water and assimilation of carbon dioxide, and suffers if leaf area is reduced (Bezbaruah, 1977). The bottom cut ends of the finished cuttings should be immersed in a bucket of water and keep them in water until planted in the nursery. When cuttings are to be transported away from the mother bush nursery, harvested shoot dipped in a bucket of water and then it should be wrapped immediately with damp sacking/fresh green plant material and transported to the shaded area of the nursery and again watered. Each cutting can be treated with growth hormone to hasten the rooting. For early root initiation, cuttings may be diped in 0.1 % zinc sulphate solution befor planting. The effect of some growth regulators on rooting has been investigated by Barman and Manivel (1986) and found triacontanol significantly better than others. Trials of a large number of pure rooting hormones and some proprietary preparations showed indole butyric acid (IBA) perhaps to be the only hormone which hastened root growth in tea cuttings and produced initially a large volume of roots (Anonymous, 1992). In south India, Rajasekar and Sharma (1989) observed that IBA at 5000 ppm in combination with zinc at 2500 ppm in solution induced higher percent of rooting and enhanced the length and dry weight of roots and shoot.

7.4.3 *Soil of cutting nurseries*

Soil has a critical influence on rooting of the cuttings. Rooting soil should be a sandy loam, porous and friable with a pH ranging from 4.5 to 4.8 though a pH upto 5.5 is not entirely harmful. High pH soil can be acidified using sulphur or aluminium sulphate or ferrous sulphate. Soil pH beyond 5 tends to promote excessive callusing at the expense of rooting.

Box 7.1 : Callus development on tea cuttings

Callus forms in order to heal the wound on the cutting mainly through the activity of the cambium. Initially the callus has the shape of a small ring which then grows and eventually covers the entire surface of the wound. The main function of the callus is to protect the cutting (plant) against invasion by pathogens and also to prevent the loss of water. But excessive formation of callus at a base of cuttings inhibits production of good roots. Soil pH, aeration in the rooting medium, type of soil, loose soil, loose planting, the condition of cutting, clone etc influence the formation of excessive callus. Plants with a heavy callus should be discarded since they may not be established well in the field.

Source: Argles (1969); Kayange (1992)

Soil with pH above 5.5 should be rejected. The organic carbon content of the soil should be in between 0.8 and 1%. Soils with a high clay content may be avoided since this soil will not drain well enough. Sub-soils upto a depth of 90 cm can be used provided well decomposed organic manure or cattle manure (I part organic manure : 5 parts sub-soil) are added. Addition of concentrated organic manures to sub-soil resulted high casualty of the cuttings (Willson, 1992). Superphosphate @ 50 g per square metre of nursery bed may be applied for development of roots. The same should be incorporated to the soil to a depth of 15-20 cm at the time of preparation of beds. In Assam, Bordoloi and Bezbaruah (1983) observed mixing of superphosphate with the sub-soil was as good as the acidic forest topsoil normally used for nursery. But a mixture of coarse sand with such soils will minimise heavyness of its texture but is expensive. Both soil and sand should be sieved separately through a 8-10 mesh sieve. In southern Africa, it was observed that mixture of topsoil with fully decomposed coffee pulp or tea waste can effectively be used for growing rooted tea cuttings (Mphangwe, 1998). Recently, Mphangwe (1999) further suggested that in absence of good top soil river sand can be used in media mixtures with fully decomposed coffee pulp or tea waste. In Sri Lanka and Kenya, the types of soil is to be used are discussed by Kathiravetpillai and Kulasegaram (1986) and Anonymous (1986d) respectively. Further, soil must be analysed (20 borings from each 0.5 ha block) to ascertain the level of nematode (eelworm) infestation. Soil having root knot nematodes less than four per 10 g of soil can be used. Application of 2 g Furadan 3G per sleeve in two splits at monthly interval helps in controlling eelworm infestation in soil.

Fig. 7.5 : Polythene sleeves after filling up with soil are covered with leafy mulch.

7.4.4 *Preparation of rooting beds and sleeves*

As soon as the beds are prepared or sleeves filled up, provision of shade should be made. If erection of shade immediately is not possible then the rooting beds/sleeves should be covered with leafy mulch (Fig. 7.5) to facilitate protection of soil from direct sun and rain. It also helps to keep down weed growth. Beds can be of 1 m width and any length depending upon availability of the area and elevate them about 15 cm. The soil of the beds should be compacted slightly by using a light roller or back of a hoe. Usually, it takes 6-8 weeks for the soil to be ready for planting. An effective open drain should be made for easy run-off of surface water. In between beds, 50-60 cm deep drains may be provided which should be connected with an outlet drain.

Polythene sleeves of the size 15-18 cm lay flat, 25-30 cm length and 150 gauge (300 sleeves in 1 kg) may be used in the nursery. The soil before filling the sleeves (Fig. 7.6), may be passed through a sieve of mesh No. 4 to remove the undecomposed organic matter and pebbles. The sleeves should be filled with the minimum of raming, 8-10 weeks before cuttings are planted. The filled sleeves should be put together within frames of about 20 cm high made of iron wire or bamboo (Fig. 7.7). This fence surrounds

Fig. 7.6 : Soil before filling the polythene sleeves.

Fig. 7.7 : Soil filled polythene sleeves put together within bamboo frame.

Fig. 7.8: Under overhead shade of hessian cloth, cutting nursery bed – plants almost ready for planting out.

each bed for holding the sleeves upright.

7.4.5 *Shade*

Bamboo lath frame and thatch in general are used for shading nursery. Coir matting, plastic net or hessian net with appropriate mesh (Fig. 7.8) can also be used. It is better to have east-west oriented nursery beds under overhead shade. In east-west beds the north side should be left open for air circulation and penetration of light. Only about 50 % or less light should be allowed to fall on the beds in a diffused manner. The usual height of the top is around 160 cm from the ground (Fig. 7.9). Two types of shade viz. low bamboo lath frame shade (Fig. 7.10) and north open overhead thatch shade with sloping roof can be done with success in north-east India. Low shade for growing cuttings should be north-south oriented. Shade made up of bamboo lath frame is provided about 22.5 cm over the propagation beds supported by side walls. The height of the lath frame should be gradually raised with the increase in height of the plants. When young plants have produced 4-5 leaves, gradual removal of the bamboo lath roof (alternately) ensures adequate hardening of plants. Heavy shading affects the growth of

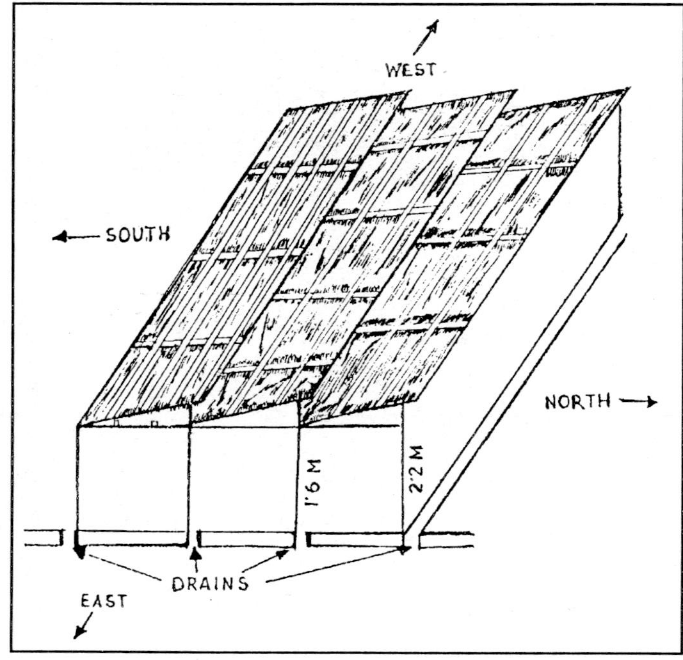

Fig. 7.9 : High overhead shade.

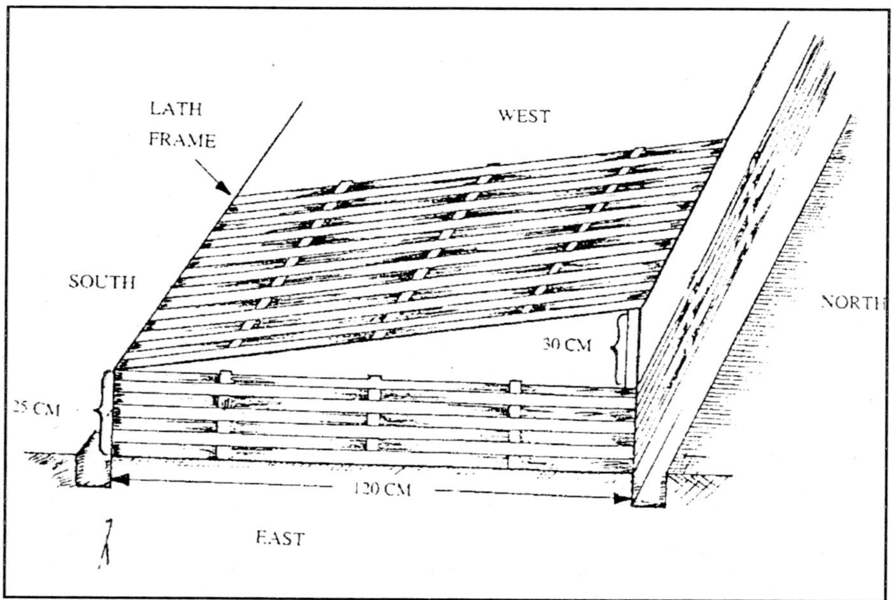

Fig. 7.10 : Lath frame shade in low level bed.

cuttings adversely. Nursery located at higher elevations (eg. Darjeeling) overhead bamboo lath frame is advised, under which each bed should be completely covered by a polythene tent (Fig. 7.11) about 45 cm above the level of the propagation beds using support of semi-circular bamboo (Fig. 7.12) or wire frame reduce drip damage and waterlogging. It allows 30 – 40% light penetration. In south India, raising of cuttings under overhead shade becomes a necessity. An overhead pandal may be raised on which double strand coir-mats of 3.5 mm^2 mesh at mid-elevations and 6 mm^2 mesh at high-elevations should be spread. It should be supported by 2.5 m high wooden or concrete posts at 3 x 3 m gap and wooden rafters of 3.5 x 5 cm size over the posts. The beds of nursery must be kept cool and moist. Cuttings raised under polythene tent should be sealed after watering. Generally, it does not require watering again before the tent is removed. This should be stretched over hoops of wire or bamboo and sealed into the ground along the sides and ends of the bed. It should be seen that the tent is humid and worm. However, tent can be opened for watering if required and remove the weeds if any. The effects of temperature on the growth of cuttings was investigated by Patabava (1987) and observed optimum temperature was 25 to 30°C. Temperature of -3 to 5 °C and 45 to 55 °C were lethal. The hardening temperatures found as 5 to 15°C and 35 to 40 °C.

7.4.6 *Planting of cuttings*

The nursery beds or / and polythene sleeves ready for planting cuttings should be watered thoroughly but slowly avoiding runoff and soil wash. The planting of cuttings should be done very carefully without damaging the cut-end and the bark. Holes may be punched in the soil with a small stick of appropriate thickness of the cuttings and then cuttings may be inserted. The node and the petiole should not be burried or touch the soil surface (Figs. 7.13 and 7.14). The soil around each cutting must be firmly pressed to avoid air pockets at the interspace of cuttings and soil. Mother leaf should be more or less in an upright position after insertion. The leaves of the cuttings planted in close proximity must not overlap. The cuttings can be planted with the tips of the leaves pointing towards west to get better success under overhead shade whereas they may be planted with tip of leaves pointing to the north under low bamboo lath frame shade (Chakravartee and Barbora, 1994). Correct method of planting cuttings are illustrated in

Fig. 7.11 : Bed covered by polythene tent under overhead shade.

Fig. 7.12 : Cutting nursery bed, plants under polythene tent.

Fig. 7.13 : Single node cuttings planted in polythene sleeves for propagation.

Fig. 7.14 : Double node cuttings planted in polythene sleeves for propagation.

Fig. 7.15. A light watering in absence of rain may be given immediately after planting. In the rooting beds, spacing of 5 cm x 5 cm or 6 cm x 6 cm is adequate for planting cuttings. The cuttings are planted in Japan at 2.5–3 cm in rows on 100–120 cm wide beds, established 12–20 cm apart.

Weeds can be removed manually (Fig. 7.16) at very early stage. Young plants should be protected from pests and diseases by spraying appropriate pesticides. When plants are 30-35 cm high with step up, disbudding or tipping may be done carefully to encourage laterals.

7.4.7 *Transplanting into sleeves from rooting beds*

During favourable season cuttings may take 4-8 weeks for callusing and 10-12 weeks for rooting. After the roots have grown cuttings should be transplanted to sleeves with a ball of earth.

7.4.8 *Manuring*

It has been discussed in Chapter 10.

7.4.9 *Storage of cuttings*

Cuttings pretreated by growing them in the nursery bed for a period of about two months until a thin layer of callus was formed over the cut end could be transported over long distance. Such cuttings retained viability up to 10 days (Anonymous, 1992). Fresh cuttings and whole shoots could be stored in sealed polythene bags under cool condition for about four days. Increase days of storage reduce survival.

7.4.10 *Budding and grafting*

Grafting to produce composite plants is a common practice in the horticultural industry. Growing of composite plants has been done with success in different tea growing countries. Crown grafting onto mature field bushes was carried out in Japan in 1902 (Katsuo, 1969), and cleft grafting onto mature root stock in the field was done successfully at Tea Research Foundation (Central Africa) for rapid production of vegetative material (Harvey, 1989). In Kenya, cleft and approach graftings have mostly been done in the field (Templer, 1971; Templer and Machaga, 1978). The latter method has also been used with young clonal plants in the nursery. However, chip-budding and splice grafting techniques were found most suitable in Kenyan nursery (Anyuka and Othieno, 1982). But Bore *et al.*

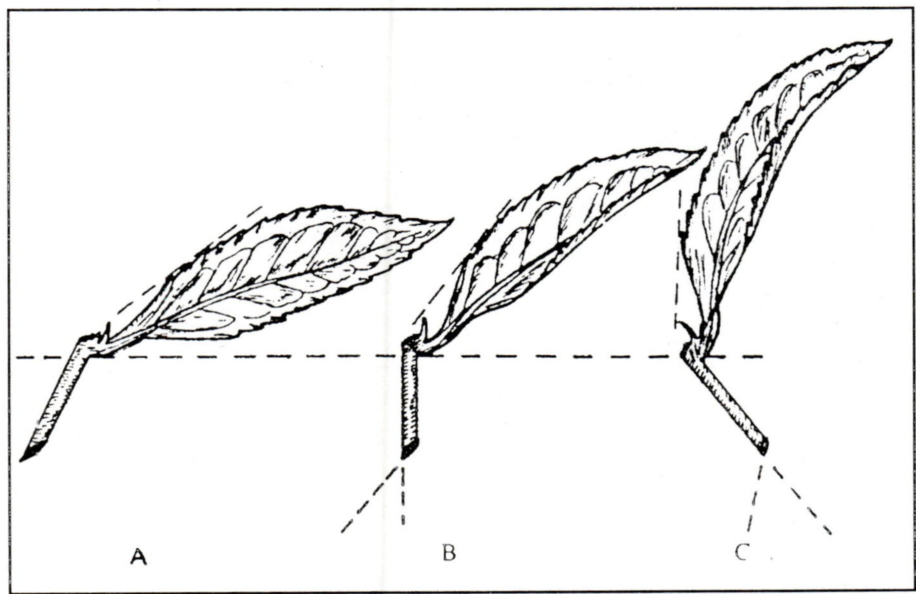

Fig. 7.15: Method of planting of leaf cuttings. Cuttings should be slightly inclined as in 'C'. If planted as in 'A' or 'B' then leaf gets splashed with soil particles exposing it to fungal infection.

Fig. 7.16 : Hand weeding in the cutting nursery.

(1995) reported no significant yield advantages by grafting in Kenya and in some cases the yield of the scion was significantly decreased when grafted onto a different rootstock. The propagation of tea through grafting was tried in Java, Indonesia in the late twenties, in north-east India from the mid-thirties (Barua, 1989) and then in southern Africa in the early 1970s (Nyirenda, 1996). In the early years, patch budding or bud grafting technique (Barua, 1968) was used and later cleft grafting (Grice, 1968) was introduced successfully. In Sri Lanka, cleft grafting has been successfully used to increase the yield of clones which already performed well by about 35 percent over a complete pruning cycle (Pallemulla *et al*., 1992). Cleft grafting and patch budding techniques are quite popular for multiplication of clonal material in north-east India. In cleft grafting, the stock must be 2-3 times as thick as scion and must be cut off horizontally within 20 cm of the soil. A cleft or slit is made in the stock; the scion, 15-22.5 cm long with a bud at its top is selected, cut into a wedge shape by trimming the opposite side, pushed into the cleft in the stock, and fixed enabling meeting of its inner bark with that of the stock. The whole graft is then wrapped over with polythene bag, tied firmly to the stock stem with a fibre or wax tape. Barua (1968) however, described a patch grafting method in tea where dormant, leafless buds are used for grafting. A rectangular patch of bark is removed from the stock and a similar size patch from scion with the scion bud at the centre is fixed exactly into the depression and firmly tied with polythene tape or wax cloth. But the type of bud required in this method is difficult to obtain in plenty as only one or two such buds develop in the *janam* scars in each shoot (Bezbaruah, 1975). Another easy and simpler method of bud grafting known as shield budding or inverted T budding has been found to be as successful as the cleft grafting or patch budding (Bezbaruah, 1975). In this method, the leaves of the budwood should be trimmed off just above the petiole for convenience of cutting out the bud (Fig. 7.17A). Then an inverted 'T' shaped cut (Fig. 7.17B) should be made on the smooth portion of the stock. Starting from the lowest bud, the bud should be cut off with a fairly thin slice with as little wood below as possible from about 0.7 to 1 cm above the bud to about 1 to 1.5 cm below (Fig. 7.17C and D). The flaps of the inverted T cut may be lifted carefully with the tip of the knife blade, then using the leaf-stock of the bud as a handle the bud should be inserted carefully upward in the flap of the stock bark (Fig. 7.17E). The graft union then firmly tied up with polythene tape (Fig. 7.17F). After 3 to 4 weeks, polythene tape should be removed and allow the bud to grow freely (Fig.

7.17G). These methods of producing plants are useful for some specific purpose such as (1) rapid multiplication of vegetative clones to establish nucleus plots of mother bushes, (2) early production of seeds on generative clones used in experimental breeding schemes, (3) conversion of vigorous seed baries of old jats into clonal baries and (4) utilizing mature, plucked bushes for clonal seed production (Barua, 1989). Further, shield budding may conveniently be used in odd bushes left failures in a cleft grafted area where errection of shade or cutting down the stock bushes for repeated cleft-grafting becomes difficult. Under such conditions the stock bushes may be allowed to produce shoots freely on which budding can be done conveniently (Bezbaruah, 1975).

A modified form of conventional method known as fresh root stock graft is developed in south India (Haridas, 1979). It reduces the time-lag between the establishment of bush and grafting. It is essentially similar to that of cleft-grafting. In this method (Fig. 7.18 A-G), the axis of the single leaf cutting selected as stock should cut about 2.5 to 3 cm above the axillary bud so as to make the cleft; scion should be taken from the rested bushes after pruning. The top cut end must be close to the leaf with the cut at a horizontal plane. The cut at the bottom which is 1.5 to 2 cm below the leaf is a tapering wedge 1.5 cm long made by an even slanting cut. The cleft on the stock axis should not be too deep and correspond to the length of the wedge on the scion. This ensures the proper alignment of cambial layers of stock and scion. The grafted portion should preferably be tied with a polythene strip. Rubber bands or special clips developed by some commercial firm can also be used for this purpose. After making the graft, the composite cutting is raised as a normal V.P. cutting (Haridas, 1979). In Malawi, inverted cleft grafting i.e. the wedge is made on the upper portion of the rootstock cutting while the split is made on the lower portion of the scion stem keeping prevention of entry of water into the cleft and easy flow of water from the union of the cleft was tried successfully (Kayange, 1991). The success will be indicated by the green colour of the stem of the scion during the growing stage. If it turns brown or black, it indicates that the graft is unsuccessful. The shoot coming out from the bud of the mother leaf of stock cutting should be cut off at its base i.e. below the lowest *janam* bud.

Although the bud grafting can be done throughout the year but the best period is when the bark or rind of stocks can easily be separated from the wood and in presence of sufficient sap. During drier months the stock

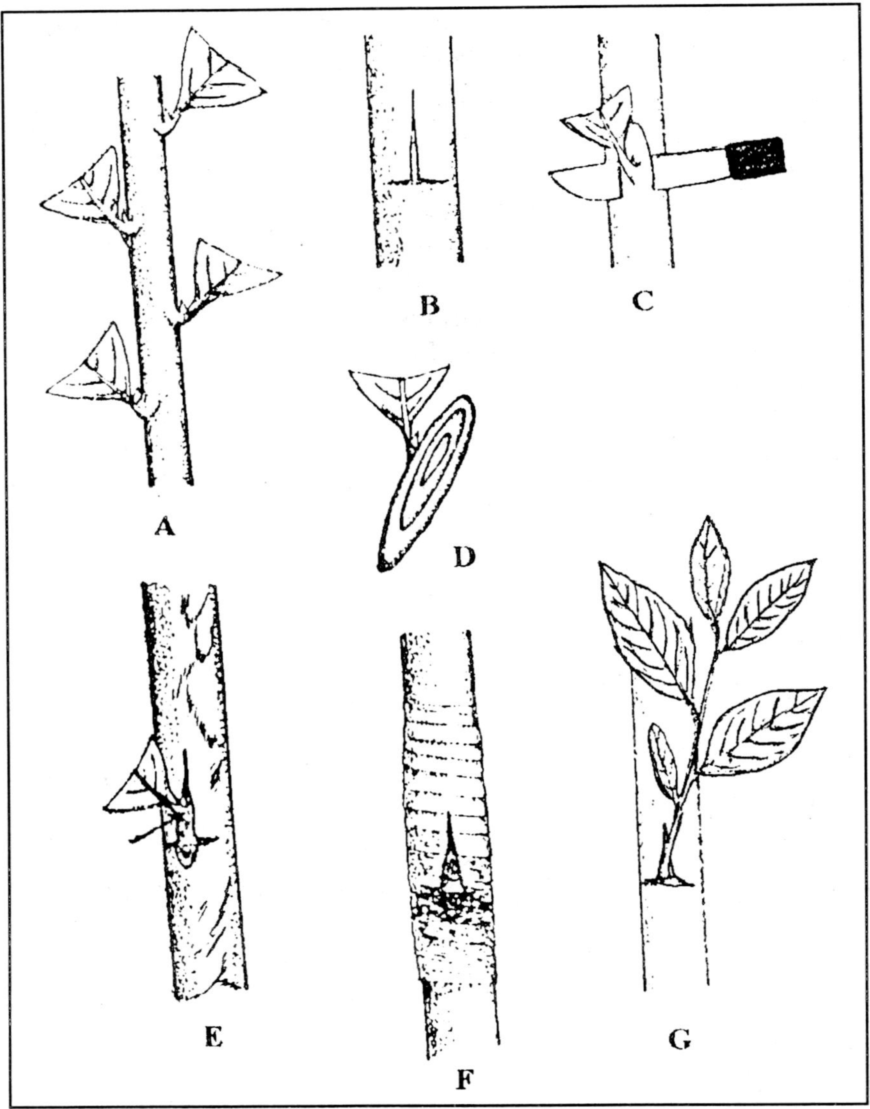

Fig. 7.17: Shield or inverted 'T' budding. A. bud wood; B. inverted 'T' cut made on the stock stem; C. lifting of bud wood; D. lifted bud wood; E. the scion inserted on the stock stem; F. polythene tape wrapped round the graft union; G. a growing scion [Source: Bezbaruah (1975)].

bushes should thoroughly be watered for a week or two to obtain such condition. Shade management during summer helps to obtain a fair amount of success (Subramaniam, 1995). For successful grafting, the stock should be more vigorous than scion. Stock influenced vigour and leaf yielding capacity of the scion but it had no effect on quality (Katsuo, 1969). The scion did not affect the depth and spread of roots of the stock plant (Anonymous, 1992).

Composite plant hold good potential in increasing the productivity

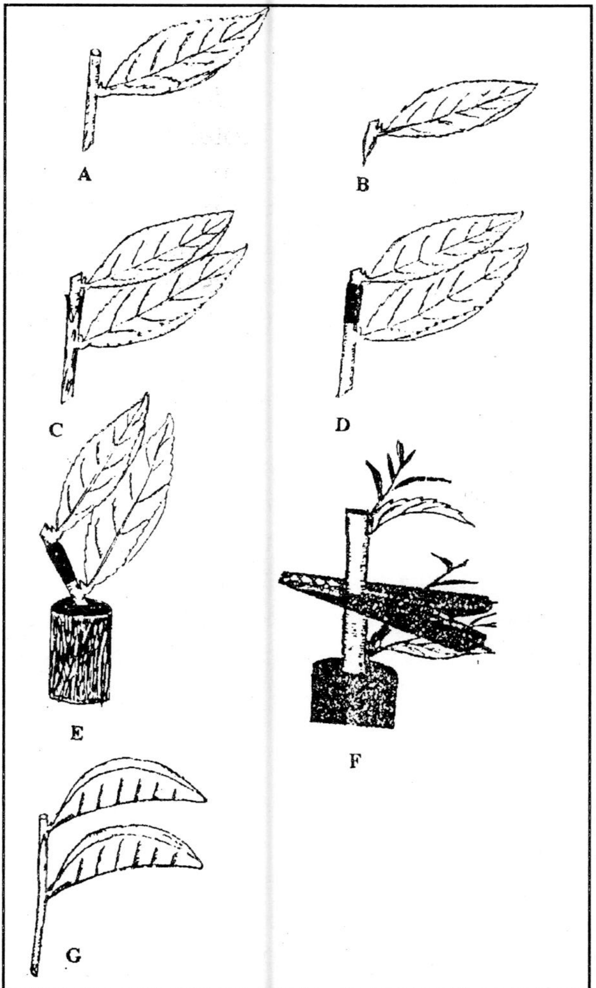

Fig. 7.18: Preparation of composite cutting. A. preparation of root stock cutting with a cleft on the stem above the leaf; B. preparation of scion cutting with a wedge on the stem below the leaf; C. the wedge of the scion cutting inserted into the cleft of the root stock cutting; D. tying of the joint by wrapping with a polythene tape; E. planting of the grafted cutting in a sleeve; F. tying of the joint with a clip; G. wedge is made on the upper portion of the root stock cutting while the split is made on the lower portion of the scion stem (inverted method).

of Darjeeling tea by combining the traits of quality and yield (Phukan, 1997). In tea, this is an established fact that the quality of scion is not affected by grafting on a vigorous rootstock. Satyanarayana *et al.* (1990) observed in a trial in south India that drought-tolerance, yield and production of phytomass of four clones used as scion varied with the rootstock used thereby indicating varying compatibility between graft partners. Considerable attention has been paid to the development of composite plant in southern Africa. It was observed if the hot dry season is extended due to a late break in the rains, much of the non-irrigated tea becomes totally defoliated, but a block of composite plants can stand out as a green island under such conditions (Whittle, 1997). This occurs because of the practice of selecting rootstocks with outstanding vigour and extensive root systems (eg. PC87 and MFS87). Several rootstocks were highly advantageous to yields of PC 114. PC 105 yields were notably increased on RS 193 (Whittle, 2000). He also reported that all clones do not benefit equally from being grafted onto all rootstocks. Therefore, care must be taken in selecting combination. Nyirenda and Mphangwe (2000) conducted a trial in Malawi and observed that single side cambium contact and alignment looked superior to both sides cambium contact and alignment. They further reported that stapling looked superior to tying with a polythene strip. Sreedhar and Ilango (1999) based on extensive trials in south India observed that the composite plant due to enhanced vigour lent by grafting the nursery span could be brought down to eight months instead of 12 to 14 months which is necessary for ungrafted plants.

7.4.11 *Propagation in vitro*

Propagation through tissue culture has the potential for rapid mass multiplication of elite planting material. Induction of plantlets in culture medium from the cotyledon has been achieved. Regeneration of shoot buds from the nodal tissue of mature field plants and seedlings and shoot tip explants have also been successful. Haploid embryoids are of importance in breeding have been developed from the pollen grains. A number of investigations on tea tissue culture carried out in India have been reported in Table 3.1 of Chapter 3. Seneviratne *et al.* (1988) claimed that first micropropagation of tea after success in inducing root formation on shoots already produced (Arulpragasam and Latiff, 1986). Haldeman and Thomas (1988) multiplied nodal sections from four to six months old aseptically grown tea seedlings. Significant progress on tea tissue culture has been carried out in India and abroad. Concerted efforts are now required to take this technology to the field since there is little pressure at the moment for application of this technology for multiplication of elite planting material using adventitious bud proliferation and somatic embryos and synthetic seeds.

References

Anandappa, T. I.1986. Seed tea. *In Handbook of Tea* (eds P. Sivapalan, S. Kulasegaram and A. Kathiravetpellai), Tea Research Institute of Sri Lanka, pp. 10-16.

Anonymous, 1986a. Fertilizers for seed bearers. *Tea Growers' Handbook*, 4th edn., Tea Research Foundation of Kenya, Kericho, Kenya, pp. 133-136.

Anonymous 1986b. The nursery. *Tea Growers' Handbook*, 4th edn., Tea Research Foundation of Kenya, Kericho, Kenya, pp. 38-40.

Anonymous, 1986c. Fertilizers for mother bushes. *Tea Growers' Handbook*, 4th edn., Tea Research Foundation of Kenya, Kericho, Kenya, pp.120-121.

Anonymous, 1986d. Vegetative propagation. *Tea Growers' Handbook*, 4th edn., Tea Research Foundation of Kenya, Kericho, Kenya, pp. 50-59.

Anonymous, 1992. Tocklai Experimental Station 1911 to 1991. Tea Research Association, Jorhat, Assam, India, pp. 42 & 49.

Anonymous, 1997. *Annual Scientific Report 1996-1997*.Darjeeling Tea Research Centre, Tea Board, Kurseong, India, pp. 25.

Anyuka, J. O. and Othieno, C. O. 1982. Techniques of grafting tea in the nursery. *Tea*, **3(2):** 31 - 36.

Argles, G. K. 1969. Root formation by stem cuttings. *Nursery Man and Garden Centre*, **148 (18):** 657-658.

Arulpragasam, P. V. and Latiff, R. 1986. Studies on the tissue culture of tea. 1. Development of a culture method for the multiplication of shoots. *Sri Lanka J. Tea Sci.*, **55(1):** 20-23.

Banerjee, B. 1992. Selection and breeding of tea. *In : Tea: Cultivation to consumption* (eds. K. C. Willson and M. N. Clifford), Chapman & Hall, London, 53-86. Barman, T. S. and Manivel, L. 1986. Triacontanol as an aid for rooting of cuttings. *Two and a Bud*, **33 (1/2):** 25-26.

Barua, D. N. 1968. Bud grafting. *Two and a Bud*, **15:** 103-109.

Barua, D. N. 1989. *Science and practice in tea culture.* Tea Research Association, Calcutta, India, p. 493.

Bezbaruah, H. P. 1975. An easy bud grafting method in tea. *Two and a Bud*, **22 (2):** 68–71.

Bezbaruah, H. P. 1977. Some observations on vegetative propagation of tea. *Proc 28th Tocklai Conf.*, November 24–26, 1977, Tea Research Association, Tocklai Experimental Station, Jorhat, India, 4–8.

Bordoloi, S. C. and Bezbaruah, H. P. 1983. Observations on vegetative propagation of tea. I. Use of sub-soil in the nursery. *Two and a Bud*, **30 (1/2) :** 12-13.

Bore, J. K. A., Njuguna, C. K. and Owuor, P. O. 1995. Chip-budding in tea (*Camellia sinensis* (L.) O. Kuntze) and its effect on yields and quality. *Tea*, **16:** 9–13.

Borthakur, B. and Banerjee, M. K. 1994. Salient features for raising a successful tea nursery. *In : Field Management in Tea* (ed. J. Chakravartee), Tea Research Association, Tocklai Experimental Station, Jorhat, Asssam, pp. 29-33.

Chakravartee, J. and Barbora, B. C. 1994. Nursery and field management of clones. *Advisory Bulletin no.3*, Tea Research Association, Tocklai Experimental Station, Jorhat, Assam, India, 1-16.

Eden, T. 1976. *Tea.* Longmans, London, p. 236.

Grice, W. 1968. Cleft grafting. *Two and a Bud*, **15:** 61- 69.

Haldeman, J. H. and Thomas, R. L. 1988. Embryo culture of tea on media amended with microbials. *The Camellia Journal*, **43(1):** 22-23.

Haridas, P. 1979. Nursery grafting techniques in tea. *Planters' Chronicle*, **74:** 128 - 131.

Harvey, F. J. 1989. Composite tea plants: a review. *Quarterly Newsletter*, Tea Research Foundation (Central Africa), Mulanji, Malawi, **94:** 12-15.

Kathiravetpillai, A. and Kulasegaram, S. 1986. Nursery and nursery practices. *In : Handbook on Tea* (eds P. Sivapalan, S. Kulasegaram and A. Kathiravetpillai), Tea Research Institute of Sri Lanka, Talawakelle, Sri Lanka, pp. 20-35.

Katsuo, K. 1969. Grafting of tea plant and its use for propagation. *J. Agric. Res. Q.*, **4(2):** 28-31.

Kayange, C. W. 1987. Propagation of binodal cuttings by misting in standard bags. *Quarterly Newsletter*, Tea Research Foundation (Central Africa), Mulanji, Malawi, **87:** 17

Kayange, C. W. 1991. Cleft grafting in the nursery. *Quarterly Newsletter*, **104:** 7, Tea Research Foundation (Central Africa), Mulanji, Malawi.

Kayange, C. W. 1992. Callus development on tea cuttings. *Quarterly Newsletter*, **106:** 5–7, Tea Research Foundation (Central Africa), Mulanji, Malawi.

Mphangwe, N. I. K. 1998. Growing media for rooted tea cuttings : Preliminary results. *Quarterly Newsletter*, **130:** 7-10, Tea Research Foundation (Central Africa), Mulanji, Malawi.

Mphangwe, N. I. K. 1999. Growing media for rooted tea cuttings. *Quarterly Newsletter*, **134:** 18-20, Tea Research Foundation (Central Africa), Mulanji, Malawi.

Nyirenda, H. E. 1996. Composite plants: nursery and field performance. *Quarterly Newsletter*, **124:** 9-15, Tea Research Foundation (Central Africa), Mulanji, Malawi.

Nyirenda, H. E. and Mphangwe, N. I. K. 2000. Production of composite plants – success rate and productivity. *Quarterly Newsletter*, **137:** 15-16, Tea Research Foundation (Central Africa), Mulanji, Malawi.

Pallemulla, D., Shanmugarajah, V. and Kathiravetpillai, A. 1992. Effect of grafting fresh cuttings on yield and drought resistance in tea. *Sri Lanka J. Tea Sci.*, **61:** 45–50.

Patabava, B. D. 1987. (Effect of temperature regime on the growth and development of tea transplants.) *Subtropichescie Kul'tury*, **2:** 58-60.

Phukan, B. C. 1997. R & D in Darjeeling : Some recommendations. *Proc. Seminar on*

Darjeeling Tea Industry, 5 September, 1997, Calcutta, Darjeeling Planters' Association & Tea Board, India, 39-48.

Rajasekar, R. and Sharma, V. S. 1989. Interaction between IBA, certain micronutrients and phenolic acids in relation to rooting of tea cuttings. *Sri Lanka J. Tea Sci.*, **58:** 25–39.

Satyanarayana, N., Spurgeon Cox and Sharma, V. S. 1990. Field performance of grafts made of fresh tea clonal cuttings. *J. Plantation Crops*, **20:** 151-156.

Seneviratne, P., Latiff, R. and Arulpragasam, P. V. 1988. Studies on the tissue culture of tea. 2. Rooting of shoots produced in culture. *Sri Lanka J. Tea Sci.*, **57(1):** 16-19.

Sharma, V. S. 1976. V. P. tea nursery. *Planters' Chronicle*, **71:** 211-213.

Sharma, V. S. 1982. Vegetative propagation of tea — a review. *Proc. PLACROSYM*, December 15-18, 1982, CPCRI, Kasaragod, Kerela, 73-87.

Sreedhar, Ch and Ilango, R. V. J. 1999. Vegetative vigour of grafted plants. *UPASI Newsletter*, **9(1):** 3.

Subramaniam, B. 1995. *Tea in India*. . Publication & Information Directorate, New Delhi & Wiley Eastern Limited, New Delhi, p.125.

Templer, J. C. 1971. Cleft grafting of tea seed bearers with clonal scions. *Tea*, **11(4):** 39.

Templer, J. C. and Machaga, J. S. E. 1978. Approach grafting. *Tea*, **18(1):** 17-20.

Venkataramani, K. S. 1963. The principles of tea clonal selection and propagation and some practical considerations in clonal planting. *UPASI Tea Sci. Dept. Bull.*, **22:** 2–14.

Whittle, A. H. 1997. Clonal tea in southern Africa. *Tea & Coffee Trade Journal*, **169(12)**: 62 - 66.

Whittle, A. H. 2000. The twentieth technical committee meeting of the Tea Research Foundation (Central Africa). *Quarterly Newsletter*, **137:** 3-6, Tea Research Foundation (Central Africa), Mulanji, Malawi.

Willson, K. C. 1992. Field operations: 1. *In : Tea: Cultivation to Consumption* (eds K. C. Willson and M. N. Clifford), Chapman & Hall, London, 201-226.

Willson, K. C. 1999. *Coffee, Cocoa and Tea*. CAB International, Wallingford, UK, p. 300.

Chapter 8

Field management

8.1 Selection of site

When site for establishment of new estates are to be selected, the climatic requirements is the first consideration because these have great influence on growth, yield and quality of tea plant. The climatic requirements of tea have been discussed in Chapter 4. Singh (1978) reported that the calculation of potential yield, requirements of drainage and irrigation of a particular site is possible if sufficient climatological data over a period of years are available. Remote sensing can indicate groundwater reserves and post-monsoon flooding prior to clearance and planting (Barbora and Bordoloi, 1993).

Ambient temperature in a year as long as possible should be within the normal range. Total rainfall and its distribution in relation to temperature suggest the necessity of irrigation. Water holding capacity and depth of soil indicates the growth and development of tea during dry season. Shelter belt is necessary for an area which suffers from excessively strong wind. Hailstorms can influence the plantation layout and forecast of crop yields (Willson, 1999). Soil requirements for tea has been discussed in Chapter 5. Some unsuitable soil such as high pH soil can be modified by different treatments described. Availability of different nutrients in the soil indicate which fertilizers may have to be applied. If planting is to be done in the slopy area then the extent of the slope should be assessed and the terraces, pattern of drainage etc should also to be worked out.

Suitable sites for construction of buildings such as factory, workshop, housing should also be kept in mind at the time of site selection. Furst and Awasthi (1981) discussed estate planning. The accessibility to the field is also important. The pattern of roadways should be designed to ensure easy accessibility to all parts of the land, transportation of green leaf, processed tea, plants, manure and personnel throughout the year. The management of estate road system, its design and establishment is discussed by Morris (1997).

8.2 Preparation of land

Virgin forest is usually the most suitable land for tea cultivation. Grassland and areas which have grown other crops can also be used. The area intended for planting is first cleared of all kinds growth. In general burning of forest growth is not advocated as it makes the soil alkaline. Trees if any should be killed in advance and possitively ring-barked or frilled. It minimises the risk of root disease. Pits created while uprooting the roots should be filled up with soil. Trees and other unwanted materials should be moved off the field. These operations generally should be carried out in dry season. When land is cleared for planting, leguminous cover crops may be sown before rains. While selecting species of the cover crops the following characteristics are to be considered :

1. It should have the characteristics of yielding large quantity of green material in a short span of time and ability in suppressing the weeds.
2. The plant should have more leafy growth than woody growth which enables rapid decomposition.
3. The plant should possess deep and fibrous root system.

8.3 Soil conservation measures

After clearing land, soil conservation works should be the major concern and must be put in hand as early as possible. Soil should be adequately protected from erosion. The process of soil erosion begins with beating of bare soil by raindrops which are then transported by run-off. Run-off could be because the soil is already too wet to hold any more water or the intensity of the in-coming water is in excess of the rate of infiltration. Also, the soil structure may be poor with very little porosity and /or the land may be too steep to allow water to infiltrate into the soil (Othieno, 1992). Many tea estates in Darjeeling hills in north-east India, the Nilgiris

Box 8.1 : Loss of soil due to erosion in tea plantations

Cachar Teela, Assam

Condition	Slope %	Cover/conservation measure	Soil loss (t/ha)
Site: Rosekandy T. E.	75	3 months old plant with poor green crop	32
Soil: Sandy loam	75	3 months old plant with satisfactory green crop	24
Altitude: 39 mt.	65	24 months old plant. No green crop and 100% weed control	10
Rainfall: 3000 mm	65	24 months old plant. No green crop and strip weed control	3
Age of bushes: 3 to 24 months	60	12 months old plant with green crop	16

Nilgiris, south India

Conditions	Canopy cover	Soil loss (kg/ha)
Altitude: 2100–2200 mt.	Excellent (about 60%)	32
Soil: Laterite	Fair (about 60%)	83
Slope: 20%		
Rainfall: 850–1390 mm	Poor (about 15%)	711
Age of bushes: 30 years		

Note: Soil loss was minimum with good ground cover and vegetation

Source. Singh et al. (1999)

and the High range in south India have recurrently suffered from soil erosion problem. Soil erosion control in tea lands has been discussed in Sri Lanka (Krishnarajah, 1985), India (Barua, 1989), Southern Africa (Grice, 1983) and Kenya (Othieno, 1992). In Sri Lanka, Krishnarajah (1985) reported loss of 20 tonnes of soil per hectare within six months in absence of any earthwork. The top soil is valuable for growth of plants and conservation is therefore necessary. The important soil conservation measures consist of terraces, contour wall, micro-catchment etc should be carried out wherever necessary before the tea is planted. Other control measures, such as proper drainage at the upstream and towards safe outlets, followed by good vegetative growth on exposed surfaces, planting of deep rooted local grasses, shrubs and trees which provide a binding force to soils exposed to the atmosphere, double hedge or staggered system of planting, practice of minimum tillage should be taken up immediately. Mulching also protects the soil from sunshine and impact of raindrop. Direct sunshine on the soil burns up the organic matter thus inherent fertility is lost with a consequent break down of soil particles whenever the soil is exposed to the sun (Grice, 1983).

8.3.1 *Terraces*

The construction of terraces in sloping land are must for minimization of rapid loss of surface soil. Terraces should follow the contour of the slope and the upper surface of the terrace may slope slightly towards the hillside to prevent soil wash. Growth of grass may be allowed along the edges and side of the terraces. There are disadvantages of terracing such as large disturbance of the soil when terraces are first constructed, increases steepness of land, exposure of subsoil where tea has to be planted, shading of plants of a lower terrace and involvement of high cost for construction and maintenance of terraces (Barua, 1989). In some places such as south India terracing nowadays is not preferred and planting is done on the natural slope. In Kenya, wide shallow terraces are recommended. They should fall at one-half percent to cut-off drains running downhill on natural drainage lines. Terraces should rise to maximum height between cut-off drains, the slope falling away on each side of the maximum. The cut-off drains should be protected from scouring by planting grasses (Othieno, 1992).

Contour planting is helpful to a terrain that cannot be mechanically terraced. Contour planting is done first marking a master row between a pair of graded contour drains and then tea rows are planted parallel to the master row. The master row is marked as (i) if a pair of graded contour drains converge in the direction of flow, the master row is made parallel to the top or higher contour drain, and (ii) if a pair of contour drains diverge in the direction of flow it is made parallel to the bottom or lower contour drain. A separate master row is marked between each pair of contour drains.

Box 8.2 : *Constituents are most liable to be carried away by surface wash*

Undoubtedly, the most important is nitrogen in the form of nitrates as converted by bacterial action for plant assimilation. The organic matter, or humus in process of decomposition which is readily carried away along with the finer soil particles by a flood or rain. Potash, phosphoric acid and other manurial constituents are also carried away, being mainly contained in the soil solution or the clay fraction of the soil.
Source: Bald (1965)

8.3.2 *Cover crops*

When land is cleared after earthworks, leguminous cover crops such as *Mimosa invisa* Mart., *Stylosanthes guyanensis* SW var. *gracilis* (HBK) Vog., *Calapogonium mucunoides* Desv., *Centrosema pubescens* Benth., *Pueraria phaseoloides* (Roxb.) Benth., may be sown before rains (Dutta *et al*., 1971). *Mimosa invisa* and *Desmodium* sp. in combination with Guatemala (*Tripsacum laxum* Nash) would be ideal (Banerjee, 1993; Manipura, 1972). The growth of cover crop can act as a barrier to protect the soil of the newly opened land against erosion. It also suppresses weed growth, improving soil structure and fertility, and helpful in keeping down the soil temperature during summer months. Sandanam and Rajasingham (1981) reported the benefits of cover crop and mulching in controlling soil erosion and increased tea yields. Cover crop can be grown in between the tea rows and in walking rows. Seeds can be sown 45 cm from tea rows. Plants can be pulled out when seeds become mature. Non-leguminous cover crops have also been used with success. *Polygonum*, *Oxalis* are of considerable value in south India, Sri Lanka and east Africa.

In Kenya, Oats (*Avena sativa* L) and buckwheat (*Fagopyrum esculentum* Moench.) is widely used as cover crop for higher and lower altitudes respectively. These crops can be established easily by direct seeding and die back after one year. The stems are tall enough to shelter young tea and roots can hold the surface soil (Willson, 1992). *Crotalaria* is also widely used as nurse crop planted between rows of young tea plants (Fig. 8.1).

In former USSR, the green crops like Soyabean or Lupines for a period of one year or so are planted. After uprooting the green crop, the land receives another round of deep cultivation, when a massive dose of phosphate is drilled in. Then only tea is planted in the prepared area (Dey, 1972).

8.3.3 *Drains*

High water table resulting in frequent waterlogging in Assam and bheel soils of Cachar tea areas of north-east India and certain swampy or river beds in a number of tea growing countries. In Darjeeling, both surface waterlogging and high water table due to lateral spring are observed. The surface waterlogging is the result of localised depression formed near the bushes. The water table rises to undesirably shallow levels and affects normal growth of tea plants due to reduced rate of mineralisation, lower microbial activity, high denitrification with lack of aeration in the soil, low soil temperature under wet conditions, and poor response to applied fertilizers.

Fig. 8.1: Recently planted tea field. *Crotalaria juncea* can be seen between the rows of young plants.

Waterlogged plants exhibit certain symptoms such as scanty shoot production, yellowing of leaves and premature defoliation, dieback in plucking points and poor recovery after pruning, infection by violet root rot and red rust diseases, loss of vigour etc. To cope with this situation it has become necessary to provide artificial drainage facilities to drain excess water below the root zone.

In south India, contour drains with lock and spill over to reduce velocity of water are generally provided. These drains are ultimately connected with catch drains running along the slope. Vertical drains paved with locally available stones having proper width and depth are also provided to draw the water safely without scouring the soil. When stones are locally not available the vertical drains can also be protected by planting weeping love grass (*Eragrostis curvula* (Schrad.) Nees) (Fig. 8.2). In steep sloppy area contour stone walls with locally available stones are sometimes constructed. In Darjeeling, contour drains of 30-45 cm deep and 30 cm wide with a bed gradient of 0.2 % (i.e. 20 cm drop in 100 m length) may be made and connected to a collector drain which is naturally formed in between two ridges. The soil escavated from the drain should be heaped on the

Fig. 8.2: Vertical drains protected by planting weeping love grass (*Eragrostis curvula*) in a south Indian tea estate [Source: K. Arunachalam, 1995. A hand book of Indian tea].

lower edge of the drain to form a small bund. The collector drains are formed in the lowest portion of the valley and are usually covered by natural vegetation. The distance between two contour drains depends on rainfall, soil texture and steepness of slope. These vegetations should not be removed which otherwise shall cause gully formation. A row of perennial citronella grass (*Cymbopogon winterianus* Jowett) can be planted at 15 cm intervals on upper and lower edge of the contour drain as well as edges of collector drains. In addition to protect the drains, the loppings of this grass can be used as mulching.

In Indonesia, soil erosion is a major problem in young tea areas. Drains are generally provided along the contour. 'Roraks' or 'lock and spill' drains are provided after every third and second contour rows in gentle and steep slopes (Bezbaruah and Barbora, 1979). In Sri Lanka, lateral drains, square in section (45 cm x 45 cm) and slope at 1 in 120 to leader or cut-off drains, are recommended. The lateral drains should have barriers at intervals of a height less than the full depth, to trap water and encourage the deposition of entrained soil (Willson, 1992).

In Malawi and Kenya, contour drains with blocked ends are constructed mainly to collect the surface run-off water during heavy rains with a view to protect the land against surface erosion and to allow the stored water seep into subsoil layers to raise the soil moisture level in the root zone.

The main purpose of drainage is to provide an optimum environment for deeper and extensive root development which enables the plants to withstand drought better. Proper drainage improves soil aeration and thereby provides a more favourable environment for multiplication of the beneficial micro-organisms and earthworms. An increased supply of nitrogen for tea can be obtained from the soil where water table is efficiently controlled at a desirable level by drainage system (a minimum of 90 cm depth below the ground level). Drainage also removes certain toxic substances, diseases organisms from soil and reduces the problem of weeds particularly aquatic type of weeds. In areas where soil pH has a tendency to rise inflowing seepage water containing dolomite (for eg. some tea estates of Jalpaiguri district of West Bengal, India located adjacent to Bhutan), the good drainage helps in maintaining the pH within the desirable limit. As a result, with scientific drainage, a remarkable increase in crop yields (10% approx.) and quality can be achieved over a period of time, provided that other inputs are maintained at their optimum. Singh (1999) discussed importance and current drainage practices in north-east India.

8.4 Planting distances

It is difficult to generalise on the number of plants ideal for a particular field area to all situation because of the involvement of number of diverse variables. Planting density is however an important determinant of yield. For high yield, complete ground cover is important. In recent years, the trend all over is to go for closer spacing in order to acheive high yield.

Generally, close spacing yield a high crop in early years and expedites recovery of the heavy cost of planting operations notwithstanding better conservation of soil and reducing the cost of weed control. The main disadvantages are suppression of overall growth of the plant (Barua, 1973; Rahman and Fareed, 1977) and restricted development of root system (Barua, 1989). There are some other factors also which have a bearing on spacing. For example, spreading habit of clone (compact or spreading frames), steepness of the slope, fertility status of the soil, machine harvesting etc. can dictate the optimum spacing. In Kenya, the influence of plant density on yield of young clonal tea has been studied by Magambo and Waithaka (1985). In north-east India, on the cost benefit angle Chakravarty and Awasthi (1981) reported that stand density should be in between 11000 to 14000 bushes ha^{-1}, although the value of this optimum will depend on climate, soil and characteristics of tea. In south India, a contour planting of single hedge (120 cm x 75 cm) with a plant population of 10800 ha^{-1} and a double hedge planting (135 cm x 75 cm x 75 cm) accommodating 13200 plants ha^{-1} and small growers in the Nilgiris plant at 120 cm x 120 cm in an up and down manner with a population 6800 ha^{-1} were adopted. Willson (1999) reported yield from mature tea is at its highest at spacing between 10000 to 15000 bushes per hectare. In Indonesia, the recommended plant density is about 10000 plants per hectare and distance between the rows is 120 cm (Semangun, 1976). Nowadays, single hedge and double hedge system of plantings are very common. The number of plants per hectare can be accommodated at different spacing are furnished in Table 1. The double hedge planting is recommended for flat or gentle sloppy area and it should be avoided in steep land where plucking would be difficult. In Japan, both single and staggered double hedges are common. Planting pattern followed is 180 x 30-45 cm to facilitate mechanical management. On the slopes, 150 x 30-45 cm spacing is practiced.

8.5 Planting

This operation basically determines the growth and yield level of a plant. The site for planting can be marked by a stick. After planting, the field should be mulched (6 to 8 cm thick) in absence of cover crop. But it must not touch the stem of the young plants.

Table 8.1 : Plant population per hectare at different spacings.

Type of planting	Spacing (cm)	Plants
Square	120 x 120	6274
	135 x 135	5315
	150 x 150	4305
Triangular	120 x 120	7770
	135 x 135	6139
	1450 x 150	4353
Single hedge	90 x 60	18518
	110 x 60	15151
	120 x 60	13888
	120 x 75	11111
	120 x 90	9259
Double hedge regular	110 x 60 x 60	19607
	120 x 60 x 60	18518
	120 x 75 x 75	13675
	120 x 90 x 90	12345
Double hedge stagger	110 x 60 x 60	20575
	120 x 60 x 60	19379
	120 x 75 x 75	14414
	120 x 90 x 90	13888

8.5.1 *Planting hole*

It should be of adequate size. Planting hole of 45 cm wide and 75 cm depth are better for establishment of young plants. When the holes are being dug the top soil should be placed on one side and subsoil on the other side. In clayey soil and drought prone areas, trench planting in place of holing will be advantageous. Trenches of 30 cm wide and 45 cm depth along the rows are dug which provide lower volume of soil for expansion of roots. Filling the holes should be done with top soil as far as possible. Sufficient surface soil has to be collected for filling the holes after removing stones and roots. A tall strong marking stick should be put in at the middle of the filled pit to indicate its position (Fig. 8.3). The fertilizer is to be mixed with the soil removed from the planting hole has been discussed in Chapter 10.

8.5.2 *Time of planting*

While planting, it should be kept in mind that the young plants must

Fig. 8.3: Marking sticks are fixed at the middle of the soil filled pits to indicate its position.

have the longest possible time to establish themselves before adverse weather sets in. In general, a cool, humid climate and moist soil are favourable for high success after planting. Planting is commenced in Nilgiris, Anamallais in south India and Darjeeling with the onset of pre-monsoon showers during the month of May. In the eastern part of Assam, winters provide sufficient moisture in the soil and cool misty weather accompanied by occasional showers makes planting possible during late autumn and early winter. The middle and western part of Assam including Cachar and tea areas in the plains of West Bengal experience droughty conditions from November to April and thereby planting is safe during May-June. In most of the tea areas of Indonesia and Sri Lanka, planting can be done at any time of the year as moderate temperature, adequate and well distributed rainfall prevails most of the year. In Japan, late March to early April is the suitable time for planting. The tea areas of east Africa experienced deficit of water in the soil, low temperature, high evapotranspiration loss due to strong wind during the early part of the year. In Central Africa, suitable period for planting is even more limited.

8.5.3 *Sleeved plants planting*

Vigorous plants of uniform height can be taken out from the nursery (Fig. 8.4) when sleeves are wet. There should be profuse feeder roots nearer the bottom-end of the sleeves at the time of planting. The main stem at the collar region should be about pencil thick and brown; it should be straight and errect. Weak and smaller plant may create problems at the time of weed control, training and susceptible to pests. Sleeved plants should preferably be carried in boxes to the planting site (Fig. 8.5) to minimise disturbance of the soil which could damage the roots. When planting seedlings, the top of the sleeve or *bheti* flush with the ground level. In case

Fig. 8.4 : Young tea plants in the polybags for field planting.

Fig. 8.5 : Sleeved plants carried in box to the planting site.

of clones the tip of the cutting should remain little above the ground. A vertical cut is made at the bottom half of the sleeve and the space around the sleeve is gradually filled with the cylinder of soil without being broken at the time of this operation. When the hole is thus partially filled the bag is slit along its full length and carefully removed. The gap around the upper part of the soil core is also filled with soil without causing any damage to the soil core. If the roots overgrow the size of the sleeve or *bhetis*, the roots at the bottom of sleeve or *bheti* should be cut to avoid twisting (Fig. 8.6). Depression if any at the collar due to poor ramming may create localized waterlogging resulting in mortality. Further, ramming the soil too hard causes compaction which is detrimental to growth. In Japan, Nakayama (1971)

Fig. 8.6 : Twisted root of mature plant.

reported that after transplanting two years old plants in the field, the photosynthetic rate of the old leaves existing from the time of transplanting was low. It was also low in new leaves at the beginning of their development and should considerably increase after about three months from the date of transplanting. Despite, a rapid decrease in carbohydrate content up to two months; gradual decrease of nitrogen content of the leaf, stem and root in percentage after transplanting was also noticed.

8.5.4 *Stump planting*

A stump with plenty of food reserve gives a vigorous shoot with a healthy leave. Stumps are at least 1 cm in diameter at the collar and have been unshaded for 5-6 months to build up starch reserve are ready for planting. Stumps after pulling out from nursery it should be cut off 10 cm above the soil level, and the root 45 cm below soil level. Thin lateral roots should also be removed. Seedling make single tap root. Special attention should also be given to ensure that the tip of the tap root is in actual contact with the soil at the bottom. Drying out of the stump never be allowed. It is always advisable to plant the stumps soon after pulling out from nursery beds.

8.6 Infilling

Tea bushes in the established field may die for a variety of reasons such as poor drainage, pests and diseases, debility, lightening etc. Infilling is the replacement of dead plants by new ones and it should be persisted with until the stand of young tea is completely filled. Vacancies encourage growth of weeds in the gaps and its presence in significant number reduces overall production. Infilling of vacancies is better done in the first two years when the young plants have not spread much so that the infills will have more space to grow freely and therefore a better chance of survival. After planting of infills, the ground should be mulched with any suitable vegetative material in order to reduce the loss of soil moisture and add organic matter to the soil. Green mulch has to be kept 10-15 cm away from the collar of infills/interplants to avoid heat injury and termite attack.

In India, the extent of vacancy in tea field could be placed conservatively at 15 percent. This is equivalent to about 60000 hectares in terms of actual area. In Darjeeling hills, the overall percentage of vacancies is 20 or even more which is equivalent to 4000 hectares in terms of area in Darjeeling (Ghosh Hajra, 1992). In south India, the vacancy varies from 10 to 30 percent and the mean figure is 14.8 percent (Hudson, 1995).

Infilling offers multiple benefits viz. (1) no loss of revenue either to the producer or to the national exchequer at any time, (2) replacement of old and unthrifty tea bushes by improved plant material, (3) increased plant population per unit area and increased productivity with minimul inputs, (4) optimum utilization of land and (5) improved quality (Kankani, 1989).

Box 8.3 : Factors responsible for stagnation or declination of yield of a tea block

- *Continuous soil erosion where top soil is lost faster than the rate of soil formation.*
- *The bush frame is severely damaged by various reasons resulting into development of cankers.*
- *Improper manuring especially under-manuring of potash over the years resulting in formation of scraggy wood and fall of maintenance leaf.*
- *Improper pruning, banjhi dal removal and faulty tipping resulting into a poor bush frame.*
- *Poor drainage of the area resulting water logging.*
- *Use of improper and sub-standard planting materials having low yield potential.*
- *Presence of more than 50 percent vacancies.*
- *Tea severely affected by root borers and primary root diseases etc.*

Infilling of individual

vacancies or in blocks should preferably be taken up in the year of heavy, medium or light pruning. Strong plants with good root system of appropriate clones or proven seed stock should be used for infilling in order to compete against the established plants around. While infilling, a large hole should be dug to ensure presence of no live roots of established tea close to the new plant. Preferably two plants per vacancy in the large pit where hard pan has been broken and adequate manure given should be planted. The side branches of the neighbouring bushes if encroaching into the vacant space to be occupied by the infills can be trimmed and infills tipped a little higher in the first two years or so. Infills should be planted at a closer spacing than the established plants. Pegs should be put around the infills which would protect plants from trampling (Fig. 8.7). Proper attention on adequate post-planting care is required for successful establishment of infills.

8.7 Uprooting, rehabilitation of soils and replanting

Replacement of old and low-yielding tea by better material has accelerated since clones with high yield potential are available. Replanting also provides opportunity to install or improve soil erosion measure. Further, spacing of the plants can be altered as per latest recommendation, which

Fig. 8.7 : Pegs have been put around the infills which would protect after transplanting.

also enable mechanical harvesting if it is relevant.

Barbora and Hazarika (1996) reported that in north-east India the productivity increases with time from planting to a peak level in about 10-12 years depending on the plant population per hectare. They further observed that by and large the productivity stagnates and starts declining from 25-30 years of age. But in tea gardens of south India are 100 or more years old and continue to respond to increased inputs with progressively increasing yields (Subramaniam, 1995). The rate of yield declination depends on the health and hygiene of the plant. However, 44 percent of the tea areas in north-east India is beyond the economic age and 80 percent of the existing tea bushes in Darjeeling is above 50 years old showing either static or declining trend in productivity. Ghosh Hajra (1992) observed that stagnancy in the yield rate is due to low yield in majority sections which are more or less 100 years old with higher percentage of vacancy. The percent area of different age group in Darjeeling, Dooars and Terai is presented below (Anonymous, 1994) :

Percent area under each age group

Age group (years)	*Darjeeling*	*Dooars*	*Terai*
1-10	5.38	14.15	20.47
11-30	7.80	23.29	25.13
31-50	6.24	16.82	17.36
Over 50	80.58	45.05	37.24

The above figures show that rate of planting has been very slow in Darjeeling and most of the tea is superannuated. The Dooars and Terai figures include Coochbehar and West Dinajpur districts respectively look healthy. The proportion of old tea over 50 years in Darjeeling is double of that in the plains.

Different methods have been used to uproot old tea. The efficacy of uprooting by bulldozer, winch and manually, and soil rehabilitation have been investigated in several countries, e.g. Kenya (Obaga and Othieno, 1986; Ng'etich and Othieno, 1993) and south India (Ponnapa, 1991). The uprooting by winch might make the soil more susceptible to erosion during heavy rain. Tabageri *et al.* (1991) reported rehabilitating tea without replanting. In Darjeeling, the traditional method of replanting is manual uprooting which is not only expensive but also leads to top soil erosion and turning over of sub-soil resulting in poor establishment of new plants. Further, mechanical uprooting is not possible because of the steepness of the slopes. The traditional method can be substituted by minimum tillage or progressive replanting technique. In Darjeeling, Kabir *et al.* (1993) reported that the old bushes can be collar pruned and killed *in situ* by applying 2, 4-Dichlorophenoxy-acetic acid (2, 4-D) at the rate of 0.8 kg a.i. in 75 litres of water (500 ml fluid stump^{-1}) or urea at the rate of 250 g stump^{-1} followed by replantation. This operation can be done in alternate strips not wider than 50 m on steep slopes and 100 m on medium slopes. No tillage replanting has also been tested in south India and Africa (Sharma, 1984; Grice, 1990).

The soil after remaining under tea for a long time is usually denuded of nutrients and some (auto) toxins (allelopathic) also get accumulated which has been reported in north-east India (George and Singh, 1990) and Kenya (Othieno, 1992). It has also been noted that the food crops such as maize and beans, despite use of fertilizers, completely failed to grow in Kenya when planted immediately after uprooting of tea. Therefore, old tea soil requires rehabilitation for restoring soil health. The recommended grass rehabilitation (with deep rooted Guatemala grass) following uprooting is a time tested practice which also takes care of crop rotational aspects to break the chain of monoculture of tea. Either the plant or microbes associated with the roots of the Guatemala detoxifies the soil by metabolising phenolics to less toxic products (George and Singh, 1990). Despite this, rehabilitation crop has various effects such as addition of organic matter and nitrogen to the soil, helping improvement in soil structure and thereby better aeration and moisture retension capacity. It also helps movement of nutrients from deep layers to top layer of soil despite increment of microorganism's activity. Satyanarayana (1971) has reported that sections that had been thoroughly rehabilitated showed a lower incidence of red rust in the replanted tea than did areas that had not been rehabilitated. Ananthacumaraswamy *et al.* (1986) investigated the economics of rehabilitation with grass before replanting in Sri Lanka. They found practice is highly profitable for the planters.

The Guatemala grass (Fig. 5.1) and Pusa Giant Hybrid Napier (*Pennisetum purpureum* Schum.) have been used extensively as rehabilitation crops in many tea areas of the world including Darjeeling hills. The Citronella grass (*Cymbopogon wintarianus*) (Fig. 8.8), Mana

Fig. 8.8: The Citronella grass planted in the uprooted tea field for rehabilitation of soil.

grass (*Cymbopogon confertiflorus*), legume (*Desmodium uncinatum*) and Weeping Love grass (*Eragrostis curvula*) have also been used widely in north-east India, Sri Lanka, Kenya and Africa respectively. All rehabilitation crop should be lopped regularly and the cut material left on site to form a mulch. For planting one hectare Guatemala, around 10000-12000 cuttings or root stocks are required. Rehabilitation crops may be manured with 10:5:10 NPK mixture @ 600 kg ha^{-1} and after each lopping 300 kg ha^{-1}. Rehabilitation period mainly depends on the built-up of organic carbon in the soil. It should be above 1 percent. The recommended rehabilitation period can be eighteen to twenty four months in north-east India. Grice

(1984) found in Central Africa that there was a benefit from soil rehabilitation under Guatemala grass grown for one year only following different tillage treatments. Two years were found not significantly better (Grice, 1987). But Kabir *et al.* (1993) observed that Guatemala did not produce sufficient green matter to rehabilitate the soil in one year in Darjeeling hills.

8.8 Shading

The young plants require protection from the bright sun and dry wind. In most of the tea areas in India, a system of shading tea plants is practised. The tradition of interplanting tea bushes with *Albizzia chinensis* (Osbeck) Merr. as shade tree started in Assam towards the end of the last century. The practice of using shade trees spread from north-east India to south India, Sri Lanka, Indonesia and Africa, where in the absence of *Albizzia chinensis* a number of local species both leguminous and non-leguminous, were tried out with varying degree of success (Hadfield, 1974a). In Indonesia, shade trees currently preferred are *Leucaena leucocephala*, especially the seedless cultivars (*L. pulverulenta*, *L. glabrata*), *Gliricidia sepium*, and *Grevillea robusta*. Large shade trees like *Albizzia falcataria* and *A. sumatrana* are rarely used. However, it was found later shaded tea in many areas had a limited response to fertilizer applications than the unshaded tea. Notwithstanding doubts about utility of shade arose with the control of blister blight epidemics in the tea plantations of south India, Sri Lanka and Indonesia following the removal of shade trees. These are probably the reasons which encouraged large scale removal of shade trees during last decade in Sri Lanka and east Africa.

Undoubtedly, shade in tea has been one of the most controversial topics and also an extensively investigated cultural practices in tea cultivation (Hadfield, 1974a and 1974b; Kulasegaram and Kathiravetpillai, 1976; Othieno, 1983; Barua, 1989).

8.8.1 *Shade and leaf temperature*

Leaf temperature of horizontal and errect-leaf tea in full sunshine may reach 40^0 to 45^0C at ambient temperatures of 30^0 to 32^0C but in shaded leaves they remain 10^0 to 2^0C above the ambient temperature (Banerjee, 1993). However, in comparison with errect or semi-errect leaves, fully exposed horizontal leaves are generally 2^0 to 4^0C warmer, but a leaf temperature of 40^0C is common to all types of leaves under unshaded

Box 8.4 : Shade in tea plantations

Advantages:

- *Regulates temperature and increases humidity at bush level*
- *Minimises loss of water through evaporation and transpiration*
- *Serves as wind break*
- *Reduces ultra-violet radiation injury and incidence of mites*
- *Helps in the circulation of nutrients and recharging of ground water*
- *Its litter and loppings contribute considerably to the soil organic matter pool and is useful in maintaining the soil fertility and recirculation of nutrients*
- *Generates additional income by way of timber and fuel*
- *Minimises soil erosion and maintains the ecological equilibrium*
- *Helps in getting better distribution of crop*

Disadvantages

- *Reduces uptake and utilization of nutrients particularly nitrogen*
- *Increases the incidence of blister blight*
- *Competes with tea for moisture and nutrients to a certain extent*
- *Serves as an alternate host for some pests and diseases*

Source: Hudson et. al. (1997) and Ranganathan et. al. (1980).

condition when the ambient temperatures are from 30°C to 32°C (Hadfield, 1968). In north-east India, this temperature range remains greater part of the harvesting season. Hadfield (1975) using laboratory grown plants, found a temperature optimum for photosynthesis at around 35°C, with a rapid fall above 37°C and no net uptake of CO_2 above 42°C. But respiration continues upto 48°C, above which leaf tissues get damaged. In some low altitude areas like north India and Bangladesh, due to high temperature in the peak growing season yield falls because tea leaf temperature exceeds the critical value and net photosynthetic rate slows down. In these areas shade trees were found to be beneficial in reducing leaf temperatures which can reach 40°C in unshaded tea and proper shade management will allow adequate quantum of light at bush level besides keeping the bush under a favourable microclimate (Raj Kumar *et al.*, 1999). Barua (1987 and 1989) opined that if the leaf temperature goes above 30°C then shade becomes necessary irrespective of location. Therefore, shade is not necessary in Kenya, Uganda, Tanzania, high and mid elevation tea areas of Sri Lanka, high elevation tea estates of Darjeeling, south India and Indonesia where ambient temperature does not often exceeds 30°C. But Tea Research Institute, UPASI recommends regulated shade in Nilgiris and under all situation not because of leaf temperature but in consideration of other factors like severe intensity of sun rays and radiation damage and leaf chlorosis were noticed in the tea plantations situated above 5000 ft. During summer in south India, draughty conditions normally experienced between December and March, at times extending over up to end of April or early May. Besides the soil moisture stress, the intensity of drought is aggravated by the low levels of atmospheric humidity during February and March and high diurnal temperature that surpasses even 35^{0}C in the mid-elevation estates during summer months. The need for shade in tea in south India is thus well recognised (Hudson and Muraleedharan, 1999). It was found that unshaded tea bushes at high altitude in Kenya produced significantly more pluckable shoots than shaded plants (Othieno and Ng'etich, 1993). In Malawi, shade trees depressed yield and reduced the efficiency of applied nitrogen (Anonymous, 1966). The conflicting results obtained from experiments with shade trees in north-east India, Sri Lanka, Africa and Indonesia can, therefore, be attributed primarily to differences of ambient temperature at the experimental sites (Hadfield, 1968).

8.8.2 *Shade and light intensity*

Shade trees absorb a considerable amount of infrared and visible radiation of different wavebands from the solar spectrum. It thereby reduces the incidence of heat and light energy on the tea bushes underneath (Barua, 1989). Light penetration into the maintenance leaf canopies of plucked tea bushes is affected by the leaf position. Generally the errect and semi-erect leaves allow a greater penetration of light in comparison with the horizontal leaves where light is restricted only to the top layer of the foliage (Hadfield, 1974a). Hadfield (1974a) further reported that in most cases, the first 10 cm layer of leaf absorbs 70 to 90 percent of visible radiation incident on its surface and visible light reaches the lower layers of the leaves in the canopy either through gaps in the canopy or by reflection from upper leaves.

In young tea, net assimilation rate (NAR) is related to log light intensity. Leaf area remains nearly steady between 50 percent intensity and full sunlight in all tea varieties. The maximum leaf area however found under 35 to 50 percent light intensity but is minimum in full sun. In contrast, the weight of the whole plant does not vary significantly between light intensity of 50 and 100 percent (Banerjee, 1993). The yield of plucked shoots is generally maximum under 50 percent light intensity—the relative values for the four light intensities being 50, 35, 100 and 20 percent (Barua, 1961).

8.8.3 *Shade and tea quality*

In another study using various artificial shading culture, it was found that shading reduced the amount of catechins in tea leaves (Saijo, 1980). It was also reported in Kenya that quality of black tea improved because of shading (Othieno, 1983). In Assam, it was observed that the strength and colour of liquor appears to improve significantly in 50 percent light though infused leaf from plants in full light was categorised as of high value (Dutta and Basu, 1955). Barua (1961) reported the overall character of the teas slightly improved at relatively low light intensity.

Box 8.5 : Are shade trees really necessary in Darjeeling hills?

It is an well established fact that there is a decrease in temperature with increase in elevation. It is ascertained that increase in altitude for every 100 metres there is a decrease in temperature between 0.6 to 0.7°C. Despite this the air turbulence in the hills is usually much more than the plains and shade trees may not be required for further reduction of leaf temperature after certain elevation. It has been observed in Darjeeling hills the ambient temperature never becomes lethally high, therefore direct effect of air turbulence in reducing leaf temperature is rather minimal. Under these conditions, it is suggested that shade is required upto an elevation of 700 m particularly on southern and south-west slopes.

Albizzia chinensis grows well and comparatively free from canker. Where establishment of A. chinensis is difficult, A. lebbek (L.) Benth., A. stipulata (DC.) Boiv. and Dalbergia sericea G. Don. can be grown. Indigofera teysmanii Miq. and Melia azedarach L. are very popular temporary shade species.

Source: Anonymous (1988); Ghosh Hajra and Kumar (1999)

8.8.4 *Silviculture of shade trees*

It is difficult to establish shade trees after tea has been planted. It would either be planted when rehabilitating old teas or during the previous year of planting young tea so that they are well established before the soil is disturbed by planting tea (Fig. 8.9). Shade trees can be raised from seeds on the plantation or direct purchase of seedlings from outside. While selecting species of shade trees, the factors like growth habit, canopy architecture, depth of root system, economic life and commercial timber value, less pest and disease incidence, tolerance to wind and frost, suitability to the soil and

Fig. 8.9: Woven bamboo guarded shade trees planted previous year of planting young tea.

climate of the locality etc. should be considered. Most of the species that have been used as shade are leguminous. Leguminous trees can fix significant amount of nitrogen from air and provide nitrogen for the tea plant.

In north-east India the permanent shade tree species found suitable are: *Albizzia chinensis*, *A. lebbek*, *A. odoratissima* (L.f.) Benth., *Acacia lenticularis*, *Derris robusta* and *Dalbergia sericea*. In south India, *Grevillea robusta* (Silver oak), a non-leguminous plant is best suitable shade tree species.

Frost is a recurring feature in the high elevation planting districts of Nilgiris and High Range. *Acacia mearnsii* de Wild. (Black wattle) and *A. decurrens* (J. Wendl.) Willd. are used in the frost prone areas. The shade provided by these is carefully regulated by pollarding the trees, where possible, and thinning out the branches as required. Shaded tea is shown in Fig. 8.10.

The young shade trees planted usually take a few years to provide effective shade to the newly planted tea. It has been observed in different places that tea was planted between the lines of green crops like *Crotalaria anagyroides* HBK. (Fig. 8.11), *Priotropis cystisoides* (Roxb.), *Desmodium*

Fig. 8.10: Tea under *Albizzia* shade in the foothills of Himalaya (Long view Tea Estate, Darjeeling).

Fig. 8.11: *Crotalaria anagyroides*, is seen between the rows of young tea in Darjeeling hills act as a temporary shade, green crop and source of mulch material.

Box 8.6 : Popular myths

- *All legumes fix nitrogen. Prior to the 1930s, it was generally believed that all legumes had nodules and the ability to fix nitrogen. This is not true as demonstrated by an extensive survey of nodulation data by Allen and Allen (1981). Species of the subfamily Papilionoideae are the most likely to nodulate (98% of those tested did) while species of the subfamily Caesalpinioideae are least like to (only 28% of those tested nodulated). Species of the subfamily Mimosoideae also are likely to nodulate; 90% tested positive for nodulation.*
- *Proven nitrogen-fixing trees (NFTs) always fix nitrogen. Just because an NFT is reported to fix nitrogen somewhere does not mean that it is always able to do so. Many factors, including poor plant vigour, absence of suitable rhizobia and environmental stress, can prevent nitrogen fixation. Any factor in the environment that affects the growth of the host plant is likely also to affect nodule development and function (Dixon and Wheeler, 1983).*
- *NFTs obtain all their nitrogen requirements through N fixation. The proportion of plant nitrogen derived from N fixation is rarely over 80%. (MacDicken, 1994).*
- *N fixation is always beneficial. In soil with good combined nitrogen levels, nitrogen fixation is inhibited by the effects of nitrate; yet carbohydrates are still provided to the rhizobia resulting in the diversion of energy that could otherwise be used for plant growth or reproduction. Under these conditions, non-NFTs with the same uses may have a physiological advantage over NFTs (MacDicken, 1994).*
- *N fixation provides nutrients other than nitrogen. Popular literature that refers to 'fertilizer' production from NFTs can give the impression that NFTs produce nutrients such as phosphorus or potassium in addition to nitrogen. This is not true. While NFTs can provide the equivalent of nitrogen fertilizer and can recycle other nutrients, help in improving physical properties of soil etc., they have no demonstrable advantage over non-NFTs in this regard (MacDicken, 1994).*

gyroides (Roxb. ex DC.), *Sesbania* sp., *Trifolium ripens* L. etc for protecting the young tea plants from direct sun rays.

Despite planting green crops, another practice which is popular nowadays in India is the planting short-duration, quick growing plants to provide shade during early years. *Indigofera teysmanii* (Fig. 8.12) *Tephrosia candida* (Roxb.) DC. (Bogamedeloa) (Fig. 8.13) and *T. vogelii* Hook. f. (African medeloa) are the temporary shade tree species extensively used both in north-east and south India. *Leucaena leucocephala* (Lam.) de Wit (Su-babul) which was introduced recently used in a very limited scale as a temporary shade. At a higher altitude in south India, *Sesbania cinerascens* Welw. ex Bak., *Crotalaria agathiflora*, *Acacia elata* A. Cunn. ex Benth.

Fig 8.12 : *Indigofera teysmanii*, is seen in the young tea field act as a temporary shade.

Fig. 8.13: *Tephrosia candida*, is seen between the rows of young tea in Darjeeling hills, act as a temporary shade, green crop and source of mulch material.

are recommended. These are initially planted at close spacing and keeping them trimmed and lopped from time to time and finally removed when permanent shade species take over.

8.8.5 *Spacing of shade*

1. Temporary shade: Initially 2 m x 2 m and later on thinned out to 4.5 m x 4.5 m.
2. Permanent shade: (based on shade provided by mature *A. chinensis*). Light shade: 15.24 m x 15.24 m (triangular) or 13.7 m x 13.7 m (Square or wider).

   Medium shade: 13.7 m x 13.7 m (triangular) or 12.2 m x 12.2 m (square).

   Heavy shade: 10.67 m x 10.67 m (triangular) or 9.14 m x 9.14 m (square or closer).

In case of steep slopy land and teelas, it is advisable that a shade spacing of the type described as heavy is used (Banerjee, 1993).

8.8.6 *Planting of shade trees*

The planting pit should be 90 cm deep and 60 cm wide. It would be ideal if the bottom soil of the pit if loosened to a further depth of 15 cm. A manure mixture composed of 10 to 15 kg dry well rotten cattle manure (or 1.0-1.5 kg of oil cake), 500 g single super phosphate (or 1 kg rock phosphate), 2.5 kg of wood ash and 1 kg slaked lime to be added in pits atleast seven days prior to planting. Sleeve grown seedling or with a *bheti* (60 cm long and 30 cm diameter) is the best method of planting. The roots found at the bottom and outside in sleeve-raised and *bheti* plants respectively should be cut off. The shade plant at the time of transplantation in mature tea areas should be 1.8 m to 2 m tall and 60 cm to 90 cm in young tea field. Woven bamboo guards help to protect young shade plants from cattle and other physical damage (Fig. 8.9).

8.9 *Irrigation*

Tea is usually a rainfed crop. Tea growing areas of few countries like Sri Lanka, Indonesia and Kenya are getting some rains throughout the year, but the major regions where irrigation is necessitated by a prolonged dry season are Bangladesh, India, Malawi, Tanzania, Zimbabwe and South

Africa (Fig. 8.14). Irrigation is also practiced for short periods in parts of China, Japan, former USSR and some other tea growing areas. In India, some parts of Assam (Cachar, North Bank, Nowgong, Golaghat), whole of Dooars and Terai and parts of south India suffer from drought and irrigation is practiced depending on the availability of water and other factors. In Darjeeling hills, plants also suffer from drought during April–May and where possible sprinkler irrigation systems are used for irrigating tea field (Fig. 8.15). This is the dominant system of irrigation in mature tea. The most common type of sprinkler used with portable system is the rotating-head sprinkler, consisting of a head with one or two nozzles. The head is rotated slowly by the action of water passing through it and water discharged covers a circular area of land around the sprinkler (Fig. 8.16). Some studies on drip irrigation have also shown encouraging results but capital cost and damage/pilferage of tubes in the field are not in favour of drip irrigation. It was reported that 60–65 percent of the total world tea area is subject to periods of drought varying from 3 to 20 or more weeks (Barua, 1989). The quantity and frequency of irrigation depends on soil, temperature, root

Fig. 8.14. The 1992 drought caused extensive defoliation in absence of irrigation in Malawi [Source: Schwarz, A. and Witt, D. The drought in Malawi and Zimbabwe. Tea International, 2: 9 – 16].

Fig. 8.15: Irrigating through sprinkler where possible in the young tea field of Darjeeling hills during May (North Tukvar Tea Esate).

Fig. 8.16: Irrigating through sprinkler in the pruned plot located in the foot hills of Himalaya (Darjeeling) during April. Water discharged covers a circular area of land around the sprinkler.

depth, saturation deficits and rate of plant growth. Irrigation requirement and water use have been studied using data obtained from rainfall, evapotranspiration and soil moisture deficit. Irrigation may be required at one or more of the three stages of development of tea–nursery, young tea for establishment and survival, and mature tea for survival / or crop (Rahman, 1999). A huge amount of water is consumed by tea bushes in the growing season. The water consumption varies with season. It was reported in China, for 1 kg fresh leaves, 400 litres of water are consumed in spring, 1000 and 1400 litres in summer and autumn respectively (Yongming, 1999). Irrigation in general in China is done when the water holding capacity of the field drops to 70%. Spray irrigation, drip irrigation besides the traditional surface irrigation are adopted in China.

Carr *et al.* (1987) reported that irrigation in the hot dry season caused contrasting responses Mulanje, Malawi and Mufindi, Tanzania. Yields of irrigated tea in Malawi did not rise to the levels recorded during the rains because of the large saturation deficits occurring during the period. However, in Tanzania, where saturation deficits were much smaller, yields of irrigated tea in October and November was as high as during the rainy season. Saturation deficit of about 2 kPa at midday has been found to be unfavourable for shoot extension (Carr, 1972). The sensitivity of shoot extension rates to the saturation deficit of the air over the range of 1.0 to 3.0 kPa was studied in Malawi by Squire (1979). Dry air led to high transpiration which reduced shoot water potential. Low shoot water potential had adverse effect on shoot extension through effects on cell turgor. Squire (1979) also observed that the linear response of shoot growth rates to mean air temperature over the range of 17 to 25^{0}C was obscured completely when the air was very dry. Tanton (1982) afterwards showed the critical saturation deficit value was 2.3 kPa. These observations have important practical implications due to one reason why irrigation can not substitute entirely for rainfall is because it can not reproduce the humid atmospheric conditions usually associated with rainy seasons (Carr and Stephens, 1992). This statement applies nearly to all crops, but particularly those which are especially sensitive to dry air conditions, such as tea.

8.10 Wind belts

Strong wind may physically damage the plant, transpiration rate becomes higher and lower the leaf temperature. For boundry/estate barrier, tall trees viz. *Casuarina montana, Eucalyptus globulus* Labill., *E. robusta, E. saligna* Sm., *Grevillea robusta* A Cunn. and *Dalbergia sisso Roxb. ex DC.* at 4.5 x 4.5 m spacing, then the medium trees such as *Cassia siame* Lam., *Acacia auriculiformis* A. Cunn. ex Benth., *Erythrina poeppigiana* (Walpers) O. F. Cook, *Melia azadirach* L., *Pongamia glabra* Vent. at 4 x 4 m spacing and dwarf trees viz. *Acacia elata* A. Cunn. ex Benth., *A. mearnsii* De Wild., *Murraya exotica* L., *M. koenigii* (L.) Spreng., *Hibiscus roseus, Dohonea viscose* L. and *Cypruss* sp. can be planted at 3 x 3 m spacing. Lines of wind belt trees should be at a right angle to the direction of the wind. In Tanzania, the effect of 4–5 m high shelterbelts (windbreaks) of *Hakea saligna* on yield and water use of tea at different distances from the windbreaks was studied by Carr (1970, 1971). Carr and Stephens (1992) observed that shelter from trees planted between fields in the direction of the prevailing wind is likely to be the best approach where high winds are considered to be a problem, but hedges may still be useful on wind-exposed ridges. They further concluded that shelter belts are probably only useful in areas where the mean air temperature is below 17–18^{0}C, or where advection of hot dry air can be expected. A list of suitable tree species with spacing for planting as shelter belts prepared by Somaratne (1986).

8.11 Training of young tea

The early management of young tea is aimed at modifying and training it to form a low bush of good spread and of appropriate height for plucking. Therefore, it is most important to establish a well developed primary frame and branch system ensure a complete ground cover as early as possible. These objectives can be achieved either by decentring and pruning, or by bending or pegging or by a combination of decentring and bending (Barbora *et al.*, 1981). Recommended procedures for different regions vary. Variations may be useful keeping the rates and habits of growth of particular clones. A few important practices followed are mentioned below.

8.11.1 *Decentring and pruning*

Young plants are centred in the field to suppress apical dominance and for the production of lateral branches which in turn helps to attain a bush habit. Decentring and pruning combination involves cutting off the leader stem at a height between 7 and 20 cm from the ground (Fig. 8.17). The cut should be clean and slope downward to the centre of the bush.

> **Box 8.7 : Test of plant food reserve in root and its availability**
>
> *The iodine solution is prepared by dissolving 1 gram of iodine crystals and 1 gram of potassium iodide in 1 to 2 ml of distilled water. Then the volume of the solution is made up to 100 ml with distilled water and store the solution in dark colour bottle. A 10 to 15 cm deep hole is dug at a distance of approximately 10 cm away from the collar (for mature plant root samples need not to be drawn so close the collar) and any root with a minimum diameter of 5 mm is severed from the plant. Two to three drops of iodine solution is applied to the smooth cut surface of the root and kept for 5 minutes. Deep blue colour all over the cut surface, starting at the centre spreading towards periphery is an indication of good starch reserve. When the starch reserve is low, the colour lacks depth or appears only at the centre of the cut surface.*
>
> *The content of crude starch in the root was found very low in September or so and begin to increase rapidly from late autumn. It remained in the maximum level during winter and then decreased with the growth of shoots. The crude starch accumulation in the main root began near the centre of the xylem, spreading to the periphery, and it decreases in spring proceeded in the reverse direction from the periphery to the centre.*

Laterals are given a cut across at 35 cm above ground level. This excision promotes more laterals from the lower portion of the stem, thus increasing the number of pruning sticks. Next pruning of the plants, about 5 cm above the last prune mark, are done within two years after the first prune for recentring. The decentering operation should be done after the plants make one flush of growth following transplanting in the field and when they are at *banjhi* stage. Further, decentering should be done in presence of sufficient moisture in the soil. Root starch is also considered important for recovery and to minimise mortality after decentring. Generally, at high elevation, say 4500 feet and above, bushes store-up more carbohydrates than burn them but in the lower elevation this precaution is all the more necessary.

8.11.2 *Bending and pegging*

The primaries/secondaries are pegged from one month after planting in case of vigorous plants. Pegs are made out of pruned tea branches, bamboo, wood etc. Young plants should preferably have at least four stems before pegging. It is better to ensure that these are formed in the nursery by removing the apical bud (Willson, 1999). Shoots that are 30 to 60 cm height are bent at angles of 30^0 to 45^0 to the ground, and are then pegged down to the ground (Fig. 8.18). All lateral shoots produced subsequent to

Fig. 8.17: Decentred young plant in the field.

the bending should also bent and pegged. In contour planting, bending should be done along with contour, preferably turned towards west or south (Banerjee, 1993). Pegged stems for the whole of their length must slope upwards. The flow of sap will stop at the highest point before a downward section which may die back. About 45^0 slope is recommended in Sri Lanka. According to Tea Research Foundation of Kenya a flatter slope is also equally effective (Willson, 1992). The area which normally have a long dry season pegging can be avoided since there are risks for sunscorch and other diseases. The pegged main stems along with laterals are first pruned 30 to 45 cm from the ground within two to three years, and then raised by 5 cm in each successive prune till permanent frame height is reached.

Fig. 8.18: Laterals of the young tea plant are pegged down to the ground to spread the branches, from which vertical shoots will form the frame of the plant.

Pegging after decentring causes an immediate crop loss but this is well-compensated later.

8.11.3 *Frame forming prune*

Planting from 20 to 26 months and at a height between 35 to 40 cm from the ground when the wood is atleast 1 cm thick, recentring and head-back of laterals to the side by pegging should be done. Subsequently pruning is done at 35 to 50 cm from the ground and then the frame is raised by 5 cm at each successive prune till a permanent frame is established.

8.11.4 *Low tipping*

Pruning is not required for first four years of planting in this method. Centring out within 10 weeks after planting and initial plucking at low level of 40 to 50 cm from the ground followed by a gradual rise or stepping up of the plucking table are to be done. The first pruning is done when the plants have attained above 85 cm height at the end of the fourth or fifth year with pruning height varying between 55 and 65 cm from the ground. This method produces good early crop but frame formation is poor.

8.11.5 *Lung or thumb pruning*

When plants are between 15 and 20 cm high, they are held between the thumb and the index finger, and the stem is broken in such a way that the tissues on one side of the stem are left intact for movement of water and nutrients. The broken portion is then bent towards the ground (Fig. 8.19) facing either south or west depending on the direction of row. The broken portion is removed after the shoots below it produced atleast 4-5 new leaves. But where the drought is fairly prolonged like in south Indian condition thumb nail pruning has not been useful. Small pruning knife can also be used for lung pruning. Lung pruning can be done even if starch content is not at its optimum.

8.11.6 *Debudding*

Buds from the leaf axils are removed from the upper portion of the plants leaving the low 20 cm untouched. This is done with the help of a small hacksaw blade or using finger. The top two and a bud should be tipped off from plants in nursery two weeks before debudding. This will help to swell the buds below. Four to five days before the plants are taken out to the field they can be debudded in the nursery. Debudding in the nursery is more thorough and economical. After about four weeks of planting in the field, removal of bud if any left out in the first operation should be done. Buds remaining below the height of 20 cm from the ground will grow and form healthy laterals. The main stem should be removed by giving a clean cut at about 20 cm with a 7.5 cm pruning knife. Thereafter normal tipping and plucking should be followed (Chakravartee and Barbora, 1994).

8.11.7 *Chemical pruning*

Certain growth retardants such as Ethephon (CEPA) at 1000 ppm,

Fig. 8.19: Thumb pruned young tea plant.

Morphactin (Chlorflurenal methyl ester) at 50 - 100 ppm, Paclobutrazol at 500 ppm and Cyclocel (Chlorocholine chloride, CCC) at 2000 ppm can be used for stimulating branches by suppressing meristamatic activity in the apical zone of leading stems. The apical growth of the plants sprayed with these growth regulators stops for 8 to12 weeks but increase the stem diameter and produce laterals at the base. The action of ethephon was found to vary under different agroclimatic conditions. However, chemical pruning is not yet very successful in tea.

In Assam, Barbora and Sarkar (1988) reported that pegging and pegging after decentring produced a large frame with more branches compared to the low tipping and step-up (LT) method. They further reported that a formative prune was essential irrespective of whether the young tea was brought up by pegging or pruning. Best results for frame formation and subsequent higher yield were obtained when formative prune was given twentysix months after planting. Banerjee (1993) reported that pegging in the initial two years has an advantage in terms of crop increase and thereafter this advantage is lost and the system's overall performance equals the combination of low tipping and frame formation pruning.

In Darjeeling hills, Kabir *et al*. (1991) while comparing the performance of decentring, pegging (since planting primaries/secondaries are pegged from 12 to 78 weeks) and application of sub-lethal concentration (500 ppm in two splits at monthly interval starting from 12 weeks from planting) of oxyfluorfen (a pre-emergent herbicide) observed that all three treatments had good impact on early spread of branches and results were almost similar with regard to crop and diameter of the stem. However, plants under pegging had an edge over other two methods especially the spread of branching was more pronounced in the clones (B157 and AV2) which otherwise have a tendency to grow more vertically under normal circumstances. Contrary to the observations in plain gardens no sunscorch damage was noticed under pegging treatment. Further, unlike in plains, the rate of growth being rather slow in Darjeeling and pegging is not difficult in young plants under double system of planting. Pegging however, requires certain degree of skill and good deal of supervision as it warrants selectivity and repeatation at least for 3-4 rounds. It also requires constant refixing of pegs loosened/dislodged by wind and soil wash by rain water.

In south India, Satyanarayana *et. al*. (1987) observed that (1) centring of young plants four to five months after planting is advantageous in promoting an early spread filling of the bush, (2) it is important to leave about ten healthy mature leaves below the point of decentring, either on the main stem or a lateral branch, (3) tall lanky plants with lesser foliage may require recentring four to six months from the first to force a low spread and (4) the plucking level may be established by tipping off three/four leaves and a bud at a height of about 50 and 55 cm in China and Assam hybrid types of tea plants respectively. Crop shoots produced above this level are plucked continually to a mother leaf up to the time of formative prune. Such a system of training has been found to result in the formation of primary frame having fewer branches, poor spread and delayed cropping in the pruning cycle (Satyanarayana and Sharma, 1991). Under these circumstances, training of young tea by two-stage tipping (first at 35 cm and the second at 50 cm) and plucking to a fish leaf up to the end of year from tipping and thereafter, to a mother leaf during the low cropping seasons and to the level during the other months is the ideal method of training young tea (Satyanarayana *et al.* 1991).

References

Allen, O. N. and Allen, E. K. 1981. *The Leguminosae.* Madison: Univ. of Wisconsin Press; London: MacMillan Publishing Company.

Ananthacumaraswamy, A., Amarasekera, A. R. and Herath, H. M. G. 1986. Economics of rehabilitating tea land before replanting: some exploration. *Sri Lanka J. Tea Sci.*, **55(2):** 89–93.

Anonymous, 1966. *Annual Report 1965 - 66.* Tea Research Foundation (Central Africa), Mulanji, Malawi, 33-36.

Anonymous, 1988. *Tea Encyclopaedia. Serial No. F1, Tocklai Experimental Station*, Jorhat, Assam, India: Tea Research Association.

Anonymous, 1994. *Tea Statistics 1993-1994.* Tea Board of India, Calcutta.

Bald, C. 1965. *Indian tea.* Thacker Spink & Co., (1933) Pvt. Ltd., Calcutta, India, p.423.

Banerjee, B. 1993. *Tea production and processing.* Oxford & IBH Publishing Co. Pvt. Ltd. New Delhi, p.336.

Barbora, B. C. and Bordoloi, P. K. 1993. Application of remote sensing in soil and water management. *In : Teatech 1993, Proc. Intl. Conf. Tea Science and Human Health,* January 11 - 14, 1993, Calcutta, Tea Research Association, India, 232-241.

Barbora, B. C. and Hazarika, R. L. 1996. Rejuvenation of old tea. *Two and a Bud*, **43(1):** 10-17.

Barbora, B. C., Jain, N. K. and Rahman, F. 1981. Bringing up of young tea. *Proc. 29th Tocklai Conf.*, December 17-18, 1981, Jorhat, Assam, Tea Research Association, India, 72-76.

Barbora, B. C. and Sarkar, S. K. 1988. Effects of different methods of bringing up young tea on frame formation. *Proc. 30th Tocklai Conf. Tea Technology-Today and Tomorrow*, March 11-12, 1988, Jorhat, Assam, Tea Research Association, India, 69-74.

Barua, D. N. 1961. Effects of shade and shade trees. *Two and A Bud*, **8(4):** 4-12.

Barua, D. N. 1973. Spacing of tea bushes. *Proc. 26th Tocklai Conf.*, November 13-15, 1973, Jorhat, Assam, Tea Research Association, India, 102-108.

Barua, D. N. 1987. Tea. *In: Tree Crop Physiology* (eds M. R. Sethuraj and A. S. Raghavendra), Elsevier Science Publishers, Amsterdam, The Netherlands, pp.225-246.

Barua, D. N. 1989. *Science and practice in tea culture.* Tea Research Association, Calcutta- Jorhat, India, p.493.

Bezbaruah, H. P. and Barbora, B. C. 1979. Glimpses of Indonesian tea industry. *Two and a Bud*, **26(1):** 1-4.

Carr, M. K. V. 1970. The role of water in the growth of the tea crop. *In: Physiology of tree crops* (eds. L. C. Luckwill and C. V. Cutting). Academic Press, New York, pp. 287–305.

Carr, M. K. V. 1971. The internal water status of the tea plant (*Camellia sinensis*) some results illustrating the use of the pressure chamber technique. *Agric. Meteorol.*, **9:** 447–460.

Carr, M. K. V. 1972. The climatic requirements of the tea plant: A review. *Expl. Agric.*, **8:** 1–14.

Carr, M. K. V., Dale, M. O. and Stephens, W. 1987. Yield distribution in irrigated tea (*Camellia sinensis*) at two sites in eastern Africa. *Expl. Agric.*, **23:** 75–85.

Carr, M. K. V. and Stephens, W. 1992. Climate, weather and yield of tea. *In : Tea Cultivation to Consumption* (eds. K. C. Willson and M. N. Clifford). Chapman & Hall, London, 87–135.

Chakravartee, J. and Barbora, B. C. 1994. Nursery and field management of clones. *Advisory Bulletin 3*, Tocklai Experimental Station, Tea Research Association, Jorhat, Assam, India, 1-16.

Chakravarty, S. K. and Awasthi, R. C. 1981. Economic evaluation of plant density. *Two and a Bud*, **28(2):** 32-34.

Dey, S. K. 1972. Tea in Russia. *Two and a Bud*, **19(2):** 73-84.

Dixon, R. O. D. and Wheeler, C. T. 1983. Biochemical, physiological and environmental aspects of symbiotic nitrogen fixation. *In : Biological nitrogen fixation in forest ecosystems: Foundations and applications* (eds. J. C. Gordon and C. T. Wheeler). The Hague: Martinus Nijhoff / Junk, pp. 295-316.

Dutta, S. K. and Basu, S. 1955. Reduced light intensity and nitrogenous manures. *Tea* Trade and Industry July 1955.

Dutta, S. K., Sharma, P. C. and Barbora, B. C. 1971. Rehabilitation of tea soils before planting. *Two and a Bud*, **18:** 1-3.

Furst, J. P. F. and Awasthi, R. C. 1981. Tea estate planning. *Two and a Bud*, **28(2):** 35-38.

George, U. and Singh, R. 1990. Biological and chemical factors affecting replanting in tea. *Proc. Intl. Conf. Tea Research Global Perspective*, January 11-12, 1990, Calcutta, Tea Research Association, India, 77-83.

Ghosh Hajra, N. 1992. Strategies for improvement of tea production in Darjeeling hills. *Indian Rev. Life Sci.*, **12:**195-210.

Ghosh Hajra, N. and Kumar, R. 1999. Seasonal variation in photosynthesis and productivity of young tea. *Expl. Agric.*, **35:** 71-85.

Grice, W. J. 1983. Soil and water conservation at planting. *Tea Research Foundation (Central Africa) Quarterly Newsletter*, **69:** 13.

Grice, W. J. 1984. The effect of cultivation and rehabilitation on replanted tea. *Tea Research Foundation (Central Africa), Quarterly Newsletter*, **75:** 20.

Grice, W. J. 1987. Effect of different cultivation and rehabilitation methods on the physical and chemical properties of the soil and the yield of replanted tea. *Tea Research Foundation (Central Africa) Quarterly Newsletter*, **86:** 4-9.

Grice, W. J. 1990. Replanting. *Tea Planter's Hand Book*, Tea Research Foundation (Central Africa), Mulanje, Malawi, 1–5.

Hadfield, W. 1968. Leaf temperature, leaf-pose and productivity of the tea bush. *Nature*, **219:** 282-284.

Hadfield, W. 1974 a. Shade in north-east Indian tea plantations. I. The shade pattern, *J. Appl. Ecol.*, **11:** 151-178.

Hadfield, W. 1974 b. Shade in north-east Indian tea plantations. II. Foliar illumination and canopy characteristics. *J. Appl. Ecol.*, **11:** 179-199.

Hadfield, W. 1975. The effect of high temperatures on some aspects of the physiology and cultivation of the tea bush (*Camellia sinensis*) in north east India. *In: Light as an Ecological Factor II* (eds G. C. Evans, R. Bainbridge and O. Rackham), London: British Ecological Society, pp.477-495.

Hudson, J. B. 1995. Transfer of technology in tea — achievements and challenges. *UPASI Tea Scientific Department Bulletin*, **48:** 67-79.

Hudson, J. B., Durairaj, J., Muraleedharan, N. and Dhankumar, V. G. 1997. Guidelines on tea culture in south India. *UPASI - KVK publication no. 30*, p. 212.

Hudson, J. B. and Muraleedharan, N. 1999. Shade for tea. *Planters' Chronicle*, **95(12):** 551-556.

Kabir, S. E., Ghosh Hajra, N. and Chaudhuri, T. C. 1991. Training of young tea in Darjeeling. *Tech. Bull.*, 6, Tea Board India, Darjeeling Tea Research Centre, 1-7.

Kabir, S. E., Ghosh Hajra, N. and Chaudhuri, T. C. 1993. Comparison of methods of replanting old tea bushes. *Trop. Agric.* (Trinidad), **70(3):** 205-207.

Kankani, N. C. 1989. Tea production. *Tea Journal*, **1:** 9-12.

Krishnarajah, P. 1985. Soil erosion control measures for tea land in Sri Lanka. *Sri Lanka J. Tea Sci.*, **54(2):** 91-100.

Kulasegaram, S. and Kathiravetpillai, A. 1976. Effect of shade and water supply on growth and apical dominance in tea (*Camellia sinensis* (L.) O. Kuntze). *Trop. Agric.* (Trinidad), **53(2):** 161-171.

MacDicken, K. G. 1994. *Selection and management of nitrogen-fixing trees*. Morrilton, Arkanasas: Winrock International, Bangkok: FAO, p.272.

Magambo, M. J. S. and Waithaka, K. 1985. The influence of plant density on dry matter production and yield in young clonal tea in Kenya. *Acta Horticulturae*, **158:** 157-162.

Manipura, W. B. 1972. Influence of mulch and cover crop on surface run-off and soil erosion on tea lands during the early growth of replanted tea. *Tea Quarterly*, **43 (3):** 95-102.

Morris, J. M. 1997. Estate road system — principles and pitfalls. *Tropical Agriculture Association Newsletter*, **17(4):** 13-18.

Nakayama, A. 1971. Changes of carbohydrate and nitrogen compounds with growth in tea plant. *JARQ*, **6(2):** 97–101.

Ng'etich, W. K. and Othieno, C. O. 1993. Performance of clonal tea plants on rehabilitated moribund tea soils. *Tea*, **14(2):** 96-106.

Obaga, S. M. O.and Othieno, C. O. 1986. Effects of different methods of uprooting old tea and soil rehabilitation on some soil physical properties. *Tea*, **7(2):** 58-65.

Othieno. C. O. 1983. Studies on the use of shade in tea in Kenya. I. Effect on nutrient uptake and yields of tea. Preliminary results. *Tea*, **4(2):** 13-20.

Othieno, C. O. 1992. Soils. *In : Tea: Cultivation to consumption* (eds K. C. Willson and M. N. Clifford). Chapman & Hall, London, pp.137-172.

Othieno, C. O. and Ng'etich, W. K. 1993. Studies on the use of shade in tea plantations in Kenya. II Effect on yields and components. *In: Teatech, 1993. Proc. Int. Symp. Tea Science and Human Health*, January 11-14, 1993, Tea Research Association, Calcutta, 282-292.

Ponnapa, P. S. 1991. Replanting - practices and economics. *UPASI Tea Scientific Department Bulletin*, **44:** 77 84.

Rahman, F. 1999. Irrigation practices in tea. *In : Global Advances in Tea Sciences* (ed. N.K. Jain). Aravali Books International (P) Ltd., New Delhi, 611-624.

Rahman, F. and Fareed, M. 1977. Studies on spacing of tea. *Proc. 28th Tocklai Conf.,* November 24-26, 1977, Jorhat, Assam, Tea Research Association, India, 8-15.

Rajkumar, S., Marimuthu, S. and Muraleedharan, N. 1999. Tea leaf photosynthesis in relation to light. *J. Plantation Crops*, **27(2):** 93-98.

Ranganathan, V., Ganesan, M. and Natesan, S. 1980. Organic matter flux in south Indian tea soils a need for conservation. *Planters' Chronicle*, **75(7&8):** 309-312.

Saijo, R. 1980. Effect of shade treatment on biosynthesis of catechins in tea plants. *Plant & Cell Physiol.*, **21(6):** 989-998.

Sandanam, S. and Rajasingham, C. C. 1981. Effects of mulching and cover crops on soil erosion and yield of young tea. *Tea Quarterly*, **51(1):** 21-26.

Satyanarayana, G. 1971. Effect of soil rehabilitation and drainage on red rust. *Proc. 25 th Tocklai Conf.*, November 16-18, 1971, Jorhat, Assam, Tea Research Association, India. 151-152.

Satyanarayana, N., Hudson, J. B. and Sharma, V. S. 1991. Response to two stage tipping of young tea in Nilgiris. *UPASI Tea Scientific Department Bulletin*, **44:** 61-66.

Satyanarayana, N., Joseph, C. P. D., Govindarajulu, V., Hudson, J. B., Surendra Mohon, M. and Sharma, V. S. 1987. Bringing young tea into bearing. *Planters' Chronicle*, **81:** 242-245.

Satyanarayana, N. and Sharma, V. S. 1991. Training of young tea by two stage tipping. *UPASI Tea Scientific Department Bulletin*, **44:** 59-60.

Sharma, V. S. 1984. New avenues to replanting of tea. *UPASI Tea Scientific Department Bulletin*, **39:** 18-21.

Seemangun, H. 1976. Introducing the Indonesian tea industry. *UPASI Tea Scientific Department Bulletin* **33 :** 26-31.

Singh, B. 1978. Prediction of drainage and irrigation requirements of tea based on meteorological data. *Two and a Bud*, **25(2):** 75–79.

Singh, B. 1999. Drainage of waterlogged tea soils with special reference to north-east India. *In : Global Advances in Tea Science* (ed. N. K. Jain). Aravali Books International (P) Ltd., 575–610.

Singh, I. D., Bhattacharya, A. K. and Bera, B. 1999. Organic tea cultivation an ecofarming approach. *Assam Rev. & Tea News*, **88(9):** 5–15.

Somaratne, A. 1986. Planting of shade trees and shelter belts. *In: Hand Book of Tea* (eds P. Sivapalan, S. Kulasegaram and A. Kathiravetpillai), Tea Research Institute of Sri Lanka, Talawakelle, Sri Lanka, pp.41-43.

Squire, G. R. 1979. Weather, physiology and seasonality of tea (*Camellia sinensis*) yields in Malawi. *Expl. Agric.*, **15:** 321–330.

Subramaniam, B. 1995. *Tea in India*. Publication & Information Directorate, New Delhi & Wiley Eastern Limited, New Delhi, p. 125.

Tabageri, L. G., Mikeldze, A. D. and Cheishvili, I. D. 1991. Results of studies on measures for rehabilitating depressed tea plantations on podzolic soil. *Subtropicheskie Kul'tury*, **4:** 136-141.

Tanton, T. W. 1982. Environmental factors affecting the yield of tea (*Camellia sinensis*) II. Effects of soil temperature, day length and dry air. *Expl. Agric.*, **18:** 53–63.

Willson, K. C. 1992. Field operations: 1. *In: Tea: Cultivation to Consumption* (eds. K. C. Willson and M. N. Clifford). Chapman & Hall, London, pp.201-226.

Willson, K. C. 1999. *Coffee, Cocoa and Tea*. CABI Publishing, Wallingford, UK, p. 300.

Yongming, Y. 1999. Agrotechnology of tea in China. *In: Global Advances in Tea Science* (ed. N. K. Jain) Aravali Book International (P) Ltd., New Delhi, 481–500.

Chapter 9

Pruning and Plucking

9.1 Pruning

As the age of tea bushes from pruning advances, size and weight of harvestable shoots decline due to the reduction in the vascular supply to growing buds. An extreme case of growth retardation is evidenced in the higher proportion of *banjhi* shoots at the plucking table. Also, the upper part of the plants below the plucking table becomes congested with stems and knots where many new stems have originated. Despite, the 'maintenance foliage', ages and losses efficiency, lower leaves dropping off, leading to uneconomic harvesting (Willson, 1999). Under continued plucking without pruning or skiffing the plucking table becomes inconveniently high and growing apices gradually lose vigour. Pruning at this stage becomes necessary by which these irregularities could be remedied. It is an important practice in tea cultivation which involves removal of certain amount of growth of the bush.

9.1.1 *Objectives of pruning*

- Maintaining convenient height to facilitate plucking operation.
- Checking reproductive growth and stimulating vigorous shoot system, rationalising wood/shoot ratio.
- Removing dead, diseased and defunct wood thus help rejuvenate bushes that have crossed the period of maximum productivity.

- Minimising formation of *banjhi* (dormant shoots) and spreading of pests and diseases.
- Renewing the growing branches and maintaining the sufficient volume of maintenance foliage to meet the physiological needs of the plant.
- Controlling crop during rush periods and to achieve better crop distribution.
- Maintaining quality in processed tea.

9.1.2 *Pre-requisites for pruning*

Starch reserve in the plant is a major determinant for pruning because the new growth relies on the starch reserves. Therefore, pruning in any form without an adequate reserve of carbohydrates could be disastrous (Banerjee, 1993) [See Chapter 8, Box 8.7]. At the time of pruning, removal of a large volume of foliage and young shoot takes place and that stops carbohydrate synthesising activities of the plants. It is the carbohydrate reserve in the roots that helps the plant to recover from pruning. Attention to the importance of starch reserves in the roots of the tea bushes for their recovery following pruning was first drawn in Sri Lanka (Gadd, 1928). In former USSR, it is believed that the carbohydrate-nitrogen balance can be taken as guide for deciding upon the pruning system (Dey, 1972).

Slow and irregular recovery from pruning has been observed if the bushes are short of nutrients. This apart, the bushes should have the efficiency for speedy uptake of fertilizer applied before pruning within a certain interval. Potash fertilizer has a great influence on the recovery of bushes from the pruning. Application of potash fertilizer before pruning is therefore a common practice. In Sri Lanka, it was investigated that poor recovery and excessive die back occurred most frequently when fertilizers were applied very close to pruning operation (Pethiyagoda, 1966). The other factors that influence pruning time are the risk of sunscorch during dry season, infestation of pests etc.

9.1.3 *Time of pruning*

This has a significant effect on the recovery of the bushes after pruning. The appropriate timing for pruning is basically considered from an assessment of the growth rate of the plant. It should be at a point when growth has sufficiently slowed down to create a favourable carbohydrate balance, so that, a layer of foliage could grow and get established well before the hot spell (Banerjee, 1993). Time of pruning also affects total yield despite its influence on the distribution of crop. In north-east India, studies using radioactive carbondioxide on the movement of photosynthates revealed that the carbohydrate start moving downwards from October until mid-February after which upward movement started again (Manivel, 1980a). Considering the starch reserve in roots, December-January is the ideal time for pruning tea bushes in north-east India. The time can be advanced upto mid-November if other factors are favourable. In the droughty areas of Cachar, Dooars and Terai it should be started in mid-December and completed by end of January. In Darjeeling hills, the time can be advanced by about 15 days as the plucking season closes earlier. Heavy pruning in mid to high elevations of Darjeeling can be started as early as August provided labour availability is not a constraint and starch reserves in the roots are satisfactory. In low elevation gardens, it should commence in October. In south India, the best time for pruning at mid-elevation is August-September and at higher elevation during April-May before the 'rush' crop is harvested. In India, lighter forms of prune or skiff are periodically applied to regulate leaf supply and provide photosynthetically active foliage.

In Sri Lanka and east Africa, the period immediately following dry season is considered ideal for pruning. Pruned bushes minimise the use of soil water in drought prone area. Ellis (1984) stressed that bushes should be pruned before they suffer from moisture stress. He discussed pruning and drought. Kilavuka (1980) reported better recovery from pruning in November/December for tea under shade than with unshaded bushes. In east Africa, the failure of tea bushes to recover from normal pruning was attributed to external damage caused by sunscorch, frost, mechanical and attack by disease-causing organisms (Templer and Machaga, 1978).

Malenga (1998) reported that the optimum pruning time for mature tea in Malawi is mid-May. Since pruning to all mature tea field is not possible to complete at this time, therefore it should be started earlier and tipping of these early pruned bushes can be done by July. Similarly, bush pruned after mid-May will be ready for tipping between early-August and end-September.

Studies conducted in Georgia revealed that semi-hard pruning at 35–40 cm from collar in 34–45 year old tea plantations (Esartiya, 1972 and 1975) or intensive (hard) pruning (Kipiani and Alkhazov, 1972) at 10–15 cm from collar in November (Chkhaidze, 1972) gave the best results in terms of growth and yield as against hard to light pruning.

In Indonesia, the main considerations for time of pruning at low elevation are sun-scorch and low content of root starch, while at high elevation it is the blister blight infestation. At low and medium elevation pruning can only be done during January to March and from October to December. If a faster bush recovery is needed then the best time for pruning is at the flushing points i.e. in April, May, October and November (Wibowo *et al.*, 1990).

9.1.4 *Types of pruning*

The pruning systems vary with the stage of the life cycle of a tea bush. These are formative pruning, maintenance pruning and corrective pruning. Formative prune trains the free growing tea plant to form a table–topped bush. The system of formative pruning are described in Chapter 8 (section 8.11). The different types of maintenance and corrective pruning and based on severity of cuts adopted on tea (Fig. 9.1) are broadly classified as follows:

a. *Light or top pruning*

Maintenance or light pruning is done at 3 to 5 cm above the previous

Fig. 9.1 : Tea after medium pruning in Darjeeling tea estate.

pruning mark when the bush under plucking becomes too high in 4–5 years. The height of pruning determines the number of leaf bearing branches. Their number is greater on longer stems while the rate of recovery is slower on bushes cut deeper (Kulasegaram and Kathiravetpillai, 1981). Light pruning is the standard method of recurring pruning in Assam and the Dooars. In lower and mid elevation gardens where tea is well manured and pruned at frequent intervals, satisfactory yield is obtained by this method of pruning The technique of light pruning involves a uniform cut-across which is carried out at various heights in different countries. In south India, it is done at 60–65 cm. In Assam with 3 to 4 year cycle, 4–5 cm of new wood is left and in Sri Lanka it is done at a lower height than in India.

b. *Medium pruning*

It is done to reduce height of the bushes, so that they do not exceed a height of 80 cm in the plains garden which enables efficient performance of the plucking operation (Venkatakrishnan and Sen, 1981). The plucking table in the slopes of the Darjeeling hills is normally lower than in the plains and the height of the bush has to be adjusted to the slopes of the terrain (Barua, 1989). The maximum height of this pruning is not well defined but it is generally done above the height of the original frame-forming pruning (Fig. 9.2). If the frame forming prune was given at 45 cm then medium prune can be done at any point above this height. In south India, it is done at a height of 45–60 cm from the ground. In *assamica* bushes medium pruning can be as high as 30–45 cm. Medium pruning helps removal of congested top hamper of weak and twiggy branches arising from continuous picking of shoots on the top of the bush. This apart, medium pruning is done to remove knots, diseased wood and to reduce the proportion of unproductive wood. Plants must be rested for six to eight weeks before carrying out medium pruning.

c. *Skiffing*

A lighter form of cut which removes knotted and congested stems permits flushing to resume after only a short interval. It is done sometimes in order to lower the plucking table. The object of skiffing, particularly in north India, is to have an early high quality first flush and total crop as well. Skiffing also helps in thickening the pruning wood and improving the general health of the bushes. In south India, skiffing is not a common practice and rarely done in well-managed estates. It is resorted to only when height of

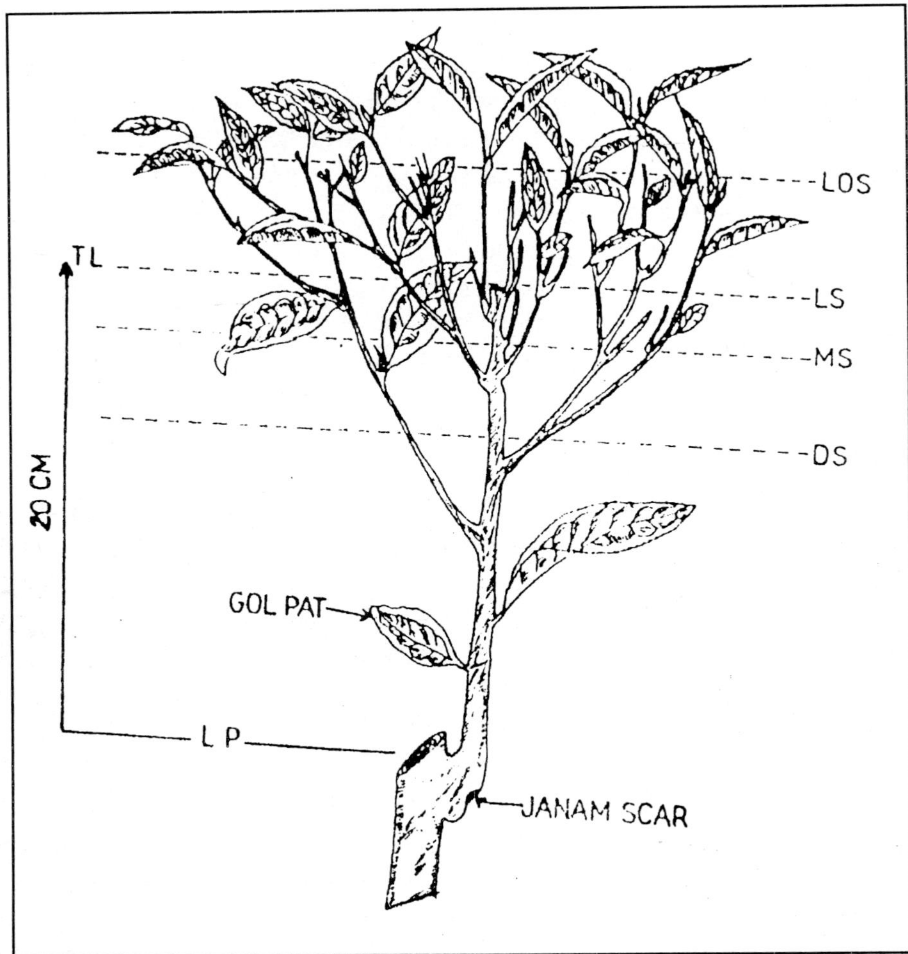

Fig. 9.2 : Different forms of pruning (diagrammatic view). LP = light prune; DS = deep skiff; MS = medium skiffing; LS = light skiff; LOS = level off skiff; TL = tipping level.

the bushes increase, making the plucking difficult and there is a necessity of postponing the pruning operation. It is done in south India at a height less than 65 cm. Skiffing is not severe enough to term a 'prune'. Different forms of skiffing are practiced which are defined as follows:

i. Deep skiff (DS)

This is normally given a year after pruning. In deep skiff, a cut is given midway between the pruning and tipping levels. For instance, if tea is tipped

Fig. 9.3: Tea after a deep skiff (DS) in Darjeeling hill. Workers are scraping the soil surface to remove weed growth using a half moon shaped flat hoe (cheeling operation).

at 20 cm, deep skiff can be given 10 cm above the last pruning level. Deep skiffing after one or more years of unpruned or level skiffs is cut midway between the pruning level and the height of the plucking table at the end of the season, provided not more than 25 cm of growth is removed by the skiff. For instance, if the plucking height at the end of the season is 30 cm above the pruning mark, tea will be deep skiffed at 15 cm above the pruning mark (Fig. 9.3).

ii. Medium skiff (MS)

The main object of medium skiffing is to remove the congestion of dead and unproductive twigs at the top, that is removal of so-called "crow's feet". Therefore, a cut often given at a height of 15 cm above the pruning mark or 5 cm below the previous year's tipping level to take care of the majority of the unproductive twigs.

Fig. 9.4 : Tea after a light skiff in Darjeeling tea estate.

iii. Light skiff (LS)

In this skiff, a superficial cut is given 20 cm above pruning level just to remove a majority of the plucking points to reestablish a level (Fig. 9.4).

iv. Level-off skiff (LOS)

This is a skiff where a cut is given 6 to 8 cm above current year's tipping level to remove the top (Fig. 9.5).

The important feature of skiffing is to retain more leaves on the bush frame. In order to keep the crop loss to the minimum, it is better to complete the skiffing before the first week of September in plains garden and mid-August in the tea estates of Darjeeling hills. In Darjeeling, where plucking of *banjhis* (dormant shoots) between flushes on unpruned or light skiffed tea in the third or fourth year of the cycle is not possible, skiffing can be done to remove the *banjhis* immediately following a round of plucking. The presence of *banjhis* could seriously affect the growth of new shoots. In terms of crop return, deep skiff would give more of second flush and main crop, whereas other forms of skiffing ensure more of first flush (Banerjee, 1993).

Fig. 9.5 : The front line chinary tea bushes in Darjeeling hill given a level-off-skiff (LOS). Note the adjoining rows are unpruned bushes.

d. *Heavy or low pruning*

In the course of years, successive up-prunings have so increased the height of the frames of the bushes, that they are difficult to pluck and when the branches have formed snags and knots, heavy pruning becomes necessary. The heavy pruning is a variant of collar pruning and common in Darjeeling when fall in yield is observed. The height of heavy pruning depends on the type of tea and condition of the frame. Generally, the bole is cut at 15 to 20 cm and 25 to 37 cm in case of Assam hybrid and Chinary kind bushes respectively. If the bushes found affected by *Aglaospora*, it is pruned to the ground level. In south India, this pruning is also resorted while doing formative pruning to young tea for spread of bushes and wherever the secondary branches are week and knotty (Hudson *et al*., 1997). This pruning can only be done after considering the health of plants and starch reserve.

Till new flush of shoots on the bushes has produced some leaves, retention of some branches as 'lungs' or 'breathers' (Fig.9.6) at the time of heavy pruning to facilitate better supply of photosynthates to the roots for

Fig. 9.6 : Retention of lung seen on the light prune bushes in Darjeeling hills.

early recovery is practised in Sri Lanka (Tubbs, 1977; Kandiah, 1984) and Indonesia (Semangun, 1976). Lungs should be left on the south-west side of the bush to reduce the impact of direct sun on the exposed bush frame. But Chakravartee and Borpujari (1994) reported that pruning is done at a time when the starch reserves are optimum and no additional benefit would be obtained by retaining a lung shoot in the year of prune or in subsequent years.

9.2 Down, Collar or rejuvenation pruning

Corrective or rejuvenation pruning is a most severe operation in which the bole of the bush is cut at ground level or even 5 cm below the ground level. Rejuvenation pruning facilitates removal of the dead and defunct wood and development of good, healthy and new primary frames. It also implies restoration of plant density by infilling and/or interplanting. It is strictly need-based operation and the factors for determining rejuvenation are the age of the bush, vacancy, drainage, yield trend and jat of tea. Fields where the frames of the bushes are diseased and unsound need to be rejuvenated. In Darjeeling, rejuvenation of bush frame by pruning at ground level is a regular practice (Fig. 9.7). In multi-stemmed chinary bushes, the regeneration of roots is a continuous process and where the roots are healthy bringing up the yield of the particular field by rejuvenation pruning following infilling and/or interplanting is much faster than the time needed to bring it through replanting. This pruning requires following essential precautionary measures to ensure success.

i. *Carbohydrate content*: It is necessary to ensure that the carbohydrate reserves in the root should be sufficient enough to support new growth before undertaking this operation. In south India, Chandra Mouli and Sharma (1993) reported that 18.5 percent total carbohydrate content in roots is adequate for satisfactory recovery from rejuvenation pruning. In the high

Fig. 9.7 : Tea plant after rejuvenation pruning in Darjeeling hill.

elevation (above 1500 m) gardens of Darjeeling hills it has been observed that bushes store-up more carbohydrates than oxidised them but in the lower elevation this precaution is all the more necessary. The carbohydrate reserves presents in the stem bark are supposed to meet early requirement of bud growth on a pruned frame to a great extent.

ii. *Resting*: Resting of bushes two to three months prior to rejuvenation will help in the build up of root starch and adequate for an early and satisfactory recovery.

iii. *Drainage*: It should be improved to keep the water table below the root zone at least 2/3 years before rejuvenation pruning. Levelling the ground to fill up depression around the collar of the old bushes if any, cleaning and further deepening the drains can more conveniently and effectively be done after pruning. On sloppy land, interceptor drains should be provided to isolate the rejuvenated area from the upper catchment.

iv. *Shade*: Diseased and very old shade trees should be ring-barked two years before rejuvenation pruning and removed from the area. The section should be reshaded by planting suitable shade trees in March–May following heavy prune. *Crotalaria* may be grown in the initial years for temporary shade (Fig. 9.8).

v. *Bush sanitation*: Repeated cut-cross at the same height lead to the formation of a lump or compound knot and branches become lub-headed; only a few vigorous branches are produced from each such knot. The development of mass-callus from the cut hinders vascular supply which ultimately lead to a decline in the yield. These bushes should be pruned low with an objective to remove maximum number of knots to get clean wood. Ferns and other obnoxious grasses around and inside the collar of the bushes if any should be removed by iron hooks. If the cut is at a height above the ground level pruned frames after pruning should be cleaned by rubbing with hessian or coir and treated with alkaline wash (Fig. 9.9). Treating the frame with 10 percent *Trichoderma* prevents *Poria*. Large pruning cuts should be given a protective coating of a fungicide paint.

vi. *Infilling*: The procedure for infilling and treatments of infills are described in Chapter 8 (section 8.6).

vii. *Plant protection*: Tea bush should be protected from thrips, jassids, mites and *Helopeltis* by spraying appropriate insecticides. Because of the

Fig. 9.8 : *Crotalaria* has been grown as temporary shade in between rejuvenated tea bushes in Darjeeling hills.

Fig. 9.9 : Tea after a heavy prune. Lime washing done on the pruned bush frame in a Darjeeling tea estate.

occurrence of abundant tender shoots in the bushes recovering from pruning, blister blight could pose a serious threat if effective measures are not taken.

viii. *Manuring*: During first two years, rejuvenated section should be treated as young tea in respect of manuring and pest control. As a remedial application to counter mineral deficiencies, extra quantities of fertilizers are often given in the year of pruning.

ix. *Tipping*: In the first year, tipping be carried out at 35 cm above the pruning height or 65 cm from the ground whichever is higher.

x. *Pruning after rejuvenation*: Rejuvenation prune fields be run for one year more than the normal pruning cycle adopted on medium prune/cut across. In the subsequent cycles the pruning should be step-up pattern.

9.2.1 *Response to rejuvenation pruning*

In south India, rejuvenation pruning in conjunction with infilling results in significant increase in productivity of the field ranging from 20 to 40 percent by the end of 8/10 years after rejuvenation (Hudson and Durairaj, 1998). They further reported that the highest yield achieved in the previous cycles is generally surpassed in the rejuvenation cycle itself. Several experiments were also laid out in Assam and West Bengal and results indicated that the rejuvenated section with infilling and/or interplanting increased crop by 30 and 47 percent in the 2nd and the 3rd cycles respectively (Sarker, 1983). It took 2–4 years to outyield the control and about 36–44 percent crop loss occurs in the year of rejuvenation. Rejuvenation, infilling and interplanting was found to be more productive over rejuvenation and infilling in Assam (Sarker, 1981). The implications of this pruning for productivity in Sri Lanka are discussed by Kandiah and Wimaladharma (1980). Malenga (1999) observed that if bushes (Clone SFS 204) are healthy, then down-pruning between 40 and 45 cm would be the best option in Malawi. Jain *et al*. (1992) found pruning in November at ground level following heavy pruning (HP) — unpruned (UP) – deep skiffing (DS) — unpruned (UP) cycle would be the most suitable practice for rejuvenating old China hybrid tea in Himachal Pradesh. Jain (1991) further reported that yield of rejuvenated *sinensis* bushes shoots upto 400 % or even more in 2nd year. In Darjeeling, Dixit (1989) reported that no loss of crop whatsoever will be incurred if rejuvenation done upto 8 percent of the total area of the garden in every two years because by diverting more pluckers to the remaining tea areas will compensate the loss of crop from the rejuvenated areas. But in a case study in south India it was observed that if 5 percent of the area brought under rejuvenation pruning annually there will be a marginal reduction in production till the 3rd year and overall production will be higher by about 16.6 percent in twenty years (Hudson and Durairaj, 1998). It would be 24 and 29 percent increase in production at the end of twenty years if 10 and 20 percent area is rejuvenated every year respectively.

Certain situations like financial constraints, set-back in uprooting and replanting programme, non-availability of land for extension programme etc warrant rejuvenation pruning and infilling and in such a situation it is beneficial in exploiting the maximum yield potential of the field in a short time. Rejuvenation pruning is not an alternative to uprooting and replanting. In east Africa, it was reported that rejuvenation pruning and interplanting with clonal plants did not result in as high a yield as complete replanting with clones (Njuguna and Magambo, 1993). Uprooting and replanting would be more advantageous than rejuvenation and infilling in the case of fields with more than 40 percent vacancy or when more than 50 percent bushes are affected by deep-bole canker extending below the ground level (Hudson and Durairaj, 1998). Fields prone to recurrent frost damage or extremely vulnerable to wind damage are not suitable for this corrective pruning. Katikarn and Swynnerton (1987) reviewed the rejuvenation of abandoned tea smallholdings in Thailand.

9.3 Pruning cycle

The interval of time between two successive prunes is called a pruning cycle. The length of pruning cycle may vary from 1 to 6 years depending on growth pattern of the tea cultivar, soil-climatological conditions, plant variety, height of pruning, quality, quantity of crop needed and availability of plucker. Further, most of these factors interact to influence productivity, therefore, without considering of these interacting process it is very difficult to generalise anything on an appropriate pruning cycle. Manivel (1980b) discussed pruning cycle in relation to movement of photosynthates. Annual pruning when bushes became dormant at the end of growing season has been a practice in north-east India hitherto, but even there the present trend is towards an extended pruning cycle. Introduction of longer pruning cycle of even 3-4 years duration without considering the age and yield level for indefinite period can not sustain productivity. Results of regional trials on pruning cycle conducted in the plains of north-east India and from commercial field data collected from the estates of Assam valley, corroborates this findings (Chakravartee *et al*., 1994).

In Darjeeling hills, as the first and second flush crop fetches better price, most of the gardens tend to keep 80 percent or more of the areas under unprune or light skiffed. Sometimes following of 8 to 9 years pruning cycle having more LS, LOS and UP. This causes damage to the bush structure and affects the yield in the long run. The effect of extended pruning cycle in Darjeeling have been investigated by Hazarika *et al*. (1965). They observed that a four-year pruning cycle is better than a six-year cycle. However, a five-year pruning cycle may also be tried at mid to high elevation. Rahman (1997) suggested suitability of a 3-year LP-UP-LS or a 4-year LP-UP-MS-LS cycle in most of the areas of Darjeeling hills. Under very slow growing conditions a 5-year cycle of LP-UP-LS-DS-UP can be considered.

In south India, after the final frame-forming prune, the regrowth is tipped for a few rounds and then plucked continuously for the rest of the pruning cycle. The pruning cycle extends over a period of 4, 5 or 6 years depending on elevation of the place. In some instances, a 'skiff' is given at a convenient height and the pruning cycle is further extended for a period of 2 or 3 years (Anonymous, 1997). However, yield trends in conjunction with the decline in the flush size, profusion of *banhij* and plucking averages give a better guidance in determining the length of pruning cycle. Yields decline earlier in the cycle in 'China' and 'China hybrid' than in Assam indicating a need for shorter cycles in that jat (Sharma, 1984). Further, pruning cycle of fixed duration for all sections in an estate is not conducive to crop maximisation and each section has to be decided on merit. (Barua, 1989).

In Sri Lanka, the pruning cycle varies from 2 years at sea level to 6 years at altitudes above 2000 m (Kulasegaram, 1986). In Kenya, the table rises at a height of 120-150 cm within 3 to 4 years therefore it must be cut back (Anonymous, 1986a). In the course of perennial growth of tea there is a continual removal of vegetative organs at short intervals by skiffing and at long intervals by pruning. In Indonesia, pruning is generally done every 3rd or 4th year, about 50 cm from the ground level depending on the elevation (Semangun, 1976) and how the plucking table is managed from one pruning period to the next. Pruning is carried out in Japan to renew the activity of old bushes and to maintain the appropriate height of bushes to facilitate plucking (Okano, 1993). In Malaysia, a three-year cycle is generally adopted and no skiffs of any sorts are given in between the prunes. Once in 15 years or so, a heavy prune is given to lower the frame heights (Dutta, 1972). Clowes (1986) reported that normally 2-year pruning cycle in China hybrid and clone SFS 204, and 3-year cycle in case of Indian hybrid and most clones are followed in Malawi.In southern highlands of Tanzania, commercially managed tea is normally pruned on a 4-year cycle. This is because yields generally decline in the fourth and subsequent years, partially because it becomes more difficult for people to harvest the shoots as the plucking table rises (Stephens and Carr, 1994).

The quality of tea is an important consideration relates to pruning system. Generally, unpruned tea is presumed, rather subjectively, to have lesser quality than pruned tea, though intrinsic factors other than pruning may also affect quality (Banerjee, 1993). Marimuthu and Ranganath (1994) reported the changes in biochemical composition and quality aspects of south Indian CTC teas arising from differences in pruning heights. They observed if bushes are pruned at a height of 60 cm, the tea shoots harvested seem to produce CTC teas with superior liquor characteristics like colour, brightness, briskness and body. But the aroma quality of the teas may-found inferior to the ones produced from shoots of 70 cm pruning height.

9.4 Mechanisation of pruning

Pruning operation, due to shortage of workers, gets delayed and resulted poor recovery of the bushes. Thus, mechanisation of pruning operation will be useful in overcoming the labour shortage, completion of pruning in time and also to cut down the cost of production. Trials conducted in Malawi with a Jonsered RS52E brush wood cutter for both down pruning and cut across pruning. The machine coped satisfactorily with both tasks. The factors which affect the economics of using machine are discussed (Grice, 1980). Pruning are mostly carried out by machine in the former USSR. However, a small percentage of the plantation is pruned manually, where hand shears are used (Dey, 1972). Japan has developed some machines for pruning and skiffing.

In south India, earlier investigations on the usefulness of motorised mechanical pruning aids with reciprocating blades developed in Japan, have shown that, a single man operated machine model KPE 200 was economic to skiff bushes at 85 cm and above. This was also useful for cutting, plucking and spraying lanes. Model, KFS 1200, operated by two persons was effective and economic for skiffing at 75 cm and above. However, both these models were found not useful for pruning at and below 65 cm (Sharma and Venkata Ram, 1980). In recent years, the performance of one man operated Kaaz Java tea cutter (model JT 430) designed by M/S Kaaz Corporation, Japan and the Agromatica Pneumatic pruning machine designed by Agromatica L Lendaro, Italy, which have a spinning discs were evaluated for their operational efficiency to prune bushes at 55, 60 and 65 cm, with clean cut ends (Sreedhar *et al.*, 1997). It was observed that the number of bushes pruned with the Kaaz Java tea cutter decreased with decreasing pruning heights, whereas the time required for pruning increased with decreasing pruning heights resulting in higher fuel consumption. In the double hedge system of planting, at a spacing of 135 x 75 x 75 cm and in the single hedge system of planting at a spacing of 120 x 120 cm, the number of bushes pruned by the Kaaz Java tea cutter was four times higher than the manual method. As a result, the number of workers required for mechanized method was 7 to 17 as against 33 to 65 workers per hectare in the manual method. In comparison, the number of bushes pruned in four hours by Agromatica Pneumatic pruning machine was found to be less than the pruned by the Kaaz Java tea cutter, because of the small spinning disc and short axis. But Pneumatic pruning machine was found to be more economical than petrol operated Kaaz Java tea cutter due to low price and less consumption of diesel. The number of bushes pruned in four hours by the Pneumatic pruner, using one lead, was two times higher than the manual method. Since four pruners can operate at a time without any additional fuel expense, this machine is very economical (cost reduction 39 to 60 percent). The performance of both the machines were also evaluated in the field under the conventional system of planting at 120 x 120 cm. Sreedhar *et al.* (1997) reported that the Kaaz Java cutter, in Assam jat, the reduction cost ranged from 29 to 37 percent, whereas in the China jat, even though the number of workers reduced to 60 percent but it was uneconomical due to higher fuel consumption. The Pneumatic pruning machine was also not found to be beneficial to prune China jat at 60 cm whereas it was economical at 65 cm due to less fuel consumption. However, these machines cannot be effectively used with irregular slopes.

9.5 Tipping

It is an essential operation aimed at forming a level plucking surface and to induce branching at a high level which will produce more plucking points in the bushes recovering from pruning. If the recovery is slow and uneven after pruning, it is advisable to defer tipping until growth is well distributed all over the frame of the bush and eventually breaking back over grown shoots to the desired level. Simultaneously, if tipping operation gets delayed it would be difficult to break the primary stems easily with fingers by the tippers. In practice, tipping is started when about 30 percent of the bushes are ready for tipping and carried out selectively at green semi-hard wood. The ideal height for tipping is still a matter of debate though the optimal height and stage of tipping must take cognizance of the growth characteristics of the cultivars (Sharma *et. al*., 1981). In north-east India, light pruned bushes under normal conditions should be plucked over five leaves, corresponds to a plucking measure of 20 cm for Assam type bushes. The measure will be less than 20 cm for China bushes. However, when pruning is done on thick wood, particularly in cycles longer than three years, the internodes are generally thicker and longer. In such situation, the average height of five leaves may go as high as 25 cm. For deep skiffed tea, the correct tipping height is the average height of two leaves which will normally vary between 7 to 10 cm. In case of medium skiffed tea, the height of one leaf which will normally vary between 3 to 5 cm. Light, level off skiffed and unpruned tea should be plucked close to *janam* or to the level (Anonymous, 1988). Barua (1989) emphasized the importance of tipping height of 20 cm for the bush of Assam kind, 15 cm for the hybrids and 10 cm above the pruning mark for China or China hybrid bushes grown in Darjeeling. Tamang (1991) investigated optimization of tipping measures for different pruning cycles and observed that 20 cm and 25 cm tipping measures in general can be advantageously adopted for the *sinensis* type and *assamica / cambodiensis* types respectively under 4-year pruning cycles. Sharma and Sharma (1992) reported the tipping height of 15 cm above the pruning mark of 37.5 cm gave the highest yield in China hybrid tea in Himachal Pradesh. Sharma and Sharma (1998) further conducted an experiment in Himachal Pradesh on rejuvenated China hybrid tea to determine the optimum tipping height above the pruning mark at 55 cm and they found that light pruned bushes should be tipped or plucked at height of 5 cm above pruning mark.

In south India, tea is tipped on the slope in between 60–65 cm from the ground after heavy pruning at mid elevations. The Assam varieties growing on low elevations, tipping height would be at 70 cm. In Assam and Assam hybrid, tipping-in material should average four leaves and a bud irrespective of height of pruning. In China hybrid, it would average three leaves and a bud on pruning levels below 45 cm and four leaves and a bud above 45 cm. However, the height of tipping is depends on the cultivar, height of pruning and altitude at which tea is grown. Banerjee (1993) opined that ideal tipping height should ensure retention of adequate number of foliage for maximum photosynthesis, apart from inducing a fast growth of mature buds.

Tipping is done several times in between 15 to 20 cm in Indonesia (Semangun, 1976) and at 15 cm above the pruning level in Malaysia (Dutta, 1972).

9.6 Shoot growth and regeneration

Tea shoots go through alternating states of growth and dormancy. The term 'flush' is applied to the growth produced by the terminal bud between two states of dormancy. In north-east India, the number of such flushes in a year is generally four and rarely five. The shoots become dormant (*banjhi*) after producing one flush of growth. The bud swells and after a variable period of time it breaks. First the outermost appendages break free and produce a *janam* or scale leaves. This is followed by the unfolding of a second *janam*. After unfolding of the second *janam* the tip of a new leaf becomes visible. Between the *janam* at the base of the shoot and the first normal leaf there is often a persistent leaf with unserrated margin known as *gol-pat* or fish leaf. The *gol-pat* is the same new leaf that is visible at the time of unfolding of second *janam* but was displaced from the original place due to elongation of shoot. Thereafter, the bud continues to grow and unfold to produce the normal leaves with the latter eventually becoming mature leaves with elongated internodes.

When a tea bush is pruned, primaries develop from the dormant buds on the stubs. When they reach a certain predetermined height, they are tipped to make a plucking table. This operation extends over a few weeks and depending on tipping height some primaries become dormant below the tipping level. Of these, some remain permanently dormant while others throw out new flush of growth above the tipping level and they are plucked. After a primary shoot has been tipped, new shoots originating from the axillary buds of maintenance leaves (first order of laterals) and these shoots are plucked when grow above the tipping level. Plucking of the laterals again stimulates production of new generation of shoots of the next higher order and the process continues till the end of the plucking season. Normally in a year, seven to eight orders of laterals are produced (Barua and Wight, 1959; Barua, 1961). The maximum quantum of yield is possible to harvest only upto the fourth order of laterals, and the fifth laterals onwards do not contribute much towards yield (Banerjee, 1993). The dynamics of leaf production in tea is dependent on the production rate of bud. However, some factors viz. vigour, water availability, intensity of light, temperature, pathogens and herbivores influence production of buds and unfurling of the leaves. The period needed for unfurling may also vary according to the characteristics of the cultivars and site qualities (Banerjee, 1993). In Assam and the Dooars, it takes three to six days while in Indonesia, Nilgiris plateau as well as higher elevations of Sri Lanka, the corresponding period is nine days. Like leaf unfurling, the time needed for a dormant bud to develop into a pluckable shoot also varies. It is eight days in Sri Lanka, twenty eight to forty two days in Assam and forty two to forty seven days in Malawi.

9.7 Plucking

Plucking is the most important harvesting operation and intricately related to pruning and tipping. In tea, the tender apical portions of shoots comprising the terminal bud, the internodes and two or three expanding leaves immediately below it, which together constitute the crop, the 'flush' is removed at periodic intervals. Plucking provides stimulus and ensures regeneration of shoots if a steady supply of assimilates to growing buds is ensured by the retention of adequate photosynthetic machinery of the bush.

It has been worked out in the past that harvesting the first two leaves together with the unfurled bud (Fig. 9.10) is the best compromise between the conflicting demands of high yield and high quality (Willson, 1992). For plucking, it is convenient to have a smooth surface at the top of the bush. The top is flat and approximately parallel to the ground in most of the tea growing countries. This is known as the 'plucking table'. In Japan, the plucking surface of the tea bush is not flat but hemispherical or dome-shaped (Fig. 9.11). In the field, bamboo-basket which is often carried in the

Fig. 9.10: Two leaves and a unfurled bud (*Camellia sinensis*).

Fig. 9.11: Tea bushes in Japan are usually trimmed as arc-shaped frame by pruning or skiffing for the convenience of mechanical plucking.

back are generally used in India (Fig. 9.12) and other countries (Fig. 9.13) for collection of plucked leaves. The baskets made of cane or plastic are also used by the pluckers. Sometimes women pluckers collect the flushes on cloth which are later emptied into large baskets.

The time of plucking of the day also influences the productivity. Gogoi (1977) showed in Assam that plucking after 11.00 a.m. gives higher (percentage of) recovery than morning (earlier) plucking and thus adjusting the plucking time itself can influence the productivity. Sinks (growing shoots) are empty in the morning and metabolise fast for routine functions leaving lesser dry matter for accumulation in the shoot. The recovery percentage is increased in shoots plucked in the afternoon than that of morning plucked ones in similar conditions. High yielding plots could be earmarked for later plucking while low yielding ones may be plucked in the morning hours.

9.7.1 *Plucking system*

A good plucking system should allow the maximum quantity of leaf of the desired quality to be harvested with a minimum period of rest to the bud without affecting the health and vigour of the bush adversely in the process. Three types of plucking are in vogue in Assam, black, standard

Fig. 9.12 : Tea pluckers in a Darjeeling tea estate showing basket on their back.

Fig. 9.13: Tea plucker in Africa showing plucking basket on his back.

and fish leaf plucking. In 'black plucking system' only *janam* and buds are left on the plucking table whereas in 'standard system', *janam*, small one-leaf-and-a-bud is also left *in situ*. In 'fish leaf plucking', standard shoots are plucked over the fish leaf instead of plucking over the *janam* (Fig. 9.14). Fish leaf plucking gives lower yield compared to black and standard plucking. Experimental results have proved that as a long term policy standard plucking is superior to black plucking in north-east India (Chakravartee and Barbora, 1994). Ghosh and Sarkar (1980) reported that irrespective of the type of pruning and/or skiffing operations, in so far as the overall yield and vigour were concerned, standard plucking was found to be superior to other systems of plucking. Keeping the same in view, standard plucking is normally be followed in north-east India. 'Single leaf' or 'step up' plucking for few rounds may be beneficial under special situations. However, black plucking can be done in extreme situations like heavy infestation of thrips and blister blight which often necessitates the removal of potential source of infestations. Black plucking over the whole season in unpruned tea or during drought may reduce yield. Fish-leaf plucking if continuously done may initially increase the yield but with falling efficiency over the period of time. Continuous fish

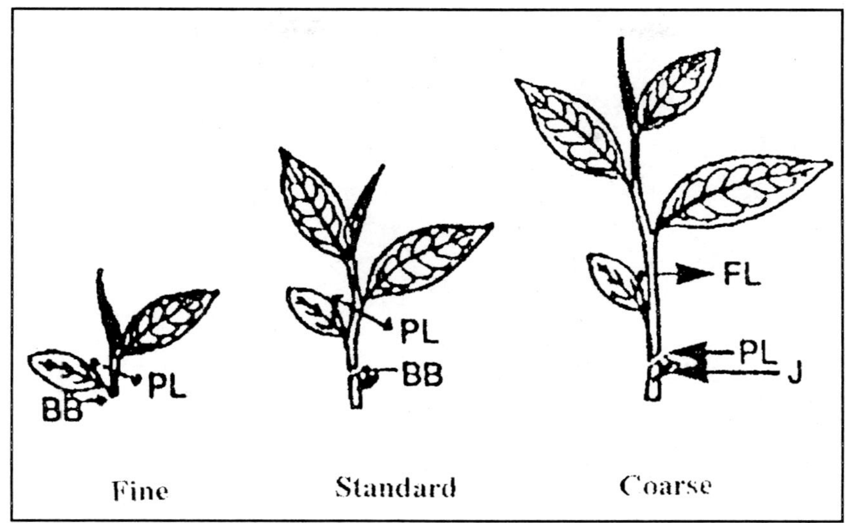

Fig. 9.14: Plucking standards (diagrammatic view); J. *janam* leaf; FL. Fish leaf; BB. Break back and PL. plucking level.

leaf plucking is unable to meet the full requirements of carbohydrate of the growing buds (Sharma, 1983). As a result, size of shoots declines faster, possibly resulting in reduced productivity. This apart, under such system in south India, the bush height increases fast becomes unmanageable from the third year onwards (Murthy, 1992).

Radioactive carbon studies of growing shoots revealed that the sink capacity gets reduced 30 percent for unfolding of every leaf from the bud (Manivel and Hussain, 1982). Thus, the two and a bud retain only 30 percent of the capacity of the bud. In order to reap the benefit of sink induced photosynthesis, growing buds, one leaf and a bud should be retained on the bush while plucking. This information led to the selective plucking of two/three and a bud rather than black plucking. Select plucking has thus contributed for sustained high productivity with quality (Manivel, 1999).

In south India, there are three types of plucking viz. light plucking, hard and level plucking. 'Light plucking' is also known as mother leaf or step up plucking. In this plucking system, which is most lenient, the shoots are plucked above the mother (or first) leaf (Fig. 9.15). In this system of plucking, axillary buds grow faster and weight of the plucked material is generally more. This system also prevents die back of axillary buds and ensures addition of maintenance foliage. But crop loss, unmanageable creep

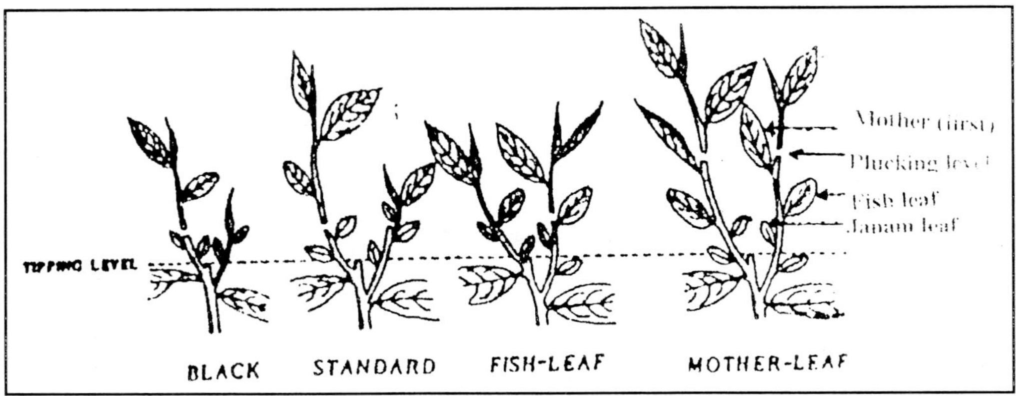

Fig. 9.15 : Generalised diagrams showing different forms of plucking.

in the height of the bushes and low plucking average are some of the disadvantages (Hudson *et. al.*, 1997). In very light plucking, the bush health improves at the cost of crop yield. Plucking is described as 'hard' when shoots are plucked below the mother leaf and above the fish leaf. It is sometimes referred to as fish leaf plucking. Hard plucking immediately increases the yield but adversely affects the bush health over a period of time. In the 'level plucking system', plucking to any leaf above pre-determined level could be done. It is generally done during high cropping season at particular interval. This system of plucking helps to achieve higher plucking average and keep the bush in low creep. There are some disadvantages of level plucking. For example, it does not add maintenance foliage, bush health to a certain extent is affected results delay in shoot development and more *banjhis* and 'crows feet' are formed despite reduction of shoot weight (Hudson *et. al.*, 1997). Considering the health of the bush and its maximum productivity, Sharma (1987) suggests that there is an optimum ratio of crop to maintenance foliage. A detail study on the rhythm of the growth in tea plant has been made in UPASI, Tea Research Station and their observations have justified the practice of 'mother leaf plucking' during the lean period (January–March and July–August) and 'level of plucking' during rest of the periods in south India (Arunachalam, 1995).

Visser (1960) reported that continuous fish leaf plucking in Sri Lanka gave higher yields than a mixture of fish leaf and single leaf plucking. Wettasinghe *et. al.* (1981) compared single leaf, fish leaf and *janam* plucking with normal estate practice (combination of fish leaf and single leaf plucking) and observed that fish leaf and *janam* plucking gave increased yields for first year. But no yield increment was noticed in the second year. The number of shoots increased but their size decreased. Continuous single leaf plucking resulted quick rising of table. The harder plucking systems reduced the addition of maintenance foliage.

In Indonesia, it was observed that the annual rise of plucking table is about 30 cm which indicates liberal fish-leaf method of plucking (Bezbaruah and Barbora, 1979). Sukasman (1987) compared three levels of plucking on bushes with 25, 20 and 15 cm depths of maintenance foliage and highest yield obtained from *janam* plucking. There were no significant differences between the yields from bushes with different depths of maintenance foliage during experimentation.

In former USSR, light plucking (the plucking of 2 or 3 leaved shoots leaving a fish leaf and one or two normal leaves) is done in the beginning of the season (May to August). Later on in the season plucking is done over one normal leaf only. October plucking is usually very coarse, and plucked leaves are generally used for making brick tea. Tskhvediani (1982) observed that highest yield was given by plucking three and four leaves and a bud. One leaf was left above the fish leaf for three months of the year and plucking was to the fish leaf for rest of the season.

Indonesia, keeping the maximum utilization of solar energy in view, the plucking is carried out in such a way that there will be minimum possibility of over shadowing of one shoot to other despite maintaining a flat plucking table. In reality, the level plucking is carried out by plucking shoot upto fish leaf especially at the centre part of the bush, but at the edges it should be adjusted to be flat (Wibowo *et.al.*, 1990).

Grice (1979) reported the increment in yield and deterioration of quality (valuation) of the processed tea with coarser plucking in Malawi. He also showed that the out turn of processed tea in the factory decreased with coarser plucking, as did the proportion of the main (size) grade. This is due to higher internal water content of the older leaves which have developed more water-holding vessels and the fibre content (Willson, 1992).

9.7.2 *Plucking standard*

In most tea growing regions, growers will specify a 'target' type of shoot that represents the most profitable compromise between the value and the quantity of the plucked leaf. For example, in southern Tanzania many growers aim to maximise the proportion of harvested shoots that have three unfurled leaves (Burgess and Carr, 1998). In China, shoots mainly of one bud and two, occasionally of one bud and three and those with tender *banjhi* buds are plucked. This is the most popular one adopted in China at present. However, in north-east India different plucking standards are in vogue. A fine plucking consists of one leaf and a bud, two leaves and a bud and single *banjhi*. Fine plucking (Fig. 9.14) is done in order to produce a very high quality and highest value tea; at times buds only have been plucked. Tea from buds was at one time referred to as 'tips'; the term now often refers to a high quality tea separated by sorting in the factory. Standard or normal plucking consists of one leaf and a bud, two leaves and a bud, single *banjhis* and three leaves and a bud. Medium consists of two leaves and a bud, single and double *banjhis*. Coarse will have two leaves and a bud, all shoots larger than two leaves and a bud and double *banjhis*. Very coarse consists of three leaves and a bud, four leaves and a bud and double *banjhis*. This kind of plucking system used when gross yield is more important than quality. Such extended plucking is amenable to mechanical harvesting. However, removal of *banjhis* and breaking back should be a part of plucking operation.

9.7.3 *Plucking rounds*

The time interval in days between pluckings is known as the 'plucking round'. This has a cardinal role in harvesting maximum crop without impairing the quality. It is also a matter of critical importance to tea estate economics. Plucking accounts for 65–80 percent (Ellis and Grice, 1976; Awasthi and Sarkar, 1983; Mitini Nkhoma, 1989; Rawlley and Beig, 1990), 30–40 percent (Burgess, 1992) of cost of cultivation in tea husbandry thus its optimization is the concern of all.

In northern India, plucking round varies from 4-14 days but in most of the estates tea bush is plucked at weekly intervals from April to December and there is a marked dormant season during winter when practically no crop is harvested. It was observed in north-east India that shorter rounds gave better quality of made tea because of a lower content of ash and fibre, though tea made from 7-day pluckings would have a balance of volatile flavour compounds (VFC), ash, soluble solids, caffeine and therefore would have an optimum quality (Baruah *et. al.*, 1986). Barbora *et. al.* (1992) studied that thermal time regulated hand plucking was more effective than the traditional system of time lag schedule. The thermal time requirement for buds to become pluckable shoots has been reported to be 389 day degrees under conditions at Tocklai, Assam.

In south India, on the other hand, the crop is harvested throughout the year. The optimal plucking interval is about 7-10 days in growing seasons and 12-15 days during lean season (cold and dry months viz. December to March). Murthy (1992) attempted to develop a system in south India to predict optimal interval between plucking rounds in different seasons based on 'leaf expansion concept'. It is more correctly defined as a phyllochron (Bond, 1945). It has been observed on an average, about 625 day degrees were found to accumulate in 52 to 68 days depending upon the months, for the axillary bud to grow into a pluckable shoot. Based on this the optimal plucking intervals during a given period could be computed.

A study conducted in Himachal Pradesh to assess the economics of different plucking intervals from 3 to 17 days (Sud *et. al*., 1996). They found shorter interval leads to lower plucker productivity and higher cost of plucking. But it fetches better price and results in higher gross and net income. In the second unpruned year, plucking interval of 9 days was found more economical. For tea planters of Himachal Pradesh with small investment capacity, plucking interval of around 13 days during unpruned years and between 7 to 9 days during pruned year was more appropriate keeping the rate of green leaf fixed by the Himachal Pradesh State cooperative tea factories.

Tea planters of Darjeeling in recent years, have been resorting to finest type of plucking to ensure better price. Sometimes 2/3 days plucking round is also undertaken for first and second flush period and there is also temptation for low plucking and dipping of hands below the plucking table. There would be invariably a sharp drop in production if such practices are undertaken. Sarkar (1989) opined that the proportion of prune and unprune/ light skiff tea be so adjusted that a regular 6–7 days round can be maintained during the first and second flush periods in Darjeeling. Further, plucking is done thoroughly so that ready or nearly ready shoots are not left on the table. Kabir *et. al*. (1996) reported that shorter plucking intervals in Darjeeling yielded higher crop, better quality compared to extended intervals. The gain in yield recorded per annum was mainly because of gain in total plucking rounds in a year. They found on an average 4/5 days plucking interval throughout the year for old Chinary tea plantation is beneficial. Das (1984) found that China jat, which had shorter leaf period gave highest yield in 5 day rounds.

In Mauritius, it was found that a short plucking round gave the highest yield from seedling and clonal tea. Yield and quality deteriorated as length of plucking interval increased from 6 to 18 days (Dumur and Naidu, 1985).

In Indonesia, under normal conditions plucking is carried out every 10 or 12 days interval to get an acceptable quality. In more favourable conditions, the interval can be shortened to 7, 8 or 9 days and in unfavourable conditions such as in dry seasons it would be 14 days or more.

In former USSR, plucking is normally carried out on a 7–8 days round excepting the months of September/October when the interval can be 10 days (Dey, 1972).

In China, the plucking interval under normal hand plucking is 4-6 days in spring, 6–8 days in summer and autumn. The interval is shortened to 2–3 days for the high grade teas. In mechanical plucking, two pluckings are taken every season and the interval is 12–24 days in spring, and 20–30 days in summer and autumn.

Plucking used to be done 2–4 times a year but plucking three times a year is a recent trend in Japan. The first plucking is done during late April to early May for the first flush. The second plucking is carried out during middle of June to early July followed by the third plucking in late July to early August. In September to October, there used to be a fourth plucking (autumnal crop) which is mostly non-economical.

In Kenya, shoots take between 10 and 14 weeks to develop to harvestable size throughout the year and there is always a good supply of shoots reaching harvestable size. In Malawi, they develop in 5–6 weeks during rains and in 10–14 weeks in the cool season (Tanton, 1992).

In Kenya, both yield (Odhiambo, 1989) and quality (Owuor *et. al*., 1990) have been demonstrated to benefit from short plucking intervals. Contrary to these observations, reduction in yield recorded in Malawi when plucking interval was too short (Palmer-Jones, 1977; Tanton, 1979). But similar trend as those obtained in Kenya where yield increased with short plucking intervals was also noticed in Malawi (Mitini Nkhoma, 1989). Owuor *et. al*. (1990) reported that quality improvement with shorter plucking intervals in Kenya was due to the increase in theaflavins, thearubigins, caffeine, sum of Group II VFC and FI levels and decrease in sum of Group I VFC (Owuor *et. al.*, 1990). This also resulted in higher sensory evaluation (Owuor *et. al.*, 1990). In southern Tanzania, the phyllochron method of planning harvest intervals has now been used in the field for over four years. It has been proved to be particularly useful between May and the end of August when there are large changes in temperature. The technique found useful for predicting harvest intervals at different altitudes within an extended estate (Burgess and Carr, 1998).

9.7.4 *Plucking and maintenance foliage*

All mature leaves, including fish leaves, left on bush below the plucking surface are known as maintenance foliage which serve a crucial dual role of meeting the carbohydrate requirements of the growing buds and replenishing the starch reserves of the bush (Sharma, 1987). These leaves must be efficient, well distributed and there must be enough of them to trap all the light which arrives at the bush (Manivel, 1978). It was reported that less than 5 percent of the incident light penetrates lower than 15 cm below the plucking table (Anonymous, 1986b). The depth of maintenance foliage in a tea bush is regulated by the tipping height. In annually pruned tea, the relationship between tipping height and yield is simple since same type of top hamper and depth of maintenance foliage is established every year. In case of longer pruning cycles, the nature of growth of top hamper and depth of maintenance foliage change from year to year depending on whether the bush is pruned, skiffed or unpruned. Willson (1992) however, reported that the maintenance layer should not be less than 25 cm in depth. During favourable growing weather, the entire maintenance canopy mobilises photosynthates to the growing shoots. Maintenance leaves reach peak efficiency within a month of their unfolding and remain photosynthetically active up to six months (Barua, 1989). The leaves in the maintenance layer will age and become less efficient. Therefore, it is necessary to add new leaves to this layer periodically in replacement of the very old and dropped leaves. In south India, fish leaf plucking system is followed and a layer of fish leaves is left on the bush at each round of plucking. Fish leaves are as efficient producers of carbohydrates as the foliage leaves (Manivel and Hussain, 1982). Since they are small, it is unlikely that fish leaves only will fulfil the demand of carbohydrates for the production of fresh shoots. Addition of fresh leaves in the maintenance layer is done by varying the depth of plucking and allowing an extra leaf to remain above the previous table level (Willson, 1992). The life of the maintenance foliage on a tipped and plucked shoot does not exceed eighteen months and its capacity for photosynthesis gradually declines after about six months (Barua, 1993).

Productivity and health of the tea bushes are governed by the interaction between the 'source' and 'sink'which have diametrically opposite functions. The canopy of foliage in tea bush is the 'source', while the proliferating and expanding organs constitute the 'sink' (Sharma and Satyanarayana, 1993) The former generates carbohydrates, while the latter use them for growth and development. Young growing shoots at the plucking table are reported to be the strongest 'sinks' (Barua, 1989) and their presence results in an upward movement of photosynthates. Photosynthates are imported into a leaf up to third leaf stage (Jain and Tamang, 1988). During winter or at other times when buds are dormant, photosynthates move downward and accumulate as starch reserves in the roots. Sink capacity drops to 70 percent of the bud in the first leaf, 40 percent in the second leaf and 30 percent in the third leaf (Anonymous, 1984). The shoot with third leaf, due to its size, is the largest sink on a plucking table. Therefore, at random removal of leaves of different ages would affect the health of the bush.

9.7.5 *Mechanical harvesting*

The enormity of labour involvement in harvesting tea could be gauged by the fact that, generally, 12500 individually hand plucked shoots of two leaves and a bud of small leaf China and China hybrid bushes in Darjeeling and 5000–7500 shoots of broad leaf Assam jat are required to manufacture one kilogram of made tea. About 70 percent of the labour employed on an estate is deployed for harvesting; thus it is the most expensive operation, accounting for about 20 percent of the cost of production (Sharma *et. al.*, 1981). This apart, the heavy rush of leaf during main cropping months is sometimes throws plucking schedule out of gear. Plucking intervals become longer and the standard of harvest falls leading to deterioration of quality of the finished product. Under these circumstances, mechanical harvesting devices which can speed up the harvesting operation will definitely be an attractive alternative. (Barua, 1998). A major difficulty in mechanising plucking process is the steep and irregular slopes on which tea is very often grown. Much of the tea areas in India is subject to this drawback. Tea grown on undulating land, can however be mechanically harvested. The hedge-trimming shears can be used on the hilly terrain. Before a plantation is converted to mechanical plucking, it is essential that plucking paths (about 45 cm wide) are established between the rows to allow for easy passage of the machine. Paths can be established during the off-season and their maintenance should be done annually. During the season, branches of bushes on either side of the path may be allowed to meet as this does not interfere with the passage of the machine. A good deal of interest on mechanical harvesting has grown in countries viz. Japan, Mauritius, former USSR, Argentina, South Africa, Zimbabwe, Uganda, Kenya, Malawi, Australia, and Papua New Guinea. Some of these countries are already using mechanical harvesters on a large scale.

During the last 3 decades or so several studies have been undertaken in Tocklai, south India and abroad to compare cost, product quality from hand and mechanical plucking. In north-east India, different plucking aids were tried viz. wheel-mounted machine, one man carrying machine, two man carrying machine, flat shear and raised shear with variable step-ups (Barbora *et.al.*, 1998). All were powered by petrol engine. The wheel-mounted machine was developed by Mutsumoto Kiko, Japan with collaboration of Tea Research Association, Assam. It was observed that July to September was found to be the most suitable time for mechanised harvesting in the plains of north-east. The wheel-mounted machine and the raised shear showed superior performance and could pluck 400 kg and 37 kg green leaf/day respectively, registering an increase of 300 % and 48 % in pluckers' productivity.

Harvesting shears have been developed for tea. The cutting action may be hand-operated or driven by a small motor. In south India, various combinations of shear harvesting and hand plucking were tried and found that all available crops can be harvested effectively during the two high cropping seasons from April to June and from mid-September to early December. It has been reported that plucking average can be increased by 50 to 100 percent and plucker utilization can be reduced by about 30 to 40 percent with shearing compared to hand plucking (Anonymous, 1999). It was further recommended that shear harvesting (Fig. 9.16) can be done after 15 months from pruning in the case of China hybrid jat and 18 months in Assam hybrid jat. Shear harvesting during the first 15 to 18 months from pruning does not enhance the productivity of pluckers because even

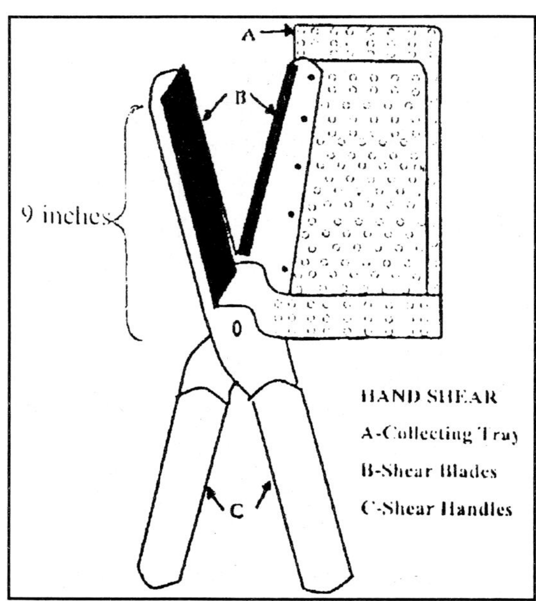

Fig. 9.16 : Hand shear machine (diagrammatic).

hand plucking on such bushes of a lower height is effective (Satyanarayana *et. al*., 1991). In Indonesia, it was reported that shear harvesting during flushing season and hand plucking in the intervening seasons caused no adverse effects on the health of the tea bush (Arifin and Semangun, 1999). UPASI, Tea Research Institute and certain progressive companies have also evaluated Japanese make single man operated (Kawasaki Ochiai), two men operated two models (Ochiai V-8 and Kawasaki T-300) tea gardens and battery operated Picky Power Harvester. Two-man operated machines (Fig. 9.17) can be covered 0.5 hectare in a day. During the peak cropping period, the machines can harvest upto 1000 kg per day. But it was observed this machine during dry season in up and down planted field with a population of 6600 bushes per hectare covered half an hectare in a day and crop harvested was only 300 kg per machine. The Picky Power Harvester can cover 100 bushes per hour. Indian models, one and two men operated Tal-Power Mobile Harvesters were also tried. Two, one man operated Agromatica Pneumatic harvesters were evaluated for their harvesting efficiency. The studies indicated that the total crop harvested with 32 cm long harvester was as efficient as the improved shears. The plucking average for the Pneumatic Harvester was 50 % higher than the

Fig. 9.17: Two-man operated Japanese harvesting machine.

improved shears. But when length of blade was increased to 72 cm depression of crop was noticed.

In Japan, plucking has been done by four methods. One person can pluck 10–15 kg by hand per day, from 60–100 kg by hand operated shears, from 400–500 kg by mechanized shears and from 1000–2000 kg by tractor mounted machinery (Takeo, 1992). However, plucking is done in Japan mostly with a powered plucking machine. In Taiwan, replacement of manual plucking by machines was discussed by Huang and Chiu (1990). Gabuniya (1991) described and assessed a number of machines developed in the former USSR. In Turkey, harvesting is mostly achieved with shears, not leaf-by-leaf plucking.

Kilgour and Burley (1991) described the machine developed at Silsoe college for use in Uganda. Mwakha (1986) reported on the performance of a Bobard Jeune type Bob M55 harvester in Kenya. It was observed that yields were reduced and quality reduced by varying degrees compared with hand plucking. Martin (1999) described that the future will see a wider adoption of mechanical plucking in South Africa and Zimbabwe in particular and the appearance of more sophisticated machinery, especially on estates which have large areas of clonal tea on flat land. Many estates, particularly in the Eastern Highlands of Zimbabwe and Malawi, however, have a variety of topography which will require the use of a range of machines. In the peak of the season, several estates in Malawi still persevere with shear plucking, with crops such as the China jat where hand plucking is difficult. In Southern Malawi, labour costs are such that mechanical plucking is not at present economic and hand plucking currently the cheaper option. Due to inherent variability in seedling tea, clones with a horizontal leaf pose are better material for mechanical harvesting. Matsumoto Kiko Co, Japan exported Valiant MCTE tea harvesters (Fig. 9.18) to Papua New Guinea which is capable to pluck 4 hectare tea area per day. Nixon (1996) studied the mechanical harvesting of tea in the Southern Highlands and reported that both wheeled and hand-held two person machines offer the possibility of substantially increasing pluckers productivity (by 4–7) times, as compared to manual harvesting.

The mechanical aid prevents or reduce the wastage of leaf, leaving plucking schedule undisturbed thereby ensuring better standard of plucked shoots and hence quality of made tea (Barua, 1998). But it has been reported that the use of mechanical aids in some tea growing areas has led to an increase in the coarse leaf content due to the extended plucking intervals and thereby deterioration in quality (Owuor, 1989). However, such adverse effects on the quality of made tea by shear harvesting was not noticed in south India (Ramaswamy, 1983). This is because the standard of shoots

Fig. 9.18: Villiant tea harvester (photograph from New Century Corporation, Tokyo, Japan).

Fig. 9.19 : Weighing of plucked leaf in a Darjeeling garden.

was maintained by adopting optimal plucking intervals (Sharma and Satyanarayana, 1993).

9.7.6 *Handling of harvested leaf*

Generally, leaf after plucking certain quantity/full basket taken to a collecting-point and weighed (Fig. 9.19) and then moved to the factory for processing. Withering in fact starts as soon as the leaf is plucked. The careless handling between the time of plucking and its arrival at the withering troughs is one of main causes of the loss of quality. Pluckers when plucking leaf in the field damage the same by grasping it too tightly in their hand, and bruising it when putting into baskets. Plucked leaves are generally packed either in coir bags or plastic containers. A variety of means of transport are used for transportation of leaf to the factory. While transporting in lorries or tractor-drawn trailers the bags should not be placed one above the other and to enable more bags to be put on a truck a removable intermediate floor is advisable.

References

Anonymous. 1984. *Annual Report 1983–1984*. Tocklai Experimental Station, Tea Research Association, Jorhat, Assam, India.

Anonymous, 1986a. Pruning. *In : Tea Growers Handbook*, 4th edn., Tea Research Foundation of Kenya, Kericho, Kenya, pp. 83–87.

Anonymous, 1986b. The maintenance foliage. *In : Tea Growers Handbook*, 4th edn., Tea Research Foundation of Kenya, Kericho, Kenya, pp. 81–82.

Anonymous, 1988. *Encyclopaedia. Serial No. H201/1,* Tocklai Experimental Station, Tea Research Association, Jorhat, Assam, India.

Anonymous, 1997. *Hand Book of Agriculture*. Indian Council of Agricultural Research, New Delhi, India, pp.1302.

Anonymous, 1999. Mechanisation in tea. *UPASI–KVK Publication Series*, **32:** 1–16.

Arifin, M. S. and Semangun, H. H. 1999. Tea industry in Indonesia. *In : Global Advances in Tea Science* (ed.N. K. Jain). Aravali Books International, New Delhi, 65–82.

Arunachalam, K. 1995. *A hand book on Indian Tea*. Arunachalam Associates (Plantation consultants), Wellington Bazar, The Nilgiris, Tamil Nadu, India, p. 271.

Awasthi, R. C. and Sarkar, A. R. 1983. A study on labour cost component in tea industry in Assam valley. *Two and a Bud*, **31(1&2):** 14–18.

Banerjee, B. 1993. *Tea production and processing*. Oxford & IBH Publishing Co. Pvt. Ltd., New Delhi, India, p.336.

Barbora, A. C., Sharma, J., Baruah, D. C., Saikia, M. C. and Baruah, S. P. 1992. Shear and machine plucking in tea (*Camellia sinensis* L.). *Proc. 31 st Tocklai Conf*., January 11–12, 1992, Tea Research Association, Assam, 108–115.

Barbora, A. C., Sarma, J., Baruah, D. C. and Baruah, S. P. 1998. Agrotechniques for mechanised harvesting of tea. *Souvenir of Seminar-cum-demonstration on mechanised harvesting of tea*. August 26, 1998, Tocklai Experimental Station, Tea Research Association, Jorhat, Assam, India, p.10

Barua, D. N. 1961. Shoot production in cultivated tea. II. The branch system. *Phytomorphology*, **11(1 & 2):** 98.

Barua, D. N. 1989. *Science and practice in tea culture*. Tea Research Association, Calcutta – Jorhat, p.493.

Barua, D. N. 1993. Crop management and nutrition. *In : Tea Culture, Processing and Marketing* (eds. M. J. Mulky and V. S. Sharma). Oxford & IBH Publishing Co. Pvt. Ltd., New Delhi, India, pp 47–54.

Barua, D. N. 1998. Mechanical plucking aid. *Souvenir of Seminar-cum-demonstration on mechanised harvesting of tea*. August 26, 1998, Tocklai Experimental Station, Tea Research Association, Jorhat, Assam, India, p.1.

Barua, D. N. and Wight, W. 1959. Shoot production in cultivated tea. I. Apical activity & radial growth. *Phytomorphology*, **9(3):** 242.

Baruah, S., Hazarika, M., Mahanta, P. K., Horita, H. and Murai, T. 1986. Effect of plucking intervals on the chemical constituents of CTC black teas. *Agric. Biol. Chem*., **50:** 1039–1041.

Bezbaruah, H. P.and Barbora, B. C.1979. Glimpses of Indonesian tea industry. *Two and a Bud*, **26(1):** 2–4.

Bond, T. E. T. 1945. Studies in the vegetative growth and anatomy of the tea plant (*Camel'ia thea* Link.) with special reference to the phloem. II: Further analysis of flushing behaviour. *Annals of Botany New Series*, **9:** 183–216.

Burgess, P. J. 1992. Initial results from the harvesting method experiment. *Quarterly Report, Ngwazi Tea Research Unit*, **10:** 3–13.

Burgess, P. J. and Carr, M. K. V. 1998. The use of leaf appearance rates estimated from measurements of air temperature to determine harvest intervals for tea. *Expl. Agric.*, **34:** 207–218.

Chandra Mouli, B. and Sharma, V. S. 1993. Rejuvenation pruning — recent developments. *UPASI Bull.*, **46:** 22–24.

Chakravartee, J. and Barbora, B. C. 1994. Plucking – important considerations under various situations. *In: Field management in tea* (ed. J. Chakravartee). Tocklai Experimental Station, Jorhat, Assam, India, pp 73–84.

Chakravartee, J. and Borpujari, N. 1994. A few aspects of pruning. *In : Field management in tea* (ed. J. Chakravartee). Tocklai Experimental Station, Jorhat, Assam, India, 64-72.

Chakravartee, J., Biswas, A. K. and Bordoloi, P. K. 1994. Effect of pruning cycle for sustained productivity and quality. *Souvenir 32nd Tocklai Conf.*, December 16–17, 1994, Tea Research Association, Jorhat, Assam, India, pp15.

Chkhaidze, G. I. 1972. The effect of hard pruning on some biological process in tea. *Subtropicheskie Kul'tury*, **1:** 21–23 (in Russian).

Clowes, M. S. J. 1986.Comparing and contrasting tea practices in south India and Malawi. *Tea Research foundation (Central Africa) Quarterly Newsletter*, 84: *Quarterly Newsletter*, **84:** 11–16.

Das, S. C. 1984. Rate of leaf unfolding in plucked and unplucked tea bushes of north east India. *Two and a Bud*, **31:** 7–11.

Dey, S. K. 1972. Tea in Russia. *Two and a Bud*, **19(2):** 73–84.

Dixit, R. K. 1989. Production of quality tea in Darjeeling. *Proc. Joint Area Sci. Committee Meeting*. November 9–11, 1989, Darjeeling, Tea Research Association, 43-48.

Dumur, D. and Naidu, S. N. 1985. The effect of plucking round length on green leaf production in tea. *Technical Bulletin, Ministry of Agriculture, Fisheries and National Resources, Mauritius*. **5:** 11–19.

Dutta, S. K. 1972. Tea in Malaysia and Indonesia. *Two and a Bud*, **19(1):** 1–9.

Ellis, R. T. 1984. Pruning and moisture stress. *Tea Research Foundation (Central Africa) Quarterly Newsletter*, **73:** 4.

Ellis, R. T. and Grice, W. J. 1976. Plucking policy and techniques. *Tea Research Foundation (Central Africa) Quarterly Newsletter*, **41:** 3-10.

Esartiya, M. V. 1972. The effect of different pruning methods on the yield of mature tea plants growing on podzolic soils. *Subtropicheskie Kul'tury*, **2:** 40–44.

Esartiya, M. V. 1975. The effect of different pruning methods on the productivity of tea plantations of different ages growing on podzolic soil. *Subtropicheskie Kul'tury*, **1:** 43–45.

Gabuniya, N. A. 1991. Perspectives of development of machines for tea growing and harvesting. *Traktory i Sel'skokhozyaistvennye Mashiny*, **3:** 34–38.

Gadd, C. H. 1928. A new view of the causation of deplodia disease. *Tea Quarterly*, **1:** 89-93.

Ghosh, T. K. and Sarkar, S. K. 1980. Effect of black and standard plucking in yield. *Two and a Bud*, **27:** 49–51.

Gogoi, B. N. 1977. Time of plucking and shoot dry matter content. *Two and a Bud*, **24:** 46–50.

Grice, W. J. 1979. The great debate. Quantity versus quality in tea production. *Quarterly Newsletter, Tea Research Foundation (Central Africa)*, **54:** 4-7.

Grice, W. J. 1980. Mechanical pruning. Quantity versus quality in tea production. *Quarterly Newsletter, Tea Research Foundation (Central Africa)*, **57:** 4-5.

Hazarika, U. N., Pradhan, S. K. and Sarkar, S. K. 1965. Effect of extended pruning cycles of varying durations on yield of tea in Darjeeling. *Two and a Bud*, **32(1&2):** 2–6.

Huang, T. F. and Chiu, T. F. 1990. Conversion of hand plucking to mechanical harvesting in high grade tea areas in Taiwan. *Acta Horticulturae*, **275:** 255–260.

Hudson, J. B. and Durairaj, J. 1998. Role of rejuvenation pruning in augmenting tea production in south India. *UPASI Tea Sci Bull.*, **51:** 22-30.

Hudson, J. B., Durairaj, J., Muraleedharan, N. and Dhanakumar, V. G. 1997.*Guidelines on tea culture in south India. UPASI–KVK Publication*, **30:** p.212.

Jain, N. K. 1991. Variations in genetic resources of tea and their evaluation*: End of the term report submitted to the Ministry of Agricultural Food and Fisheries Govt. of Japan. 1991*.

Jain, N. K., Chakrabarty, D. N., Tamang, M. B., Singh, R. D. and Sood, P. K. 1992. Studies on time and level of rejuvenation pruning in China hybrid tea in Himachal Pradesh. *J. Plantation Crops*, **20 (Supplement):** 29–34.

Jain, N. K. and Tamang, M. B. 1988. Physiological basis of tea management practices. *Proc. Intl. Cong. Plant Physiol.*, February 15–20, 1988, New Delhi, India, 469–479.

Kabir, S. E., Ghosh Hajra, N. and Kumar, R. 1996. Effect of plucking intervals on yield and quality of Darjeeling tea. *J. Plantation Crops*, **24(Supplement):** 758–762.

Kandiah, S. 1984. Rim lung pruning reduces mortality of heavy pruning. *TRI Science J. Tea Services*, Tea Research Institute of Sri Lanka, Talawakela, **55(2):** 94–102.

Kandiah, S. and Wimaladharma, S. 1980. Studies on the physiology of pruning tea. 3. The implications of renovating ageing plants by 'rejuvenation' pruning and infilling. *Tea Q.*, **49(2):** 13–19.

Katikarn, K. and Swynnerton, R. 1987. Rejuvenation of abandoned tea smallholdings. *Thai J. Agric. Sci.*, **20(3):** 217–256.

Kilavuka, C. I. 1980. An observation on the recovery of pruned tea under shade. *Tea*, **1(1):** 4–6.

Kilgour, J. and Burley, J. 1991. Tea mechanisation in Uganda. *Agricultural Engineer*, **46(2):** 52–55.

Kipiani, E. Ya. and Alkhazov, Yu. G. 1972. The effect of pruning on the growth and development of tea shoots. *Subtropicheskie Kul'tury,* **2:** 26–28.

Kulasegaram, S. 1986. Pruning. *In : Handbook on Tea* (eds. P. Sivapalan, S. Kulasegaram and A. Kathiravetpillai), Tea Research Institute of Sri Lanka, Talawakelle, Sri Lanka, 82–87.

Kulasegaram, S. and Kathiravetpillai, A. 1981. Effect of severity of pruning on growth and yield of high country seed tea (*Camellia sinensis*). *Tea Quarterly,* **50(1):** 16–25.

Malenga, N. E. A. 1998. Why tip tea as early as possible. *Tea Research Foundation (Central Africa) Quarterly Newsletter,* **129:** 8.

Malenga, N. E. A. 1999. Effect of height of pruning mature SFS 204 on bush survival and tea yield. *Tea Research Foundation (Central Africa) Quarterly Newsletter,* **133:** 7–9.

Manivel, L. 1978. Importance of maintenance foliage in tea. *Two and a Bud,* **25(2):** 74–75.

Manivel, L. 1980a. Role of maintenance foliage. *Two and a Bud,* **27:** 52–55.

Manivel, L. 1980b. Time of pruning bushes in relation to movement of photosynthates. *Two and a Bud,* **27(1):** 8–10.

Manivel, L. 1999. Physiology of tea productivity. *In: Global Advances in Tea Science* (ed. N. K. Jain). Aravali Books International, New Delhi, 463–480.

Manivel, L. and Hussain, S. 1982. Photosynthates in tea. I. Contribution of photosynthates to pluckable shoots by maintenance leaves. *Two and a Bud,* **29:** 13–16.

Marimuthu, S. and Ranganath, S. R. 1994. Effect of different pruning heights on chemical composition of green leaf and quality parameters of CTC teas. *J. Plantation Crops,* **22(2):** 115–118.

Martin, P. 1999. Mechanical harvesting of tea in the central and southern African region. *Tea & Coffee Trade Journal,* **171(9):** 112–119.

Mitini-Nkhoma, S. P. 1989. The effect of round length on the yield, pieces per kilogram and relative value. *Tea Research Foundation (Central Africa) Quarterly Newsletter,* **95:** 12–13

Murthy, S. B. K. 1992. Harvesting in tea – precepts and practices. *UPASI Tea Sci. Dept. Bull.,* **45:** 43–62.

Mwakha, E. 1986. Clonal tea response to frequency and height of mechanical harvesting. *Tea,* **7(2):** 48–57.

Nixon, D. J. 1996. Mechanical harvesting of tea in the Southern Highlands – a summary of findings and conclusions. *Quarterly Report, Ngwazi Tea Research unit,* **25:** 8–26.

Njuguna, C. K. and Magambo, M. J. S. 1993. Effect of rejuvenation pruning and increasing plant density on yields of old tea bushes interplanted with clonal tea plants. *Tea,* **14(1):** 13–20.

Odhiambo, H. O. 1989. Nitrogen rates and plucking frequency in tea. The effect of plucking frequency and nitrogen fertilizer rates on yield and yield components of tea (*Camellia sinensis* (L) O. Kuntze) in Kenya. *Tea,* **10:** 90-96.

Okano, K. 1993. Tea industry in Japan. *Proc Intl. Symp. Tea Science and Human*

Health. January 11–14, 1993, Calcutta, Tea Research Association, 52–59.

Owuor, P. O. 1989. Black tea quality : Effects of some agronomic practices on tea quality. *Tea*, **10:** 134–146.

Owuor, P. O., Odhiambo, H. O., Robinson, J. M. and Taylor, S. J. 1990. Variations in the leaf standard, chemical composition and quality of black tea (*Camellia sinensis*) due to plucking imtervals. *J. Food Agric*., **52:** 63–67.

Palmer-Jones R. W. 1977. The effect of plucking policies on yields in Malawi. *Expl. Agric*., **14:** 43–49.

Pethiyagoda, U. 1966. The importance of the timing of pruning in relation to recovery. *Tea Quarterly*, **37:** 80–85.

Rahman, F. 1997. Suggested remedial measures for augmenting production in Darjeeling. *Proc. Seminar Darjeeling tea industry*. September 5, 1997, Calcutta, Tea Research Association, India, 24–32.

Ramaswamy, S. 1983. Quality of tea in relation to harvesting. *UPASI Tea Sci. Dept. Bull*., **38:** 107–116.

Rawlley, N. K. and Beig, M. V. 1990. Comparison of plucking systems and leaf standards between north-east India and south India. *Planters' Chronicle*, **85(10):** 479–484.

Sarker, S. K. 1981. Rejuvenation of old tea. *Proc. 29 th Tocklai Conf*., December 17–19, 1981, Tea Research Association, Jorhat, Assam, India, 21–25.

Sarkar, S. K. 1983. Rejuvenation of old tea. *Tea Times (Sandoz)*, **1/83**, 1983.

Sarkar, S. K. 1989. Factors influencing productivity of tea in Darjeeling. *Proc. Joint Area Sci. Committee Meeting*. November 9–11, 1989, Darjeeling, Tea Research Association, 1–8.

Satyanarayana, N., Spurgeon Cox, Govindarajulu, V., Surendra Mohon, M., Hudson, J. B. and Sharma, V. S. 1991. Implications of mechanical harvesting in tea. *J. Plantation Crops*, **18 (Supplement):** 332–341.

Semangun, H. 1976. Introducing the Indonesian tea industry. *UPASI Sci. Dept. Bull*., **33:** 26–31.

Sharma, D. K. and Sharma, K. L. 1998. Standardization of tipping/plucking height for a light pruned section of China hybrid tea (*Camellia sinensis* (L) O. Kuntze) grown in Himachal Pradesh. *In : Developments in Plantation Crops Research* (eds. N. M. Mathew and C. Kuruvilla Jacob), PLACROSYM XII, November 27–29, 1996, RRII, Kottayam, India, 228–229.

Sharma, K. L. and Sharma, D.K. 1992. Standardization of tipping height in relation to pruning levels of China hybrid tea (*Camellia sinensis*) grown in Himachal Pradesh. *Indian J. Agric. Sci*., **62(11):** 758–761.

Sharma, V. S. 1983. Plucking styles. *UPASI Tea Sci. Dept. Bull*., **38:** 33–48.

Sharma, V. S. 1984. Pruning of tea – precepts and practices. *UPASI Tea Sci. Dept. Bull*., **39:** 63–67.

Sharma, V. S. 1987. Harvesting in tea. *Planters' Chronicle*, **81:** 261–266.

Sharma, V. S., Haridas, P. and Venkata Ram, C. S. 1981. Mechanization of harvesting in tea. *UPASI Tea Sci. Dept. Bull*., **37:**106–114.

Sharma, V. S. and Satyanarayana, N. 1993. Harvesting in tea. *Proc Intl. Symp. Tea*

Science and Human Health. January 11–14, 1993, Calcutta, Tea Research Association, 242–255.

Sharma, V. S., Satyanarayana, N. and Haridas, P. 1981. Tipping in relation to pruning heights and its effect on the yield of tea (*Camellia* spp.). *J. Plantation Crops*, **9:** 112–118.

Sharma, V. S. and Venkata Ram, C. S.1980. Motorised mechanical aids to prune/skiff tea (*Camellia* L. *spp*.). *Proc.* PLACROSYM III: 283–288.

Sreedhar, Ch., Victor, R., Ilango, J. and Sasidhar, R. 1997. Mechanisation of field operations in tea culture.*UPASI Tea Sci. Dept. Bull.*, **50:** 8–19.

Stephens, W. and Carr, M. K. V. 1994. Responses of tea (*Camellia sinensis*) to irrigation and fertilizer. IV. Shoot population density, size and mass. *Expl. Agric.*, **30:** 189–205.

Sud, R. K., Singh, C. M. and Jain, N. K. 1996. Economics of different plucking intervals in China hybrid tea (*Camellia sinensis*) in Himachal Pradesh. *J. Plantation Crops*, **24(1):** 29–33.

Sukasman 1987. Plucking and yield with quality of pluckable shoot of tea. 2. Effect of maintenance foliage thickness and systems of hard plucking on yield and quality of pluckable shoot of tea. *Menara Perkebunan*, **55(2):** 25–29.

Takeo, T. 1992. Green and semi-fermented teas. *In: Tea: Cultivation to consumption* (eds. K. C. Willson and M. N. Clifford). Chapman & Hall, London, 413–457.

Tamang, M. B. 1991. Structure, development and productivity of the tea bush. *J. Plantation Crops*, **18 (Supplement):** 108–121.

Tanton, T. W. 1979. Some factors affecting yield of tea. *Exptl. Agric.*, **55:** 187–191.

Tanton, T. W. 1992. Tea crop physiology. *In: Tea: Cultivation to consumption* (eds. K. C. Willson and M. N. Clifford). Chapman & Hall, London, 173–199.

Templer, J. C. and Machaga, J. S. E. 1978. Preplanting tea land to Guatemala grass. *Tea*, **18(1):** 16–17.

Tskhvediani, I. Sh. 1982. Effect of plucking methods in tea on the yield of a Kolkhida plantation and other cultivars. *Subtropichieskie Kul'tury*, **1:** 47–49.

Tubbs, F. R. 1977. The length of the pruning cycle. *J Tea Res. Inst. Ceylon*, **10:** 22–23.

Venkatakrishnan, N. S. and Sen R. N. 1981. Time, motions and ergonomic study in plucking. *Proc. 29 th Conf., Tocklai Experimental Station*, 46–52.

Visser, T. 1960. Plucking practices in relation to maintenance foliage. *Tea Q.*, **31(2):** 38–52.

Wibowo, Z.S., Sukasman, and Sukasmono.1990. Some key developments of tea cultivation in Indonesia. *Proc. Intl. Conf. R & D in Tea*, January 11–12, 1990, Calcutta, Tea Research Association, Jorhat, Assam, India. 84–90.

Wettasinghe, D. T., Nathaniel, R. K. and Kroon, D. D. 1981. Effect of severity of plucking on the growth and yield of low grown tea. *Tea Q.*, **50(3):** 123–130.

Willson, K. C. 1992. Field operations: 2. *In : Tea: Cultivation to consumption* (eds. K. C. Willson and M. N. Clifford). Chapman & Hall, London, 227–267.

Willson, K. C. 1999. *Coffee, Cocoa and Tea.* CABI Publishing, UK. p.300.

Chapter 10

Mineral nutrition and fertilizer management

10.1 Introduction

Fertilizer is one of the major agro-inputs contributing to the cost of production and productivity in tea plantation. Tea is normally grown as a long term monoculture. For proper maintenance of the health of tea bushes and to obtain high yield, a well balanced fertilization and manuring is necessary at certain intervals throughout the year. Judicious use of fertilizers, in association with other appropriate cultural practices make possible sustained increased production of tea over a long period.

Recommendations on the levels of fertilizer application are mostly based on the results of long or short-term field experiments carried out by the different Tea Research Institutes, analysis of soil or leaf or the presence of mineral deficiencies symptoms or toxicities. But in general physiological attributes viz. nutritional needs of different tea cultivars, effect of different pruning, standard of plucking, age and production of dry matter etc have not been considered (Banerjee, 1987). However, the most logical approach may be what is removed by the plants from the soil during its growth is given back to the soil thus maintaining the *status quo* of the soil with respect to the nutrients. Further, nutrients inherently present in soil and also the fixation, leaching and volatilization losses of nutrients should also be considered. In the past, manuring on replacement basis was popular in many countries but later it was virtually abandoned and yield of different sections/blocks in a tea estate is still the major consideration in the allotment of fertilizers. If fertilizers are applied in a fixed proportion, a low yielding section in the tea estate gets less nitrogen, phosphate and potash as well. The complexities which affect the determination of optimum fertilizer application have been discussed by Tolhurst (1960).

10.2 Essential plant nutrients

Green plants are autotrophic since they manufacture their own food from CO_2 and H_2O in presence of sunlight. But for inorganic matter they depend on the outside world. Plants obtain a wide variety of inorganic elements from the soil. These inorganic elements which are usually grouped together under the class "mineral elements or mineral nutrients" and the nutrition of mineral nutrients is called mineral nutrition. The source of these elements is mainly the ions present in the soil solution and absorbed by the root system.

Among the essential elements, oxygen, hydrogen, carbon are obtainable from air and water and need no specific consideration. The other thirteen mineral nutrients are essential for the growth and development of all plants including tea. For reasons of convenience and tradition, these are usually categorised as follows :

Macro		Micro	Beneficial
Primary	Secondary		elements
Nitrogen (N) Phosphorus (P) Potash (K)	Sulphur (S) Calcium (Ca) Magnesium (Mg)	Zinc (Zn), Iron (Fe), Manganese (Mn), Copper (Cu), Boron (B), Molybdenum (Mo), Chlorine (Cl)	Aluminium (Al), Silicon (Si)

The role in plant metabolism of Al and Si is not clear. However, silicon is well known for its oxydising power making native phosphate easily available. Aluminium has also been found to be an essential nutrient for tea (Konishi *et.al.*, 1985). All these above mentioned nutrient elements carryout specific functions in the plant. Though they are equally essential but they are required and absorbed by the plant in widely different amounts. For example, 100 kg processed tea may contain 4 kg N but only 4 grams of Zn

(1000 fold difference) and yet if the tea bushes do not get these 4 grams of Zn, they neither make efficient use of other nutrients nor produce high yields (Tandon and Ranganathan, 1993). The requirement of mineral nutrients of plants obtain largely from the soils, fertilizers and manures. A soil may be rich with nutrients for low yield level and may become deficient for a high yield level. In south India experience shows that a yield level of 800 kg processed tea ha^{-1} or less could be supported by the soil sources without application of nutrients externally. However, as yield go up increasing number of nutrients become important and enter the fertilizer schedule. The soil limiting factors in relation to productivity of 4500 kg processed tea ha^{-1} are presented in Table 10.1.

Table 10.1 : Emergence of soil limiting factors in relation to productivity of tea in south India.

Productivity of made tea (kg ha^{-1})	Limiting factor
Below 800	Nil
800 – 1000	N and K
1000 – 2000	N, P, K, Zn + liming
2000 – 3000	N, P, K, Zn + liming with material containing $MgCO_3$
3000 – 4000	N, P, K, Zn, Mg, Si, B + liming; transport process within the soil
Above 4500	N, P, K, Zn, Mg, Si, MO, B + liming; transport process within the soil

Source: Tandon and Ranganathan (1993).

10.3 Nutrient composition and loss

Before considering nutrients required to produce a given level of yield of processed tea, it is worth having a look at the mineral composition of the various plant parts and their native proportions in the total dry matter (Table 10.2 and 10.3). These show the relative abundance of various elements and their distribution among different parts of the tea plant. Tea being a vegetative crop, the concentration of N is the largest of all elements. The concentration, next to N is K which is followed by Ca, P, and Mg (Verma, 1993). It is seen that the Assam CTC tea had higher mineral content than that of Darjeeling orthodox tea. This may indicate the better nutrient absorption capacity of Assam tea compared to the Darjeeling tea or more solubility of minerals in the CTC tea compared to orthodox tea.

Table 10.2 : Nutrient contents (to produce 100 kg marketable tea) – macro and secondary nutrients

Plant part	Proportional Dry matter		Nutrient contents in the various parts							
			kg					g		
	kg	%	N	P_2O_5	K_2O	*MGO*	*CAO*	*Al****	*Cl*	*Na*
1. Plucks (marketable tea*)	100	12.5	4.0	1.15	2.4	0.42	0.8	100	6	8
2. Foliage on the bush	120	15.0	3.9	0.98	1.1	0.60	2.1	120	15	9
3. Fallen leaves (litters)	80	10.0	2.6	0.66	0.7	0.40	1.4	80	10	6
4. Stems/ branches	320	40.0	3.2	1.90	0.7	0.80	1.1	160	10	24
5. Roots	180	22.5	3.2	0.98	3.8	0.71	1.3	411	NA	27
Total	800	100.0	16.9	5.68	8.8	2.92	6.7	871	-	74
Removed by 1 & 4**	420	45.0	7.2	3.05	3.1	1.21	2.0	260	16	32

* 12.5% in high yielding fields under intensive cultivation systems; in low/moderate yielding fields it varies between 8 to 10%.

** Removed crop (1) and for fuel at the time of pruning (4).

*** Al content in mature foliage can go several times (up to 20 times) the figures shown above. NA - Not available.

Sources: Magambo and Othieno (1977); Ranganathan and Natesan (1988); Ling et. al. (1989); Natesan and Ranganathan (1990); Annual Report 1968 to 1989, UPASI TRI, Coonoor, India.

However, the mineral content in the tea brewed from Assam orthodox tea is comparatively lower than that of Assam CTC tea (Kalita and Mahanta, 1993). Tea flush normally contains 5 percent nitrogen, 1 percent phosphorus and 2 percent potash. These three major nutrients when their levels are exhausted may affect the entire dynamics of crop yield (Ranganathan, 1979).

Table 10.3 : Nutrient contents (to produce 100 kg marketable tea) – micronutrients

Plant part	Proportional dry matter		Nutrient contents in the various parts g				
	kg	%	Fe	Mn	Zn	Cu	B
1. Plucks (marketable tea*)	100	12.5	20	85	4	4	3
2. Foliage on the bush	120	15.0	40	180	4	12	9
3. Fallen leaves (litters)	80	10.0	27	120	3	8	6
4. Stems /branches	320	40.0	24	72	11	6	6
5. Roots	180	22.5	130	22	16	8	2
Total	800	100.0	241	479	38	38	26
Removed by 1 & 4**	420	45.0	64	157	15	10	9

* 12.5% in high yielding fields under intensive cultivation systems; in low/moderate yielding fields it varies between 8 to 10%.

** Removed crop (1) and for fuel at the time of pruning (4).

Sources: Magambo and Othieno (1977); Ranganathan and Natesan (1988); Ling et. al. (1989); Natesan and Ranganathan (1990); Annual Report 1968 to 1989, UPASI TRI, Coonoor, India.

In a tea field, nutrients are removed in a various ways. The loss had been more where soil husbandry was poor. It has been estimated that one quintal of crop removes about 6 kg nitrogen, 1 kg of phosphate and 2 kg of potash and in absence of pruning litter the level of losses increases (Banerjee, 1993). Wood *et. al.* (1964) observed about 4 kg nitrogen, 1 kg phosphate and 2 kg potash are removed by way of 100 kg dried leaf. Othieno (1979) reported that production of 1000 kg dry tea by clone no. 6/8 in Kenya removed 40 kg N, 4 kg P_2O_5 and 19 kg K_2O. The details of losses reported by Eden (1952) are presented in Table 10.4. The nutrients are continuously removed at each plucking round and thus in a year considerable amount is removed from the soil. Wood is taken as fuel at the time of pruning and thus nutrient retained in the wood are also removed from the soil. Despite removal of nutrients by the harvested crop and prunings, the nutrients are

Table 10.4 : Losses of major elements in the production of 1000 kg dry tea.

	Nutrient loss in kilograms		
	N	P_2O_5	K_2O
Young shoot	40.2	8.5	16.0
Stems	23.6	7.0	18.7
Old leaves	27.2	4.6	13.0
Total	91.0	20.1	47.7

Source: Eden (1952)

also locked up in the shoot and roots, lost in leaching, added to the soil by the fallen litter of leaves and the decomposed roots. Further, substantial quantities of nutrients are absorbed by the weeds in the tea field and these nutrients are temporarily not available to the tea plant and may not be available permanently if the weeds are removed during weeding. Willson (1969) examined the problem related with loss of nutrients in a field of mature tea in east Africa over a pruning cycle of three years. The summary of his findings is depicted in Table 10.5.

Table 10.5 : Loss of nutrients in tea field over a three-year period (kg ha^{-1}).

		Nutrients		
		N	P_2O_5	K_2O
	In plucked shoot (5050 kg)	252	40	108
A Loss	In prunings (22400kg)	785	135	570
	Leached	52	1	16
	Loss as gas	150		
	Total	1239	176	694
	Fertilizers applied	510	102	102
B Gain	Nutrients in leaf fall	56	9	22
	Stored in mulch	100	62	2
	Total	666	173	126
	Net loss (A – B)	573	3	568

Source : Willson (1969)

Since consideration of removal of nutrients by prunings at the time of calculation of fertilizer requirements of tea field will introduce a huge error, fertilizer requirements based on crop response is the sound method of computation. This system automatically allows for the loss of fertilizers from tea field (Barua, 1989).

10.4 General role of mineral elements

Mineral elements as such can not participate in any physiological process of the plant, but are only effective when present in an ionic form. On the whole mineral elements perform various functions, the most vital of which are described bellow :

Constituents of protoplasm and cell walls

Some of the mineral elements enter into the composition of the cell. Carbon, hydrogen and oxygen form important component of cellulose, lignins and various reserve foods and major part of the protoplasm in plants. Beside these the role of sulphur in protein, magnesium in chlorophyll, phosphorus in nucleoprotein, calcium in calcium pectate have been well established.

Catalytic effect

Some elements act as catalyst in many metabolic processes of the plants. Some of them act as enzyme precursors, while some others like manganese, magnesium etc act as activators or inhibitors in the enzymatic system.

Balancing effect

Some elements counteract the toxic effect of some other elements by maintaining an ionic balance. This behaviour of one ion in reversing the usual effect of another ion is also called antagonism. Ca, Mg and K belong to this group of elements and are called as balancing elements.

Influence on acidity and buffer action

The mineral elements have a great influence on the pH of the cell sap. Some of the important buffer systems in plants, phosphate and carbonate have been found to be originated from some of the elements absorbed by the plants from the soil.

Influence on permeability of cytoplasmic membranes

Some elements have a great influence on the permeability of the cell membranes. Thus, calcium and other di-or trivalent cations usually have a inhibiting effect on the permeability of the cytoplasmic membrane, whereas the monovalent ions have an enhancing effect.

10.5 Effect of individual nutrients

10.5.1 *Macro elements*

1. NITROGEN

Although nitrogen is the earth's most abundant element (the atmosphere is 78% nitrogen gas) and an essential component of all life, plants only use nitrogen 'fixed' with hydrogen and oxygen in the form of inorganic compounds. Soil microbes and lightning fix around 90 to 140 million tonnes of nitrogen per year. But humans are now adding atleast 140 million tonnes annually through the combustion of fossil fuels, power generation, other industrial processes and the production and use of inorganic nitrogen fertilizers.

Of the various essential elements, tea being a leaf crop, nitrogen probably is the key element in plant nutrition and investment on nitrogen occupies a major portion in the manurial bill. It accounts for about 4 to 5 percent of the dry weight of the shoot plucked for processing tea. Nitrogen is essential for growth and development of all living tissues, promotes branching, leaf production, enlargement of leaf surface and yield. Between 40 to 50 percent of the protoplasm is made up of nitrogenous compounds and 18 to 20 percent of dry protein is in the form of nitrogen. Therefore, nitrogen is of considerable importance in the vegetative growth of plants. In the soil the amount of this element in available forms is small, while in quantity withdrawn annually by crop is comparatively large.

The establishment of exact limit to plants' response to nitrogen is very difficult because of the operations of various interacting factors viz. characteristics of soil, cultivar, age of plants, shade, pruning cycle, plucking standard etc and extent of those factors seems to vary in different tea growing areas. The effects of cultural practices, soil characteristics, application of fertilizer and nitrogen uptake were studied by Karasuyama *et. al.* (1985). The main reason for variation in nitrogen response in different tea soils may be related to the characteristics of soils particularly this aspect related with process of mineralization and rapid conversion of nitrate by nitrification (Banerjee, 1993).

Nitrogen is added to the tea soils generally either as urea or as sulphate of ammonia. Tea plants assimilated nitrogen either as ammonia (NH_4^+) or as nitrate (NO_3^-) ions. When urea is applied to soil, it is split into NH_4^+ and CO_2 by the enzyme urease. Ammonia ions thus formed from urea or available from ammonium sulphate applied to soils is partly oxidised to nitrate by soil microbes. Nitrate, after assimilation by plants gets transformed to nitrite by the enzyme nitrate reductase, which is subsequently reduced to ammonia by another enzyme nitrite reductase. Ammonia thus liberated or absorbed directly from soil is assimilated in the roots of the plants with the help of enzymes glutamate dehydrogenase and glutamine synthetase/ glutamate oxoglutarate amino transferase systems to form amino acids, amides and finally proteins (Dev Choudhury *et.al.*, 1988). In tea the interaction of the nitrate and ammonia on the uptake of each other is not known though in comparison with nitrate, ammonia at high concentrations is more detrimental to root growth. The uptake of nitrate is influenced by availability of oxygen, temperature and level of soil moisture (Canvin and Atkin, 1974). The nitrogen uptake mechanism by the tea plant has been studied and found that the tea plant absorbed ammonium nitrogen more effectively than nitrate nitrogen (Ishigaki, 1978).

Synthesis of amino acids and proteins is accelerated in presence of adequate potassium. Potassium also helps in the translocation of plant metabolites from the roots to the shoots and thereby increases the efficiency of nitrogen assimilation. Potassium and nitrogen when applied together have a synergistic effect on tea. It has been observed that the best response of nitrogen is obtained when adequate amounts of potassium applied. Potash and phosphate also have been shown to interact favourably.

Paul and Sen (1984a) have conducted a long term experiment at Tocklai, India and reported scouring of feeder roots and tendency of downward movement of main root to avoid high accumulation of soil nitrate in the upper layer of higher nitrogen treated (147 kg ha^{-1}) plots. In Tocklai, another study has been conducted to examine all possible biochemical components of tea shoots and root following application of 100, 200 and 300 kg ha^{-1}, with 50 kg P_2O_5 and 100 kg K_2O as single dose applied during April. Biochemical studies had shown that uptake of nitrogen in tea shoot was not significantly enhanced with 200-300 kg N doses as compared to 100 kg N treatment. This was found to be caused by accumulation of nitrogen in the feeder root as amide (mainly theanine) and depletion of root starch. High dose of nitrogen produced more ammonia which was toxic to plants and was retained in the root as amide to evade ammonia toxicity in plant cells. The enzyme nitrate reductase was partially inhibited. With adequate potassium, the accumulation of theanine was considerably reduced, enzymes were activated and starch reserve was built up to supply carbon source and energy (Dev Choudhury *et. al.*, 1983; Paul and Sen, 1984a, 1984b).

In the high yielding clonal plots (China jats) at the Anamallais, south India, Manivel *et. al.* (1997) reported that nitrate reductase activity in feeder roots increased upto 200 kg N ha^{-1} $year^{-1}$ and it declined. Amino acids, caffeine and chlorophyll contents in the tea shoot showed a gradual increase with N level. They further observed an increasing trend in the value of colour index upto 300 kg of applied N. However, the values reduced at 600 kg N ha^{-1}.

In Sri Lanka, Kularatne and Bhavanandan (1971) in a sand culture experiment using N^{15} found that young tea plants absorbed twice as much as nitrogen when applied in the form of ammonium nitrogen compound compared to plants receiving nitrate nitrogen. After absorption, the plants appear to be able to assimilate both forms equally efficiently for synthesis of amino acids, protein etc.

In tea, nitrogen is generally applied in the form of a straight nitrogen compound. Among the various forms of nitrogen fertilizer, sulphate of ammonia showed to be the most efficient, safe fertilizer for tea and was mainly used for several years. Its long continued use increases acidity of the soil (Gokhale, 1957; Tolhurst, 1966) and create problems of base deficiencies and manganese toxicity (Bonheure and Willson, 1992). Ammonium nitrate, less acidifying than sulphate of ammonia and calcium ammonium nitrate (CAN), almost neutral materials have been included in the various experimental comparison by different researchers at different locations. It has been observed in east Africa that CAN at 80 lb per acre produced a significant reduction in yield (Anonymous, 1965). An experiment on the efficacy of split and basal application of organic and inorganic fertilizers in the optimisation of yield conducted at Darjeeling Tea Research Centre, Kurseong. The five years' yield trend reveal that single basal dose (during April/May) of CAN:DAP:MOP @120:45:80 per hectare gives highest return. The organic treated plots showed visible signs of improvement in the soil properties and the uptake of N, P, K nutrients is also comparable with inorganic fertilizer plots. The differences in yield between the organic and inorganic plots were also not very high (Anonymous, 1995). In south India, Verma (1993) reported greater efficiency and economy are achieved if 20 percent of total annual nitrogen is applied as ammonium sulphate, 65 percent as urea and 15 percent as CAN. Further, the efficiency of different sources could be increased by manipulation of $N:K_2O$ ratio for urea 1:1, ammonium sulphate 2:1 and CAN 4:3, upto yield level of 3000 kg ha^{-1}, while above 3000 kg processed tea ha^{-1}, the $N:K_2O$ ratio recommended is 1:1 irrespective of source of nitrogen. Ranganathan *et. al.* (1987) found that yield responses decreased in order of sulphate of ammonia, CAN, urea on soil pH 5.0-5.5. Yield depression following continuous use of CAN was reported at Tocklai, Assam (Anonymous, 1970). Urea was found to be 5 percent less efficient as a source of nitrogen than ammonium sulphate (Anonymous, 1966).

Meskhidze (1985) found high yield response in sulphate of ammonia than ammonium nitrate or ammonium carbonate. In east Africa, Willson (1967) concluded that sulphate of ammonia was most likely to produce a yield response. He further concluded that the significant loss of ammonia will occur if urea is applied as a soil dressing in unfavourable weather conditions. Urea is ideal for foliar application. Obatulu (1985) observed no significant difference between urea and sulphate of ammonia when it was used to manure one year old tea cuttings.

In Sri Lanka, Pethiyagoda and Krishnapillai (1971) observed in a pot culture study that the total growth and content of elaborated nitrogenous compounds were better when a mixture of nitrate and ammonium forms was applied. But results were obtained when 60-80 percent of the nitrogen was available in the form of nitrate. They further suggested that the mature tea may behave differently.

The development of a plant or its increase in complexity during its approach to maturity is govern by the internal ratio of carbohydrates to nitrogen within the plant tissue. If carbohydrate foods are relatively more abundant than organic nitrogen compounds proportionately more cell wall material and less protoplasm will be synthesized. Tissue composed largely of such cells are usually compact and more or less woody. Nitrogen manuring, therefore, plays a vital role in a crop like tea where simulation of vegetative growth is the ultimate aim of the cultivar (de Jong, 1950).

In general, 2 to 3 times more nitrogen is applied than what is actually been used by the plant because of low recovery of nitrogen. It was observed in south India that 13 to 14 percent of the applied nitrogen is made use of by the tea plants (Ranganathan, 1969). The experiments conducted in Assam have shown that only half of the applied nitrogen is utilized by the tea plant while the other half is lost by leaching or denitrification (Jain, 1983). Chen (1987) reported that the utilization rate of nitrogen is generally lower than 50 % when it was investigated by isotope labelling method. Paul (1992) reported that 33 percent of the applied nitrogen may be lost through volatilization and leaching. Such losses are lower in soils with high organic matter status. Singh (1989) reported that not more than 10 percent of the applied nitrogen is harvested in the shoots and only about 45 percent is accounted for by the whole plant.

It has been found that the increase in yield in case of unshaded tea is directly proportional to the amount of nitrogen applied. The linear response is found for N applications upto levels of 150 kg N ha^{-1}. Any increase above the application of optimum N level may cause reductions in dry weight contents, increase in fibre content (Cloughley, 1979) and increase in the rate of internodal extension (Banerjee, 1993) resulting stem components gaining relatively more weight than leaves. The application of higher rate of nitrogen (above 185 kg ha^{-1} $year^{-1}$) improved the dry weight of harvested shoots and also increased fibre contents.

The application level of nitrogen have been raised subsequently in recent years. In some experiments conducted at Tocklai, Experimental Station, India response to nitrogen on high yielding tea was obtained upto 235 kg ha^{-1} (Anonymous, 1992). Rojoa *et. al.* (1979) found in Mauritius, application of 375 kg N ha^{-1} $year^{-1}$ given highest yield. In Russia, higher yields recorded from 300 kg N ha^{-1} than from 450 or 600 kg N ha^{-1} (Rikhter and Lyashko, 1979). A linear response to nitrogen upto 250 kg N ha^{-1} in Malawi was reported by Malenga (1985) and he further reported that response fell below above 250 kg N ha^{-1} (Malenga, 1987). In south India,

Manivel *et.al.* (1997) reported the effect of different levels of N and N K_2O ratio on yield of three different cultivars which is presented in Table 10.6. Results indicated that there was yield response to increasing rates of nitrogenous fertilizer application upto 300 kg ha^{-1}. There was a depression in productivity in the case of Assam and China jats at 600 kg N ha^{-1} $year^{-1}$ and 500 kg N in the case of Cambod jat. Sharma (1991) at Tea Experimental Station, Himachal Pradesh, north-west Himalayas conducted a manurial trial in a 130 years old Chinary hybrid tea and found an increase in yield to the extent of 50.6 and 50.9 percent over control with the application of nitrogen 90 and 180 kg ha^{-1} respectively. The crop return at these levels was 5.30 and 3.11 kg processed tea per kg nitrogen respectively. Willson (1975a) also reported diminishing crop return with the increase in nitrogen application rate due to the limited availability of other materials. However, the response to nitrogen as high as 270 kg ha^{-1} has been reported in Himachal Pradesh by Marwaha *et. al.* (1977). Yield responses are often dependent on time of application. Aoki (1987) found in Japan that application of nitrogen before bud break at the end of the winter increased the photosynthetic rate substantially for limited period. The response at

Table 10.6 : Effect of N levels and NK ratio on tea productivity*.

Treatments	Yield of processed tea (kg ha^{-1}) (cycle average)		
	Assam	China	Cambod
NK_2O ratio			
2 : 1	4566	4719	5557
2 : 2	4613	4828	5755
N level (kg $ha^{-1}year^{-1}$)			
50	4292	4396	5169
100	4463	4609	5509
200	4699	4847	5785
300	4914	5143	6139
600**	4578	4875	5700
C.D. at P = 0.05			
NK ratio	170	312	202
N levels	268	382	288

* Mean of data recorded between 1990 to 1995

** Cambod jat received 500 kg N $ha^{-1}year^{-1}$

Source: Mcnivel et. al. (1997).

application of higher levels is reduced and remains positive. But its continuous decreasing ultimately would be uneconomic at application of very high dose of nitrogen. The best response of nitrogen at any level is dependent on adequate availability of potassium and other nutrients.

The response of shaded tea to nitrogen is complex type, level of shade have a vital role. It has been generally observed that unmanured shaded tea yields at higher rate in comparison with unmanured unshaded tea, but the response to applied nitrogen is much higher in unshaded tea (Bonheure and Willson, 1992). Efficiency of applied nitrogen was reduced by shade, although minimum yield was obtained by applying nitrogen under moderate shade (Anonymous, 1992). Manivel *et. al.* (1997) while conducting experiments in south India using clone UPASI-10 with or without shade trees (*Grevillea* sp), found that the productivity of shaded tea was higher than the unshaded tea. Further, they observed that shade in plantations helps efficient assimilation of nutrients. The elemental concentrations of maintenance leaf collected from the tea plants grown under shade are comparable with unshaded tea. Moderate decline in potassium, iron, zinc contents and accumulation of manganese were evident in unshaded tea.

Virgin soils are higher in nitrogen than those of top soils of old tea areas. It has been estimated that average nitrogen content of virgin soils is 5200 kg ha^{-1} against 3600 kg ha^{-1} for old tea area soil (Banerjee, 1993). In Assam, it was observed on an average 1600 kg N ha^{-1} is lost during a 40-year period of planting. The losses in nitrogen in 30 to 60 cm sub soil are nearly two-third of the loss in 0 to 30 cm soil layer (Dey *et. al.*, 1968).

Malenga (1996a) in Malawi, investigated the response of four high yielding good quality tea clones to nitrogen on fertile virgin soil. He observed that the yields of all the clones were maintained without additional nitrogen during first two years after planting but clone PC81 responded to nitrogen in the second year after planting. The other clones examined all responded to nitrogen in the third year after planting, and the yields of all the clones were substantially increased by the higher rates of nitrogen applied in the early years after planting, though the responses to nitrogen differed between seasons. But Ellis *et. al.* (1983) found that seedling and clonal plants grown on virgin soil showed no response to nitrogen fertilizers until the tea was nearly ten years old.

The response to nitrogen in a three-year pruning cycles, the average yield increment for third year applications was five times more than the first year (Eden, 1965). In a four-year pruning cycles in Kenya the response often fall in fourth year as reported by Willson (1965). However, the uptake of nitrogen was highly stimulated after heavy pruning (Hoshina *et.al.*, 1979).

Carbohydrate reserve in roots can be used as an index of plant health. Carbohydrate content increased from 100 kg N ha^{-1} $year^{-1}$ to 300 kg N but it declined at 600 kg N ha^{-1} $year^{-1}$ (Manivel *et. al.*, 1997).

Plants also have the ability to absorb nitrogen through leaves. Nitrates or ammonia salt or urea in suitable concentration in water solution are satisfactory materials for spraying on the leaves of plant. A study has been carried out to evaluate foliar application of urea combined with muriate of potash (1% each, 2 kg urea and 2 kg muriate of potash in 200 litres of water per hectare per round) as a palliative measure to combat drought and sustain the health and productivity of the tea plants. Effects of NK foliar application on crop is pronounced during the stress period recorded yield increment upto 10 percent. Foliar applied potash regulates stomata and retains turgidity (Manivel *et. al.*, 1997). Further, Marimuthu and Raj Kumar (1998) found that urea improves the permeability of the cuticle and thus favours the diffusion of foliar applied nutrients. Urea is readily absorbed and metabolized within the plant system and serves as a carrier for absorption of potash.

Bonheure and Willson (1992) reported when urea applied as foliar spray, most of the nitrogen was retained in the leaves as glutamine and arginine.

Marimuthu and Raj Kumar (1998) reported that under dry weather condition 55 percent of the foliar applied N-15 urea (1%) was absorbed at the end of four hours and no significant differences in the nitrogen absorption between four and twentyfour hours was observed in the field grown young and mature plants of the clone UPASI-3. The remaining part of the foliar applied N-15 urea might have been lost due to driff, volatilization and drip the ground.

2. PHOSPHORUS

Phosphorus plays a key role in DNA synthesis, encourages the formation of new cells, promotes root development particularly of the lateral and fibrous rootlets, hastens leaf development, flowering, fruiting including formation of seed and above all is indispensible for growth. It counteracts the harmful effects of excess nitrogen application. Phosphorus also plays a vital role in energy transformations and takes part in the metabolic processes of plants. It is involved in the basic reactions of photosynthesis and also in the activation of a number of enzymes particularly in the dark reactions in photosynthesis.

Tea soils are fairly rich in phosphorus (Mann and Gokhale, 1960). Phosphorus is strongly fixed in acid soils as insoluble aluminium phosphate, iron phosphate or calcium phosphate (if free calcium is available) and thus remains unavailable to the plant till a point is reached when repeated applications of phosphate make the soil well-saturated with phosphate. Tea plant is able to utilize phosphorus compounds efficiently by surface contracts and through micorrhizal associations. Adsorption of applied phosphate was observed to take place within 17-24 hours after application. This happened in all soils but more phosphate got adsorbed in fine grained than coarse grained soils. The absorption of phosphorus is a continuous process except in winter. Soil receiving phosphatic fertilizer over a period of time released more phosphate for absorption by plants than freshly applied phosphatic fertilizers. In tea approximately 60 percent of the phosphate absorbed by the plant is removed and 40 percent is returned to the soil or utilized with perennial growth (Ranganathan, 1976a). The plant need to absorb 40 kg of P_2O_5 to produce one tonne of marketable tea. However, the uptake of phosphate depends on the influence of soil and growth conditions, genotype etc. The phosphorus absorbed by tea clones from the soil solution in which it exists as $H_2PO_4^-$, HPO_4^{--} and PO_4^{---} ions. The relative abundance of these ions is largely dependent on pH.

The availability of inorganic phosphorus is largely determined by (1) soil pH; (2) soluble iron, aluminium and manganese; (3) presence of iron, aluminium, and manganese containing minerals; (4) available calcium and calcium minerals; (5) amount and decomposition of organic matter; and (6) activities of micro-organisms (Brady, 1984). It was estimated only 10 percent of the applied phosphorus is immediately used by tea. The remaining phosphorus stays in soil and is made slowly available in subsequent years (Ranganathan, 1976b). The role of lime in increasing the availability of phosphate is well documented (Mandal, 1975). The availability of phosphorus in the soil is at its highest when the soil has a pH between 5.5 and 7.0 (Bonheure and Willson, 1992). Ahsan (1994) reported in

Bangladesh tea soils contain over 85% aluminium in the exchangeable sites and phosphate in the fertilizer will readily be fixed with such a high aluminium unless lime is used.

Phosphorus requirement of tea and the quantity of phosphorus removed in the crop are comparatively much lesser than that of nitrogen and potassium. In the past, the experiments have been carried out in north-east India indicated application of 22 kg P_2O_5 ha^{-1} to mature tea caused significant and consistent increase in yield (Barua, 1989). But Gogoi (1993) observed in north-east India that optimum dose of phosphate was 50 kg P_2O_5 ha^{-1} for mature tea yield. This concentration was optimal for the activity of most enzymes such as nitrate reductase, nitrite reductase, glutamine synthatase, glutamate synthatase and glutamate dehydrogenase and decrease at higher phosphate doses. Sen (1967) reviewed the results of the experiments of phosphate manuring in north-east India. Response was found to vary with regions. The main reason for variability in response may be related to the variation in the inorganic and redundant soluble phosphates, as well as organic and total phosphates in the soils (Bhattacharya and Dey, 1978). The responses of added phosphate in the south Indian soils is low, not more than 15 percent and often statistically insignificant (Ranganathan, 1971). Traditionally, phosphate application in Assam was restricted to once in three years and that too @ 20 kg P_2O_5 ha^{-1} during pruning year (Rahman, 1977). Barua (1989) opined that supply of 30 to 80 kg ha^{-1} of phosphate appears to meet the requirement of most soil. A field trial has been laid out in Kurseong , Darjeeling to ascertain the optimum dose phosphatic fertilizer (rock phosphate) along with various combinations of potassic fertilizer (muriate of potash) keeping the dose of nitrogen fertilizer (urea) constant. It was observed in the first two years that N:P:K @ 120:45: 80 kg ha^{-1} (basal) gives significant high yield (Saha and Bisen, 1995). They have also observed that the percent phosphorus content in the flush and third leaf component were high where the dose of phosphate was maximum (45 kg ha^{-1}). This is followed where dose of phosphate was 30 kg ha^{-1}. On an average, it was observed that the ratio of percent phosphorus in the flush component to third leaf is maintained at 10:1.

In Sri Lanka, Eden (1944) reported significant improvement in yield when old established tea was supplied with 34 kg phosphate ha^{-1}. Increase of the dose of phosphate to 64 kg had no additional effect.

Dicalcium phosphate was found superior to single super phosphate in releasing P for uptake by plants. At low doses of phosphatic fertilizer, availability of phosphate from rock phosphate was found to be poor (Anonymous, 1992).

Since rock phosphate is cheap and sparingly soluble in acidic soil, gives a continuous supply of soluble phosphate to the crop and unlike superphosphate it is not prune to quick fixation, its use is therefore recomended. In a seven years field trial on tea clone (S15/10) which had been replanted after uprooting old tea in Kenya, east Africa, Othieno *et. al.*, (1997) observed that application of triple superphosphate (TSP) significantly increased pH values of the soil toward the end of the trial compared to rock phosphate (RP). RP significantly increased the extractable P especially at high rates compared to TSP. The content of mature leaf phosphate was significantly higher in the TSP treatment compared to RP while mature leaf N, K, Ca, Mg, Mn and Al contents were not affected significantly by the rate and source of fertilizers. The mean annual yield was not significantly affected by the fertilizer source but increasing the rate increased the yields in a quadrate manner. Tea research in the north-eastern India has confirmed the usefulness of Mussoorie rock phosphate for tea. In trials at the Nagrakata sub-station of the Tocklai Experimental Station, Assam, application of Mussoorie rock phosphate mixed with pyrites in 4:1 ratio gave 35% increase in green leaf yield as compared to bushes which received P through single super phosphate (Anonymous, 1984). However, the rock phosphate should not be applied at the rate exceeding 100 kg P_2O_5 ha^{-1} at a time. A crop depression for a year or two due to fluoride toxicity occurs if application exceeds 100 kg P_2O_5 ha^{-1} (Ranganathan, 1981).

Adequate potassium is a prerequisite to get the response from phosphate application. Suitable application methods often help in maximising the efficiency of phosphate uptake. In different trials in south India, the placement of phosphate fertilizer at 15 to 25 cm depth of the soil in alternate years has given 7 percent increase in yield while the shallow placement by eyebrow method has given 5 percent increase over broadcasting (Ranganathan *et.al.*, 1978). In Russia, Shuvalov *et. al.* (1986) reported highest translocation of radio-active phosphorus to aerial mass when tagged fertilizers were placed at 30 to 40 cm depth. But no gain from placement of phosphatic fertilizers below soil surface was reported in north-east India (Anonymous, 1982). However, it has been recently observed in north-east India that forking into a depth of 5-7 cm after broadcasting of rock phosphate @ 50 kg ha^{-1} showed higher yield than other method of application (Anonymous, 1998). Ikegaya (1975) in Japan reported broadcasting between rows followed by surface incorporation was found to be the most effective method. It may be due to the availability of the maximum concentration of white, feeding roots of tea near the surface of the soil (Barua and Dutta, 1961).

There is some evidence that the response of phosphate increases due to the application of organic manure in the form of compost and cattle manure (Koyamu and Nambiar, 1978).

Willson *et. al.* (1975) working in Kenya reported positive yield response of mature tea when there was undisturbed mulch on the soil surface. They have also noticed presence of more available phosphate and active feeder roots on the surface layer of soil. It can be stated that greater absorption of phosphate was encouraged due to increase in production of active feeder roots following mulching.

In most of the tropical and subtropical tea soils, the low available phosphorus is induced by the fixation of phosphate rather than the low natural phosphate contents in soils. Investigation showed that the phosphatase promotes the conversion of conjugated P to available P. The phosphatase activity unfortunately is rather low in tea soil in comparison with other soils. Therefore, the application of phosphatase preparation or the micorrhiza preparation into tea soil was proved to be beneficial in increasing the P availability (Chen, 1995).

The symbiotic association between plant roots and fungi mycelia is termed mycorrhiza (fungus root, plural mycorrhizae). Vasicular arbuscular mycorrhiza (VAM) is one of the common type of micorrhizae which have important effects on plant phosphate uptake. Greater soil exploration by micorrhizal roots as a means of increasing phosphate uptake is well established. In phosphate deficient soils immobile phosphate ions develop a phosphate depletion zone around the roots. The hyphae reach beyond this zone and directly translocate nutrients from the soil to the root cortex (Hayman, 1983). Experiments with ^{32}P-labelled phosphate indicate that VAM hyphae obtain their extra phosphate from the labile pool rather than dissolving soluble phosphate (Raj *et. al.*, 1981). Sparingly soluble rock phosphate is better utilized by the hyphae by closer physical contact with ions dissociating at the particle surface (Bagyaraj and Ravindra, 1993). Rajagopal and Ramarethinam (1998) investigated the influence of VAM on P nutrition of tea in south Indian soils observed uptake of phosphate was accelerated and directly related to the colonisation of tea roots by Glomus fasciculatum.The rate was found lower from rock phosphate amended soil than that of super phosphate. However, research on the details of the root of VAM, P uptake, variable source of applied phosphate, selected fungal strains capable of growing acidic tea soils etc remains to be explored.

3. POTASSIUM

Among the minerals, nitrogen and potassium are the two elements which are required in large quantities than other elements. Though potassium is not a constituent of any organic component of plants, it is required in large quantity. Potassium is mainly present in the meristematic tissues and quite absent in the nucleus and plastids. It plays many essential roles in plants. It is absorbed as the k^+ ions. Potassium takes part in many biophysical functions like osmoregulation (Hanson and Koeppe, 1975; Lauchli and Pfluger, 1978), stomatal opening (Nagarajah, 1979) and biochemical processess like enzyme activation (Wyn Jones and Pollard, 1983), metabolic regulations (Wyn Jones, 1975) etc. Potassium is most abundant cation in the cytoplasm, and salts of potassium are greatly responsible for the maintenance of osmotic potential of cells and tissues (Mondy and Munshi, 1993). It determines metabolic status of a plant by acting as a co-factor atleast for about 40 enzymes (Evans and Sorger, 1966) and thus, play an important role in the growth of the plants and yield. Because of its case of transport across plant membranes, it is highly mobile within the plant. Its relatively high concentration in plant cells helps to regulate the opening and closing of stomates in the leaves and the uptake of water by root ceils (Brady, 1984).

Potassium is essential for photosynthesis, formation of starch and translocation of sugars. It encourages formation of strong root and stem systems and enhances the ability of the plants to resist diseases and insect attacks, cold and other adverse conditions. It offsets the harmful effects of excess nitrogen in the plants. Application of potash fertilizer has been observed to increase the tolerance of the tea plant to water stress and also increase water use efficiency of the plant during water stress condition (Dey, 1977). Potassium also helps the plants in overcoming specific fungal infection to an extent.

Response of potash depends on number of factors viz. soil temperature during main growing seasons, severity and type of pruning to which tea is exposed, rainfall and soil moisture. Since the availability of soil potash is greatly reduced under unfavourable moisture conditions therefore presence of optimum moisture is an important factor for uptake of potash. The uptake depends to a large extent on the degree of development of root system. The absorption of k from the soil solution is plentiful when aeration is adequate. When CO_2 is excess, the absorption of nutrient elements decreases in order: K>N>P>Ca>Mg. Inadequate cultivation, the compactness of soil and waterlogging decreases O2 concentration and hence reduce the uptake of K by plants (Ramamoorthy and Velayuthan, 1976). The uptake of potassium by tea is inhibited by excess calcium in the soil (Zaid and Guerrier, 1983). The uptake of potassium and calcium by the roots of young tea plants has been investigated by Chamuah (1986). He observed that calcium and potassium ratio in the bush tops was significantly correlated with brown to white roots ratio. Plants having more active white roots have a greater capacity for uptake of calcium relative to the potassium uptake. Since aluminium in highly acidic soil, interferes with potash uptake the pH of the soil influences the uptake of potassium. In south India, the higher efficiency of potassium was found between pH of 4.5 and 5.0 (Verma, 1993). In east Africa, Willson (1975b) reported that responses occurred only when pH was below 5.2 and best response of potassic fertilizers observed at pH 4.9 which led to maximum benefit out of the nitrogen applied.

Plant can take up potassium in much larger quantities than what is needed for optimum growth. Unlike the situation with the respect of phosphorus, however, much potassium is lost by leaching if the plants do not use it efficiently. Further, under ordinary field conditions and with an adequate nutrient supply, potassium removal by crop is high (2-4 kg per 100 kg crop). The problem of maintaining soil potassium is outlined diagrammatically in Fig. 1.

The incidence of scorching following foliar spray of potassium bearing fertilizers was investigated by (Rahman *et. al.*, 1978). Though the potassium sulphate caused less damage in young and mature tea in comparison with potassium chloride (muriate of potash) it is always desirable to avoid foliar application of potassic fertilizer.

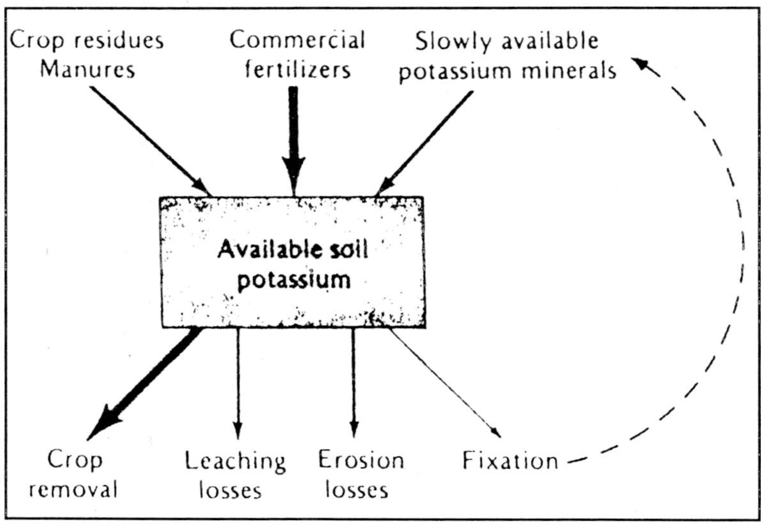

Fig. 10.1: Gains and losses of available soil potassium under average field conditions. The approximate magnitude of changes is represented by the width of the arrows. For any specific case the actual amount of potassium added or lost undoubtedly may vary considerably from the above representation. As was the case with nitrogen and phosphorus, commercial fertilizers are important in meeting crop demands.

It has been found in north-east India that there is clearly a positive response to potash upto 150 kg K_2O ha^{-1} but in south India there is around 250 kg ha^{-1} when combined with appropriate levels of nitrogen and phosphorus on the basis of crop losses.

Broadcast on soil surface is the best method for optimal utilization of potassium since it freely moves down to the root-zones with percolating water.

Potassium and nitrogen when applied together have a synergistic effect on tea. It has been shown that the best response of nitrogen is obtained only in presence of adequate amounts of potassium. The influence of nitrogen-potash interaction in augmenting yield increase of tea have been reported from north-east India (Sharma *et.al.*, 1977), south India (Ranganathan and Natesan, 1985), Sri Lanka (Bhavanandan and Manipura, 1969), east Africa (Wanyoko, 1979). Since then emphasis has always been given on the application of adequate potassium in relation to nitrogen to maintain high yield. Ranganathan *et. al.* (1988) reported that the potassium level and the ratio of potassium to nitrogen have significant effects on yield. They further observed that the yield of tea over a cycle reveals that application of higher level of K_2O in pruned year and adoption of 1:1 N:K_2O ratio in the other years resulted in higher productivity. Potash and phosphate also have been shown to interact favourably. Zinc also has a positive effect on the uptake of potash. Potassium efficiency may be enhanced roughly 4 to 6 percent with the correction of zinc deficiency (Ranganathan, 1979). There has been a marked effect and very important role of potassium on the recovery of tea from pruning and on bush frame size (Willson, 1975).

4. SULPHUR

Although sulphur has long been known to be essential for plant growth, it has received only recently the attention it deserves. However, it is an essential nutrient for tea. Sulphur is mainly absorbed by roots as sulphate (SO_4^{--}) ions, but in some cases very small amount may enter through moist mesophyll cells of leaves from the atmosphere in gaseous form as sulphur dioxide (SO_2). sulphur is a constituent of large number of amino acids like cystine (27%S), cysteine (26%S) and methionine (21%S) in plants which are ultimately built up into proteins. Sulphur it seems to be involved in the formation of chlorophyll, activation of enzymes and takes part in oxidation-reduction reactions in respiration. It forms an important constituents of certain vitamins like thiamin. Sulphur increases the nodule formation in legumes (shade trees) and development of roots. Sulphur also is thought to be connected with the conversion of the sunlight energy into chemical energy in the photosynthesis process. The harvestable crop contains sulphur between 0.15 and 0.20 percent (Othieno, 1992) and 0.08 and 0.3 percent (Verma *et.al.*, 1998a) on dry weight basis.

Sulphur is lost due to leaching, fixation and immobilisation. To compensate for leaching losses about three to four times the quantity of actual removal (1.53 to 2.04 kg sulphur for every 100 kg processed tea) has to be applied (Ranganathan and Natesan, 1987a). Due to continuous removal of sulphur in various ways it may be seriously depleted from the tea growing soils. Further, regular application of sulphur to the soils of the tea estates is getting limited due to replacement of ammonium sulphate and single super phosphate by urea and rock phosphate respectively. Despite this, due to the dramatic increase in productivity in recent years there is substantial removal of sulphur from the soil. When sulphur carrying fertilizers were used in tea, the need of additional sulphur application was never thought of. The situation in tea has gradually been changed with almost complete switch over to non-sulphorus fertilizers resulting sulphur deficiency in soil and hence in the plant.

Assam tea soils appeared to have lower soil available sulphur compared to Dooars soil. Soil available sulphur is also higher in West Bengal as compared to Assam (Chakravartee, 1996). The critical level of available soil su;phur in north-east India and Darjeeling have been found to lie between 35 and 40 ppm (Chakravartee, 1996) and 20-30 ppm (Saha and Bisen, 1996) respectively.

The sulphur deficiency can be corrected by application of fertilizer containing sulphur. Data generated over six years at Tocklai Experimental Station in Jorhat, Assam indicated that the soil application of sulphur produced significant effect on yield (Table 10.7). The highest mean crop response (1991-93) was recorded at 20 kg S ha^{-1}, followed by 40 kg S ha^{-1} which was significant over control (Barbora, 1995). On contrary, Sinha *et. al.* (1992) recorded the highest crop response after two years of annual application of 40 kg S ha^{-1}. In Darjeeling, the highest yield of made tea (1992-95) was recorded by annual application of 20 kg S ha^{-1} during April/ May through ammonium sulphate in a single basal dose (cost-benefit ratio 1:1.95) (Saha and Bisen, 1996). They have also reported that the optimum leaf nitrogen and sulphur ratio for Darjeeling tea are 9/10:1 (N: S) and is a better index for detecting 'tea yellows'. At Tocklai Experimental Station, Barbora (1995a) reported that different sulphur sources significantly

Table 10.7 : Effect of applied rate of sulphur on yield of processed tea.

S rate (kg ha^{-1})	Mean processed tea (kg ha^{-1})	% over control
0	2154	100
20	2387	111
40	2318	108
60	2210	103
CD at P = 0.05		60

Source: Barbora (1995a)

influenced the average yield of a pruning cycle. Calcium sulphate (Gypsum), sulphate of ammonia, aluminium sulphate and iron pyrites were found at par and significantly increased the yield over control. Sparingly soluble and insoluble sulphate such as gypsum, single super phosphate are good sources of sulphur for neutral and alkaline soils. In acid soils they form acid pockets. The uptake of nutrient by the ramifying roots is impaired in these acid pockets and in extreme situations, the nutrients come out of the roots (Verma, 1993). Soluble sulphates such as ammonium sulphate, potasium sulphate and magnesium sulphate are good sources of sulphur for acid soils. All common fertilizers except pyrite and elemental sulphur contain sulphur in the sulphate (So_4^{--}) form which is readily available to the plants. Sulphur pyrite and elemental sulphur have to undergo chemical and microbial transformation into the sulphate ions. This conversion is accompanied in three to eight weeks in Indian soil conditions (Tandon, 1995).

There is no fixation of sulphate ion in soils hence it does not accumulate in the soils of south India even after continuous use of ammonium sulphate for several years, thus stressing the need for application of sulphur regularly. In tea, the requirement of sulphur is 1.01 kg for producing 100 kg of marketable tea, out of which 0.50 kg is recycled. Although the net removal of sulphur is 0.51 kg for every 100 kg marketable tea, the actual application has to be made at three to four times the quantity of the actual removal (1.53 to 2.04 kg sulphur for every 100 kg of marketable tea) in order to compensate leaching losses, limitations imposed by transport processes in soil, and loss by reduction to sulphides during monsoon (Verma, 1993). The requirement of sulphur in south India has been calculated for various yield levels upto7000 kg of processed tea and enumerated in Table 10.8.

Considering the efficiency of ammonium sulphate and soil characteristics, the sulphur requirement of tea in south India could best be met by way of applying 20 % nitrogen as ammonium sulphate annually. According to this schedule, a balance N:S ratio in south India is maintained at 1:0.24 in the manuring schedule for sustainable productivity of tea (Natesan, 1989). Verma *et.al.*, 1998a) further reported that if situation warrants to economise on cost of manuring, in a four-year cycle, ammonium sulphate may be applied in the pruned year and third year. Alternatively, 10 percent of total nitrogen may be applied annually in the form of sulphate of ammonia instead of annual application of 20 percent of total nitrogen.

Table 10.8 : Sulphur requirement and nutrition through sulphate of ammonia (SOA)

Processed tea	Sulphur required	20% nitrogen soil application as SOA (kg ha^{-1})		
(kg ha^{-1})	(kg ha^{-1})	N*	SOA	S
1000	13.2 – 17.6	20	100	24
2000	26.4 – 35.2	40	200	48
3000	39.6 – 52.8	50	250	60
4000	52.8 – 70.4	60	300	72
5000	66.0 – 88.0	70	350	84
6000	79.2 – 105.6	80	400	96
7000	92.4 – 123.2	90	450	108

* For medium organic matter status

Source: Verma (1997)

In sulphur deficient fields at Malawi, the highest crop response was obtained at 25 kg S ha^{-1} (Table 10.9). Grant and Shaxson (1970) observed that application of sulphate of ammonia once in every 3 to 6 years in Malwi would prevent recurrence of deficiency. Brandram-Jones (1963) reported the short-term benefit of a small amount of gypsum in the planting pit. But application of higher rates of gypsum were not so effective.

Table 10.9 : Response of sulphur in Malawi (Mimosa)

S (kg ha^{-1})	Processed tea (kg ha^{-1})	% of increase over control
0	1692	100
25	1913	113
104	1855	110
156	1910	113
312	1734	103

Source: Mkwaila (1992)

Soil organic matter status under well drained and well aerated soil was found to have influenced in building up of available sulphur both in Assam and in the Dooars. There exists a linear and positive correlation between the soil organic matter and soil sulphur availability. On the other hand, soil rich in phosphate appeared to have retained less sulphate in the available form (Chakravartee, 1996).

Banerjee (1987) reported that the tea cultivars with semi-erect leaves respond better to sulphur than those with horizontal leaves. The reason is not yet known.

Barbora (1995a) found existence of significant interaction between sulphur and nitrogen on the yield of processed tea. Application of 20 kg S ha^{-1} along with 135 kg N ha^{-1} produced the highest yield. This was followed by 20 kg S ha^{-1} along with 90 kg N ha^{-1}. Thus for efficient utilization of nitrogen it should be adequately balanced with sulphur.

Banerjee (1993) opined that sulphur bearing fertilizers may provide sulphur needs of tea but the differential responses of the cultivars suggest sulphur in combination with other nutrients might increase overall yield of tea.

It was observed in a pot culture experiment conducted in China that with the supply of sulphur regardless of the sources used (urea, $(NH_4)_2SO_4$, MgO, $MgSO_4$, K_2CO_3, K_2SO_4) the photosynthetic rate and chlorophyll content in sulphur treated plants were much higher than those in non-sulphur treated plants with equal amounts of the same cation (Xun and Jianyun, 1994).

5. CALCIUM

It is a constituent of cell wall in the form of calcium pectinate and may not be replaced by other bases. Calcium is required for cell division and cell enlargement, in translocation of carbohydrate and an activator of several enzymes. It is also required for formation of seed in case of seed bari plants. Calcium influences the water economy of plant, synthesis of proteins, neutralisation of acids in plants, root development and absorption of nitrogen. Plants take this element in ionic form as Ca^{++}.

The harvestable crop, on average, contain between 0.5 and 1% calcium dry weight basis. Othieno (1992) reported that 10 to 20 kg calcium per ha^{-1} is removed for production of 2000 kg processed tea ha^{-1} $year^{-1}$ and it would be in between 25 and 50 kg for high yielding fields producing 5000 kg processed ha^{-1} $year^{-1}$.

It was found from nutrient culture studies that the uptake of nutrients, specially cations, has highest at the pH range of 4.5 and 5.0 which is also suitable for optimum calcium nutrition of tea (Ranganathan and Narayanan, 1975). Since most of the phosphatic fertilizers (except ammonium phosphate) contain large amounts of calcium so tea soils received calcium along with phosphates. But continual use of nitrogenous fertilizers either ammonium sulphate or urea, sometimes soil pH goes below 3.9 and chemical degradation of soil occurs. Owuor *et. al.* (1988) reported the reduction of the content of leaf calcium by application of various nitrogenous fertilizers. Therefore, positive response have been observed in terms of yield to calcium application in acidic soils low in calcium concentration with pH below 4.8. The highest efficiency of nitrogen and efficiency of potash at reasonable level are maintained in soil pH around 4.8. Liming is advantageous to maintain pH of the tea soils. But at the time of application of calcium in tea soils, despite soil acidity, some factors viz. depression of potassium uptake by calcium, increase of leaching loss of nitrogen by stimulating nitrification have to be taken into account. Krishnapillai and Pethiyagoda (1980) reported acceleration of nitrification of ammonium and quick growth of young potted sand cultured plants by application of calcium carbonate in nutrient solution. They also found that the calcium was not absorbed by the plants and it was due to pH of the soil.

6. MAGNESIUM

It is one of the essential mineral constituents of chlorophyll molecule where magnesium combines with iron containing precursor to a molecule of chlorophyll. Magnesium activates many enzymes concerned with carbohydrate metabolism and is predominant in the citric cycle. It also catalyses many phosphorylation reactions relating to nitrogen metabolism in plants. It plays an important role in the phosphorus metabolism of plants and also linked up with the synthesis of nucleoprotein, fats etc and respiratory mechanism. Magnesium is extremely mobile and quantity required by tea and that lost in the crop is relatively smaller amount than calcium. Magnesium fertilizers directly increases the efficiency of phosphatic fertilizer. The loss of magnesium through leaching is very high. It comes next to that of calcium but is greater than that of potassium. It is absorbed by plants in the form of the Mg^{++} ions for photosynthesis.

Some magnesium from solution is lost through fixation by the clay mineral, possibly Mg^{++} replaces A^{+++} in 2:1 clay minerals through isomorphic replacement. But the fixed magnesium can gradually become available to plants (Kanwar, 1976a).

Magnesium fertilizer on soils with pH in the range of 4:1 to 5.0 had no effect on yield of tea (Willson, 1975c). But significant increases in yield in Georgia from application of magnesium fertilizer has been reported by Godziashvili and Peterburgsky (1986). In south India, the effect of soil applied magnesium on yield of mature tea over two pruning cycles has been studied by Pilani and Verma (1993). They obtained significant higher response in the second cycle which may be attributed to the high levels of N, K and Mg applied in second cycle. In both the cycles, significant interaction between K and Mg have been observed. While there was crop depression or marginal increase due to application of Mg at low levels of N and K. However, the crop response was significantly higher at high levels of N and K due to application of Mg in both the cycles. Magnesium is applied as magnesite or magnesium sulphate or magnesium silicate if silicon is needed. In south India, the levels of magnesium in tea soils are generally in low to medium category and adequate to maintain high yield of tea. The application of dolomitic lime (3.6 % Mg) once in a pruning cycle, a regular practice to correct acidity of the soil. This also prevents the occurence of magnesium deficiency to a certain extent. But it may not fulfil the entire need of magnesium of tea fields. Widespread deficiency of magnesium has been noticed in clonal as well as in high yielding fields of south India inspite of the application of dolomitic lime. Thus, in addition to liming with dolomitic lime, it is recommended to apply 8 to14 kg of magnesium sulphate per hectare as foliar spray on the basis of yield levels (Table 10.10) in four to six foliar sprays on the following day of plucking. Since magnesium sulphate is compatible with zinc sulphate, manganese sulphate and boric acid therefore they may be applied together without any additional cost of spray (Verma et al., 1998b).

Magnesium sulphate @ 200 and 400 kg ha^{-1} or magnesium carbonate @100 and 200 kg ha^{-1} for tea yielding 3000-4000 kg ha^{-1} and for those above 4000 kg ha^{-1} respectively is recommended for field application in south India (Ranganathan and Natesan, 1987b). Magnesite should be used when 20% annual nitrogen is applied as ammonium sulphate, while in absence of ammonium sulphate, magnesium sulphate should be used by mixing nitrogen and potash fertilizers. But magnesite may be mixed only with urea-based mixture (Verma, 1993).

Table 10.10 : Requirement of magnesium for foliar application.

Processed tea (kg ha^{-1})	*Magnesium sulphate* (kg ha^{-1})*
Below 3000	8
3000 – 4000	10
4100 – 5000	12
Above 5000	14

* One / two percent concentration; Spray volume: 200 l ha^{-1}

Source: Verma et. al. (1998b)

10.5.2 *Micro elements*

1. ZINC

It is considered to be an essential element for tea whose deficiency is widespread. Deficiency of zinc was first reported from Sri Lanka (Tolhurst, 1962). Zinc is important for the synthesis tryptophane which is the precursor to growth hormone indole acetic acid (IAA). Zinc also promotes starch formation. It is also found to activate the enzymes (dehydrogenase and tripeptidases) and causes the synthesis of diphosphopyridine nucleotidase (DPNase)-an enzyme which split DPN at the nicotinamide riboside linkage.

Zinc is taken up by the plants as Zn^{++} ions or in the form of a complex with chelating agents, e.g. EDTA (ethylenediamine tetraacetate). Salts and complexes of zinc can easily be absorbed directly through leaves. Therefore, application of zinc mostly as foliar spray for correcting deficiency has become a routine practice. Chelated sources of zinc in place of zinc sulphate were more effective both in soils and foliar applications (Borpujari, 1982).

The availability of zinc is least between pH 5.5 and 7.0, but its availability increases at a lower pH. Zinc availability at pH above 7 becomes a complex problem, as the positively charged zinc ion gets converted into a negetively charged zincate complex whose availability tends to be reduced in alkaline soils.

In most tea growing countries, zinc is widely applied as foliar spray, as the sulphate in India and Sri Lanka and as the oxide in African countries. Dosage of zinc also varies. In south India for example, zinc is applied as zinc sulphate at one percent concentration @ 8 to 12 kg ha^{-1} per annum in four to five splits (2 pre-monsoon, 2 or 3 post-monsoon) (Verma *et. al.*,

1997). In north-east India, the present recommendation for zinc sulphate application is 12.5 kg ha^{-1} for all areas (Chakravartee and Sinha, 1994). Zinc is a heavy metal and several tea consuming countries have imposed restriction on its content in the processed tea. Keeping this restriction and also from the information generated on enzyme activity in view, the annual dose of zinc sulphate should be restricted to 12.5 kg ha^{-1}. In north-east India, the ideal times of zinc application are biginning of first flush, end of first flush and biginning of autumn flush for unpruned/light skiffed teas and biginning of autumn flush in pruned/deep skiffed teas. Zinc application should be avoided during second flush and main cropping period (Tea Encyclopaedia, TRA, Assam, Serial no. 176/1). Foliar application of zinc sulphate not exceeding 12.5 kg ha^{-1} at 1 to 2 percent concentration was found to enhance nitrate reductase activity, nitrogen uptake and protein content in the shoot without any adverse effect on the quality of CTC tea (Chakravartee and Sinha, 1994). In Darjeeling, highest significant yield of

Box 10.1 : Chelates

Chelate is applied to compounds in which certain metallic cations are complexed or bound to an organic molecule. In complexed form the cations are protected from reactions with inorganic soil constituents that would make them unavailable for uptake by plants. Iron, zinc, copper and manganese are among the cations that form chelate complexes.

The effect of chelation can be illustrated with iron. In the absence of chelation, when an inorganic iron salt such as ferric sulphate is added to a calcareous soil, most of the iron is quickly rendered unavailable by reaction with hydroxide.

$$\underset{\text{(available)}}{Fe^{3+}} + 3OH^- \rightleftharpoons \underset{\text{(unavailable)}}{FeOOH} + H_2O$$

In contrast, if the iron is added in the form of an iron chelate, such as FeEDTA, the iron remains mostly in the chelate form, which is available for uptake by plants.

$$\underset{\text{(available)}}{FeEDTA} + 3OH^- \rightleftharpoons \underset{\text{(unavailable)}}{FeOOH} + EDTA3^- + H_2O$$

The magnesium by which micronutrients from chelates are absorbed by plants is still obscure. The chelating agents in most cases are absorbed by growing plants, but the rate of their absorption is lower than that of the metals they carry. The primary function of the chelate is therefore to keep the metals available in the soil. Further, there is evidence that some of the benefit from the chelating agents is through increased translocation of the metals once they are absorbed by the plants.

Source: Brady (1984)

processed tea can be obtained by applying 10 kg zinc sulphate ha^{-1} $year^{1}$ in two split doses of 5 kg ha^{-1} each in the month of June and August. Four split doses @ 1.25 kg ha^{-1} in the month of April, June, August and October also increases the productivity of processed tea on a relatively lesser scale. In Malawi, Malenga (1979) found significant responses to aerial application of zinc to mature seedling tea. Phiri (1985) reported significant and economic responses to foliar application of zinc oxide on young clonal tea in Malawi. Zinc oxide contains 80 % zinc and not soluble in water. A suspension of finely ground zinc oxide is used as foliar spray. Borpujari (1982) reported that nearly 4.6% of the total zinc of the soils of north-east India remained in the available form which correlated positively with the silt and clay content of the soil and negatively with the available phosphate. Zinc is not readily absorbed by tea from the soil. It is possible that zinc applied to the soil gets quickly fixed with the soil particles, though pH, organic matter content, phosphate availability, silt and clay percentages may also influence zinc uptake from soil. It is equally possible that organic colloids act as a sink for zinc. However, if zinc is continuously applied to the soil its absorption is completed in about three weeks. It has been seen that roots accumulated the largest amount, the leaves the least and the stem occupy an intermediate position. Chinary cultivars were the most efficient in uptake of zinc, while Assam varieties were the least. The cation exchange capacity (CEC) of roots affected the zinc uptake significantly (Borpujari, 1982). The uptake of zinc enhanced by nitrogen, but depressed by phosphorus, manganese and iron (Borpujari, 1982). Further, if the soil is rich in phosphorus, zinc is quickly converted into zinc sulphate which is stable compound with less moveability and hence the slow upward movement (Banerjee, 1993). Soil applied zinc accumulates in the collar region of the bush and is not translocated to growing shoots. Moreover, zinc is required at the growth point because it is a precursor of hormone. Radio tracer studies of ^{65}zn in potted plants indicated that within an hour about 51 % of the foliar applied one percent zinc sulphate solution had translocated to growing shoots and it increased to about 70 percent at the end of two hours (Table 10.11). Therefore, to obtain better efficacy a minimum period of two rainless hours is required for absorbtion and translocation of foliar applied zinc. However, addition of NAA (synthetic auxin) at 10 ppm to one percent zinc sulphate solution has increased (6 to 8 %) the translocation of zinc in comparison to application of zinc sulphate alone. Shuvalov (1985) studied the distribution of zinc in tea plant.

Table 10.11 : Translocation of zinc (potted plants of UPASI-3).

Treatments	Period (hr)	^{65}Zn translocated
$ZnSO_4$ (1%)	1	51
$ZnSO_4$ (1%) + NAA*	1	59
$ZnSO_4$ (1%)	2	70
$ZnSO_4$ (1%) + NAA*	2	76

* *NAA (Naphthalene acetic acid) 10 ppm (Agronaa / Planofix 45 ml / 200 litre)*

Source: Marimuthu and Manivel (1992)

The experiments in south India have confirmed the usefulness of spraying mother bushes with 2 percent zinc sulphate solution, one or two weeks prior to harvesting of cuttings in reducing overcallusing and in hastening rooting. Incorporation of manganese at 500 ppm and boron at 100 ppm increases the efficiency of zinc absorption. Further, speedy growth of young plants in the nursery by application of zinc as foliar spray has been observed (Ranganathan, 1981). A minimum of 100 litres of spray fluid per hectare is required to cover the maintenance foliage effectively. Zinc sulphate is compatible with urea, ammonium sulphate and muriate of potash and can be used in combined sprays. Large increase in mite population after zinc sulphate application has been noticed by Sarma *et.al.* (1985). However, the spray of a mixture of minor nutrients did not have this effect.

2. IRON

It is an essential part of several enzymes. It has a role in the formation of iron porphyrin, a precursor of chlorophyll. Although cholorophyll does not contain iron but synthesis of chlorophylls is dependent upon its pressure. It serves as catalyst for photosynthesis, respiration and reduction of sulphates and nitrates. It plays a role in the metabolism of nucleic acids.

Iron is abundantly available in acid soils because it is a constituent of most of the minerals. Iron deficiencies are not common and its amount in plants is very small.

3. MANGANESE

It serves as a catalyst in various physiological and enzymatic reactions, and is involved in the reduction of nitrates during respiration. It seems to be associated with iron in the chlorophyll formation.

> **Box 10.2 : Need for maintenance of nutrient balance**
>
> - *Excess copper or sulphate may adversely affect the utilization of molybdenum.*
> - *Iron deficiency is encouraged by an excess of zinc, manganese, copper, or molybdenum.*
> - *Excess phosphate may encourage a deficiency of zinc, iron, or copper, but enhance the absorption of molybdenum.*
> - *Heavy nitrogen fertilization intensifies copper and zinc deficiencies.*
> - *Excess sodium or potassium may adversely affect manganese uptake.*
> - *Excess lime reduces boron uptake.*
> - *Excess iron, copper, or zinc may reduce the absorption of manganese.*
>
> *Source: Brady (1984)*

Manganese is abundantly available in acid soils. Soils with pH lower than 5.5 may contain large amounts of their manganese in water-soluble and exchangeable bivalent form, but with the rise of pH, the amounts of these forms decrease (Kanwar, 1976b). Therefore, deficiencies in tea soils are uncommon and toxicities are also rare but found on highly manganiferous soils. Tolhurst (1962) reported the appearance of manganese deficiency symptoms on some tea bushes of Sri Lanka.

Manganese is absorbed by the plant as manganous ion (Mn^{++}) and also in molecular combinations with certain organic complexing agents e.g. EDTA. The manganese uptake from soil was improved by chelating with EDTA. The uptake is also stimulated by phosphate (Prasad and Dey, 1979). Foliar application of manganese was more efficient than soil application. Manganese is used either as sulphate or as chloride. In Assam, Kalita (1992) reported that manganese availability was negatively correlated with soil organic matter status, application of manganese sulphate was most cost effective and manganese fertilizer decreased the optimum time of fermentation and improve the quality of made tea. He further reported that the soil application of manganese @ 20 kg ha^{-1} was optimum dose for yield. Investigations have however shown that foliar application of manganese sulphate at 1 to 2 percent (W/V) substantially increases nitrate reductase activity and protein content of tea shoot (Chakravartee and Sinha, 1984).

4. COPPER

It is an essential plant nutrient and important for tea. Most of the deficiencies of copper have been reported in the case of organic soils, sandy soils and highly leached soils. In certain highly organic soils of Malawi, deficiencies of copper has been observed, where tea shoots do not oxidise properly after rolling causes poor fermentation. Clowes and Mitini-Nkhoma (1987) demonstrated the relationship between poor fermentation and a low content of copper. They have reported that foliar application of a copper compound gave a lasting benefit. In Sri Lanka, Ramaswamy (1960) investigated copper in tea. He had observed increment in the content of copper in tea during manufacture which was due to the application of copper in the field.

Copper mainly absorbed as cupric ion (Cu^{++}). Copper is essential constituent of the enzyme polyphenol oxidase which is vital for fermentation, important for photosynthesis, respiration, protein and carbohydrate metabolism and probably nitrogen fixation. It takes part in the synthesis of chlorophyll and some vitamins. Copper is also involved in the utilization of iron.

It has been proved many times that the absorption and fertilization of one element is profoundly affected by the concentration of other elements present both in the soil and in the plant. Copper concentration has been known to affect or be affected by several other elements. On account of the high copper fixation capacity of the soils, its toxicity except sandy and gravely soils is rarely seen on plants. In north-east India, copper sulphate sprayed at high concentration (12 %) suppressed growth of tea bushes (Barua, 1989).

The common copper fertilizers are copper sulphate ($CuSO_4.5H_2O$), copper ammonium phosphate $Cu(NH_4)PO_4$, copper chelates, copper frits and even oxides of copper. Generally, one percent or less solution of copper sulphate (25% copper) is used for spraying on tea plants to correct copper deficiency. Copper levels below 5 ppm in the leaves indicate copper deficiency in Malawi. It is advisable to apply the spray in the early morning or late evening during hotter and drier periods of the year. Copper in the tea field is applied indirectly as copper oxychloride (50 % copper) to control blister blight disease in tea.

In Darjeeling, foliar application of copper sulphate @ 0.5% increased the crop yield upto 9.6% over the control. This dose did not affect the shade of colour of processed tea and further no residual metallic impact was noticed during the tea tasting. However, polyphenolic contents in copper treated teas were found to decline whereas the content of total chlorophylls increase (Anonymous, 1995).

5. BORON

Since a small quantity of boron is required by the tea plant and adequate boron is available in most of the tea soils, therefore boron deficiencies are rarely seen in the tea garden. In Japan, Ishigaki (1984) demonstrated, using tea plants in sand culture, that only a small quantity of boron was necessary. However, boron is involved in the carbohydrate transport by increasing membrane permeability, increases the absorption of water and takes part in the formation of pectin in the cell walls. A large number of sterility and malformations of reproductive tissues observed in the boron deficient plant. It also plays a role in the development of roots and root nodules in leguminous plants. It has a effect on uptake and utilization of calcium, and also to control the balance between phosphorus and calcium. Boron deficiency is in general accelerated under extremely dry conditions. During drought periods plant roots are forced to exploit only lower horizon of the soil where boron availability is quite low. But during rains roots can again absorb boron from the top soil where its content is comparatively higher than sub-soil (Brady, 1984).The availability and utilization of boron are determined to a considerable extent by pH. Despite pH, organic matter, salinity, alkalinity, calcium carbonate, texture, quality of irrigation water, manurial treatments etc., have also been found to affect the availability of boron in soils.

When boron in solution is added to the soil, it is adsorbed by the soil complex very readily and some of it can not be displaced with water. In north-east India, the boron uptake was studied by Prasad and Dey (1979). They reported that the uptake of boron was much quicker from foliar application in comparison with soil application. Boric acid (H3BO3) and borax solutions can be sprayed on the leaves of the plant to correct the deficiency of boron. Boric acid proved to be a better source of boron than borax. The optimum dose of boron for soil application was found to be 2 kg ha^{-1} (Kalita, 1992). The concentration of boron between 110 and 120 ppm in leaf is possibly the threshold value of boron toxicity of tea (Banerjee, 1993). Boron is absorbed in one or more of its ionic forms viz. $B_4O_7^{--}$, $H_2BO_3^{-}$, $H_2BO_3^{--}$, or BO_3^{--}. In plants the highest concentration of boron found in leaf and the lowest in root, stem concentration occupying an intermediate position.

6. MOLYBDENUM

It is a constituent of the enzyme nitrate reductase (responsible for the reduction of nitrates in soils and plants) and nitrogenase. Molybdenum applied to soil @ 1 kg ha^{-1} increased nitrate reductase activity in Assam tea soils (Anonymous, 1992). No deficiency symptoms of molybdenum has been observed in tea soils. It is however, felt that high yielding tea may benefit from molybdenum application. It may be applied at the rate of 15 grams molybdenum sulphate ha^{-1} $year^{-1}$ along with zinc foliar spray. The liming of acid soils will usually increase the availability of molybdenum. Molybdenum is essential for nitrogen fixation (both symbiotic and non-symbiotic) and nitrogen assimilation.

7. CHLORINE

It stimulates the activity of some enzymes, influences carbohydrate metabolism and water retention. Chlorine is a part of certain organic acids. It is applied through muriate of potash used as potash fertilizer. Chlorine is added to soil in considerable quantities each year through rain water. Its addition through fertilizer and in other ways helps to prevent the deficiency of chlorine under field conditions.

10.5.3 *Beneficial elements*

1. ALUMINIUM

It is freely available in acid soils and tea plants absorb large quantities of aluminium. Soil pH is a major factor that controls the uptake of aluminium from soil into the tea plant (Dong *et.al.*, 1999). Chenery (1955) reported only a small quantity of this element is essential for tea. Aluminium influences the assimilation of phosphate and potash but high levels of aluminium decrease uptake of potash. The same can be corrected by regular potash application (Sivasubramanian and Talibudeen, 1971). They also found a significant linear relationship between aluminium and phosphorus in the plant. In a sand culture experiment Ishigaki (1984) in Japan found accumulation of aluminium in roots and old leaves.

Aluminium stimulated the absorption and utilization of phosphorus, nitrogen and potassium. In presence of aluminium, strong absorption of calcium has been observed by Konishi *et. al.* (1985). Aluminium remarkably alleviated excessive toxicity of magnesium (Konishi, 1992). In plant tissues, aluminium is recognised to bind directly and toxically to DNA, calmodulin and other cell constituents (Haug and Caldwell, 1985).

2. SILICON

The deficiency of this element has not yet been reported. Silicon is not classified traditionally as a nutrient but considered important for tea at high yield levels. Application of 250-300 kg ha^{-1} of silicate is suggested once in two to four years in the very high yielding tea fields (above 4000 kg ha^{-1}). The interval may be reduced to one or two years in the case of exceptionally high yield tea (Ranganathan and Natesan, 1987).

10.6 Organic fertilizers

Organic fertilizers have been time-tested materials for improving the fertility and productivity of the soil. They also have a corrective effect on the adverse soil conditions caused by the continuous and excessive use of inorganic fertilizers. Apart from promoting soil aggregation leading to better water holding capacity in coarse textured soils and drainage in heavy soils, organic manures cause favourable changes in soil reaction and enrich the nutrient status of the soil. Humus, derived from the decomposition of organic manures, has chelating properties as well as nutrient buffering capacity. By virtue of these properties, humus increases the availability of both the added and native nutrients. The fixation of soil phosphorus and the leaching losses of secondary and micronutrients will be comparatively less. Erosion is reduced and the infiltration of water improved. It stimulates root growth. The nutrient contents of the organic fertilizers are comparatively low (Table 5.9).

10.6.1 *Green manure*

Green manure is a cheap alternative to the use of fertilizer nitrogen. The process also makes a positive contribution to the maintenance of soil organic matter content at a satisfactory level. Green manuring is a low cost and effective technology in minimising investment cost of fertilizers and safeguarding the productive capacity of the soil without any impoverishment

Table 10.12 : Effect of mulch on soil acidity and leaf nutrient content (%).

	Without mulch	With mulch
Soil pH	4.00	4.30
Nitrogen	2.96	3.69
Phosphorus	0.28	0.23
Potassium	1.00	1.51
Magnesium	0.08	0.12

Source: Grice (1984)

(Ramkrishnan and Ranganathan, 1993). Trees, shrubs, cover crops, legumes, grasses, weeds, ferns etc provide green manure, an expensive source of organic fertilizer to built up or maintain soil organic matter and fertility. Several locally available weeds could also be used as green manure. The importing of green manure is not commonly practiced in the tea estates. The usefulness of mulching is dealt in chapter 5. In north-east India it has been calculated that 5000 kg of vegetative mulch can provide 2.5 cm thick surface cover per hectare of young tea (Barbora, 1995b). In south India, Swaminathan (1992) reported that effective mulching of young tea fields could be achieved with 15 to 25 tons of plant materials per hectare. Though the nutrient content of mulch materials may vary but its application will contribute to overall improvement of physical and chemical conditions of soil. It also shows positive effect on soil acidity and leaf nutrient content and yield (Grice, 1984; Barbora, 1995b) (Table 10.12 and 10.13).

10.6.2 *Manure from field plants*

Leaf-fall from tea plants, shade trees and leguminous plants planted as temporary shade provide some green manure. Leaves and stems (trimmed frequently) of the trees planted as windbreaks also form green manure. It

Table 10.13 : Effect of mulch on yield of young tea (kg ha^{-1} of processed tea).

	1963/64	1964/65	1965/66
No mulch	155	347	826
With mulch	360	600	1259

Source: Grice (1984)

Table 10.14 : Nutrients in prunings in south India.

	Organic matter (tonnes ha^{-1})	Major nutrients (kg ha^{-1})		
		N	P	K
Leaves	7.2	252	30	72
Twigs	3.6	85	10	21
Secondary stems	4.2	44	13	13
Primary stems	9.6	101	28	28

Source: Ranganathan (1973)

has been calculated that the annual deposition of organic matter from leaves and stem of tea were 2088 and 2664 tonnes ha^{-1} $year^{-1}$ and Grevillea robusta were 7487 and 2125 tonnes ha^{-1} $year^{-1}$ respectively (Goodchild, 1960).

The distribution of nutrients among the various parts of the prunings and different methods of pruning were measured by Ranganathan (1973) and Barbora (1995b) in south and north-east India respectively (Table 11.14 and 10.15). The large quantity of nutrients recycled through the soil to the plants has a significant effect on the soil and on the nutrition of the plants. Removal of the prunings from the field takes a large quantity of nutrients out of plant-soil system. In Malawi, removal of prunings in the cycle after a down prune produced a 4 % reduction in the crop yield (Grice and Malenga, 1987). Continuous removal of prunings for eight years reduced the crop yield upto 19% in north-east India (Dutta and Sharma, 1966).

Table 10.15 : Nutrients and dry weight of pruning litters under different methods of pruning in north east India.

Pruning/ Skiffing	Dry weight (kg ha^{-1})	Major nutrients (kg ha^{-1}) in pruning litters		
		N	P	K
LP	7960	159	40	119
DS	3708	74	19	56
MS	1035	21	5	16

Source: Barbora (1995b)

10.6.3 *Manure from other organic materials*

The application of tea waste (53% organic carbon) rapidly produced a substantial increase in the growth rate of young plants. But application of excessive quantities, however, toxic symptoms (Krishnapillai, 1981). The other waste such as neem, castor, mohua, groundnut oil cakes are of common use in agriculture. Use of neem oil cakes has some additional advantages besides providing nutrients viz. nitrogen (4.15%), phosphate (0.9%) and potash (1.2%). It has some antifeedant and repellant action on soil borne pests. It also retards nitrification and thereby reduces the loss of soil nitrogen. Ranganathan (1977) reported increase in long term yield by application of neem cake at low levels but at higher rates there was depression of yield. It has also been observed that other concentrated manure like rape seed oil cake had slightly greater stimulatory effect than sulphate of ammonia (Watanabe and Ikegaya, 1985).

10.7 Diagnosis of fertilizer requirements of tea by soil analysis

The importance of soil analysis for diagonising the fertilizer requirements of tea is well recognized. Though only a minute fraction of soil is used for analysis whereas a large mass of soil and sub-soil is used by the plants still soil testing gives a measure of the availability of nutrients to crops. This apart when nutrients are applied as fertilizer they may be used by the plant, fixed in the soil or leached away. In a situation like this leaf analysis may only show the absence or presence of the nutrient. On the other hand, soil analysis could give some indication on the fate applied fertilizer (Owuor and Wanyoko, 1983). Prior to undertaking planting of tea, representative soil from the area should be analysed for fertility status. This practice is particularly important, if the area is outside the tea growing tract. If the growth of the plants is satisfactory the recommendations based on initial analysis may be followed during the first few years.

10.7.1 *Method of collecting soil samples*

For the purpose of analysis it is necessary to take some precautions at the time of collecting soil samples from the tea estate. The sample collected should be truely representative of the area sampled. Soil samples should not be collected from the field within 90 days of the application of mixed (N, P & K) fertilizers and within 30 days of application of nitrogenous fertilizers. If there are marked differences in the nature of soil, lie of the land, manurial history, age and growth of the plants, separate samples should be collected from different areas. While collecting soil samples areas like roadside, labour line, cattle shed, below shade trees, composite pit, swamps, vacant patches and eroded areas should be avoided. The field should be divided into units across slope and never along the slope. The size of the unit may be taken as 1 to 5 acres depending on the variations to be considered for sampling purposes as indicated above. After removing the surface litre and mulch cut a thin vertical section of soil from the top to depth of 25 cm using sharp-edged chisel. Samples can be taken easily in tea field by means of two-inch screw auger. The auger is screwed to a depth of nine inches, withdrawn by a straight pull and the adhering soil rubbed off into a container. Pool all the samples of 0-22.5 cm (0-9 inches) depth from the different pits and mix well. Gohain *et. al.* (1999) reported that atleast sixteen random borings from one hectare area is necessary in order to make a composite and representative sample to estimate pH, organic carbon, phosphorus, potassium and sulphur. If the size of the composite sample is large, reduce by quatering. For this purpose, spread the well-mixed soil into a thin-layed square on polythene sheet or brown paper. Divide the square into four equal squares and discard the soil in the diagonally opposite squares. Repeat the process until about 500 grams sample of soil is obtained. Prepare composite sample from 22.5-45 cm (9-18 inches) in similar manner. Dry the samples under shade and pack them in clean cloth or polythene bags and never in manure contaminated gunny bags. Label each sample as mentioned below and should be tied to the outside of the bag. Write label with pencil or ball pen and never in ink. Label should not be put inside the packet/bag along with soil.

Soil information label

1. Name of the tea estate; 2. Name of the division / section; 3. Depth of sampling; 4. Date of sampling; 5. Age of tea; 6. Jat of tea; 7. Pruning cycle adopted; 8. Yield of tea per hectare; 9. Manuring system followed; 10. Parameters to be tested; 11. Object of testing etc; 12. Check the following:

Texture of the soil:	sandy, loam, sandy loam, clay loam
Slope:	level, steep, moderate
Drainage:	satisfactory, moderate, water-logged

13. Any other useful information

10.7.2 *Foliar analysis as a diagnostic tool*

Despite soil, mineral analysis of leaves can also be useful tool for assessing the nutrient status of the bush. The advantage of leaf analysis shows that the quantity of nutrients getting from the soil at the time of sampling. The Tea Research Foundation of Kenya and Tea Research Foundation (central Africa) in Malawi, now provides a diagnostic and advisory service based on leaf analysis (Othieno, 1988; Grice, 1990). The uppermost mature leaf in Kenya (Othieno, 1988) and third leaf in Malawi (Willson, 1970) are recommended for sampling. Figure 2 indicates the leaf to be

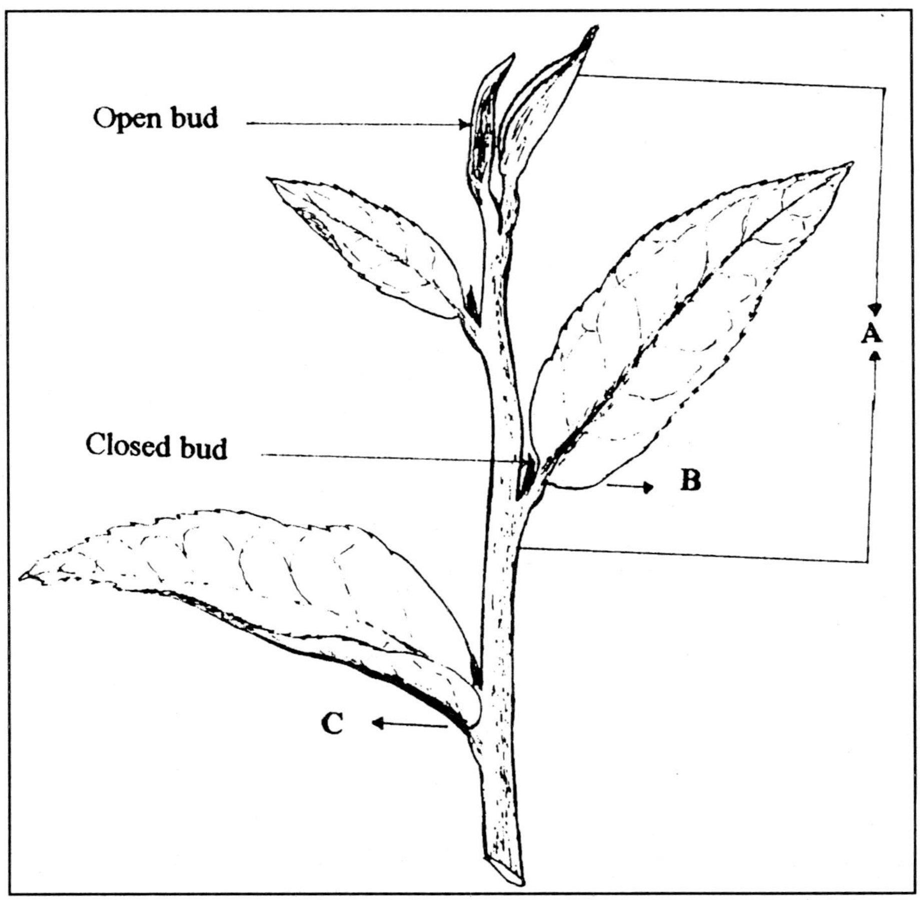

Fig. 10.2: Active tea shoot showing leaves recommended for foliar analysis by different authors. A. Three leaf flush (Filippova, 1979); B. Third leaf (Willson, 1970; Barua, 1989); C. Mother leaf (Hasselo, 1965; Wickremasinghe and Krishnapillai, 1986), Uppermost mature leaf (Othieno, 1988).

sampled for chemical analysis. Grice (1990) further suggested that sample leaf taken off a flushing shoot from every bush in the two intersecting lines. It will be necessary to collect samples on the day of, or on the day before plucking to meet the third leaf requirement. The highest cropping months of January to February under Malawi conditions were considered to be the best time for drawing leaf samples. The minimum sample size should be of 100 leaves. On a bright, sunny day, samples can be sun dried but it would be best to dry the samples for 4 hours at a temperature between 80-100 ^{0}C

The critical levels of different nutrients established for third leaf of an active shoot are listed in Table 10.16 and have been used satisfactorily in different countries. Further, the critical levels of N, P, K, Mg and Zn for uppermost mature leaf set by TRF, Kenya are lower (Table 10.17) than those used by K. C. Willson.

Table 10.16 : Critical levels of nutrients in the third leaf of a tea shoot.

	N %	P %	K %	Ca %	Mg %	Na %	S %	Fe ppm	Mn ppm	Zn ppm	Cu ppm	B ppm	Mo ppm	Cl %	Al ppm
Deficient															
	3.00	0.35	1.60	0.05	0.05			60	50	20	10	8			
Subnormal															
	4.00	0.40	2.00	0.10	0.10		0.05	100	100	25	15	12			
Normal															
	5.00	0.50	3.00	0.35	0.30	0.10	0.50	500	5000	50	30	100		0.10	
Excess															

Source: Bonheure and Willson (1992)

In Sri Lanka, the mother leaf i.e. the leaf from whose axil a pluckable shoot has developed, has been shown to be the most suitable (Hasselo, 1965; Wickremasinghe and Krishnapillai, 1986). The leaf nutrient status has not been critically evaluated against fertilizer application and crop yields.

In a study, carried out at Tocklai in north-east India, it has been observed that third leaf of growing shoots to be the most sensitive indicator in determining N, K, Mg and Zn status of the tea bushes. However, the second leaf in comparison with third leaf appeared to be little better in determining P status. In case of Ca, each one of first, second and third leaf was equally satisfactory. Though wide seasonal variation in leaf nutrient content was

Table 10.17 : Critical levels of nutrients in the uppermost mature leaf set by the Tea Research Foundation of Kenya.

Nutrient	Deficient (%)	Borderline (%)	Adequate (%)
Nitrogen	Below 3.0	3.0 – 3.5	Above 3.5
Phosphorus	Below 0.15	0.15 – 0.17	Above 0.17
Potassium	Below 1.20	1.20 – 1.50	Above 1.50
Magnesium	Below 0.10		
Zinc	Below 10 ppm		

Source: Owuor and Wanyoko (1983)

recorded at Tocklai still best time for drawing leaf samples in north-east India observed to be in highest cropping period of July and September (Anonymous, 1980).

In the Krasnodar region of former USSR, the optimal N, P and K levels for three-leaved flushes has been established by Filippova (1979).

In south India, Venkataramani (1987) reported that they did not have the basic information to use leaf analysis as diagnostic tool. Keeping the different difficulties for selection of leaf in mind, third leaf of a growing shoot is the best representation of the nutrient status of a tea bush. The leaf samples should be drawn during highest cropping period. If errors for sampling could be minimised, the top most maintenance leaf would be even better (Barua, 1989).

Plant analysis for predicting fertilizer needs is not very useful since chemical composition of tissues is little affected except in extreme case of deficiency for long duration. However, composition with the average figures shown in the Table 10.2 and 10.3 can prove useful in modifying current fertilizer schedules where the content of any particular nutrient(s) shows a decreasing trend over the years, and to correct imbalances. For example, the ratio of K in growing shoots to that in older leaves gives an indication of withdrawal of that nutrient from mature foliage as a result of inadequate K supply, this was needed extensively in east Africa for correcting K imbalance in the past (Ranganathan, 1992). In south India, the K content of mature leaves, and their ash has been used for identifying K deficiency. Healthy tissues contain > 0.8% K and their ash > 8% K. This has also been used in

Indonesia, former USSR and Taiwan.

For P, the sensitive tissues are, flush in Indonesia and south India, mature foliage in Sri Lanka, and second, third and fourth internodes in south India. P varies between 0.22 % and 0.50% in these tissues and decline with age.

If the Na content of leaf tissues exceeds 100 ppm, crop yield reduce. Soil containing more than 60 ppm exchangeable Na than large quantities of K should be used to reduce the uptake of Na.

The Ca content of the maintenance foliage should be kept between 0.6% to 0.8%. When it exceeds 0.8%, crop depression occurs and when it falls below 0.6% toxicity of heavy metal (particularly Mn) observed (Ranganathan, 1992).

The critical levels of different nutrients listed in Table 10.16 have been satisfactorily used in different countries.

It can be seen from the foregoing account that though some progress has been made but lot of research on standardising the procedure for sampling of levels for nutrient analysis remains to be done.

10.8 Forms of fertilizers

Knowledge of common fertilizer used in manuring tea field is useful in order to apply them judiciously. The nutrient content of some common fertilizer are presented in Table 10.18. The fertilizer used in tea gardens are described below.

10.8.1 *Nitrogenous fertilizers*

These fertilizers are classified into four groups according to manner in which their nitrogen is combined with other elements. These are nitrate, ammonia and ammonium salts, chemical compounds containing nitrogen in the amide form, and plant and animal by-products.

1. AMMONIUM SULPHATE

This has been the traditional nitrogenous fertilizer in tea. It is a white crystalline salt and non-hygroscopic in nature. It is easy to handle and it stores well under dry conditions. But in comparison with other nitrogen sources such as ammonium chloride and urea, the unit cost of nitrogen is the highest. It is relatively slow acting as the ammonium has to be converted

Table 10.18 : Important commercial fertilizers and their nutrient contents.

Fertilizers	Nutrient content (%)
Ammonium sulphate	N – 20.6, S – 24
Ammonium sulphate nitrate	N – 26, S – 15
Urea	N – 46
Calcium ammonium nitrate (CAN)	N – 25, Ca – 8
Ammonium phosphate (20 : 20)	N – 20, P_2O_5 - 20
Ammonium phosphate (16 : 20)	N – 16, P_2O_5 - 20
Ammonium phosphate sulphate	N – 20, P_2O_5 – 20, S – 15
Di-ammonium phosphate	N – 18, P_2O_5 – 46
Single super phosphate	P_2O_5 – 16, S – 12, Ca – 21
Dicalcium phosphate	P_2O_5 – 42 (acid soluble), S – 1, Ca – 23
Triple super phosphate	P_2O_5 – 43, Ca – 1
Rock phosphate (Mussoorie, India)	P_2O_5 - 20 to 24 (acid soluble), S – 2.3
Rock phosphate (Udaipur, India)	P_2O_5 - 30 to 40,
Muriate of potash (Potassium chloride)	K_2O – 40 to 60, Cl - 47
Potassium sulphate	K_2O – 50, S – 16 to 22
Zinc sulphate	Zn – 23 to 35, S – 11 to 18, Cl – 0.5 to 2.5
Zinc chelate	Zn – 9 to 14
Magnesium sulphate	Mg - 9.88, S – 13
Magnesite (80 % $MgCO_3$)	MgO – 38
Manganese sulphate	Mn – 26 to 28, S – 15 to 17
Molybdic acid	MO – 40
Borax	B – 11.3
Boric acid	B – 17
Calcium sulphate (Gypsum)	S – 13 to 15, Ca – 16 to 19
Phosphogypsum	S – 14, Ca – 21 to 22
Iron pyrites	S – 22 to 24, Fe – 20 to 22
Elemental sulphur	S – 85 to 100,
Dolomite (magnesium limestone)	CaO – 34, MgO – 18 to 21

N = Nitrogen; P_2O_5 = Phosphate; K_2O = Potash; S = Sulphur; Zn = Zinc; Mg = Magnesium; Mn = Manganese; Mo = Molybdenum; B = Boron and Ca = Calcium

to nitrate before absorption by the tea plant. This fertilizers is recommended for application (1) during pre-monsoon or hot weather when soil moisture is not optimum, (2) when soil pH is high, (3) in fertilizer mixture, and (4) once in three years to combat possible sulphur deficiency. During rainy season, it sometimes forms lumps and should be powdered before application to the field. Its long continued use increases acidity of the soil by removing calcium therefore it is advisable to use this fertilizer along with bulk organic manures to safeguard against ill effects.

2. UREA

It is a fertilizer of white granules with very hygroscopic in nature and has a poor storing quality. However, the unit cost of nitrogen is the cheapest. During monsoon when significant moisture is available in the soil the urea should be applied. Since it is highly soluble in water therefore subject to rapid leaching. When applied to the soil, its nitrogen is rapidly changed into ammonia. If application is followed by dry spell then there will be a volatilisation loss. Urea contains a chemical called biuret which is supposed to produce toxic effects on tea if its proportion in the urea exceeds 1.5 to 2.0 percent.

3. CALCIUM AMMONIUM NITRATE (CAN)

It is a fine free-flowing grey or light brown granular fertilizer and hygroscopic in nature. Keeping quality under storage is poor. The unit cost of nitrogen is in between ammonium sulphate and urea. It is commercially prepared from ammonium nitrate and ground lime stone. Since nitrate form is liable for leaching therefore application during monsoon is not advisable. It can be applied in pre-winter or winter. It is almost neutral and can be safely applied even to acidic soils.

4. AMMONIUM CHLORIDE

It is a white crystalline compound, non-hygroscopic and keeps well on storage but does not contain sulphur like ammonium sulphate. It could be used as substitute for urea. The unit cost of nitrogen is higher than urea and cheaper than ammonium sulphate. But it continuous application will increase acidity of the soil since it has a powerful acidifying effect. It is not commonly used in tea plantation. High application levels may lead adverse effect on tea for its chloride content.

10.8.2 *PHOSPHATIC FERTILIZERS*

These fertilizers are divided as natural phosphates, treated or processed phosphates, and by-product phosphates and chemical phosphates.

1. ROCK PHOSPHATE

It occurs as natural deposit of rock. It is a brown powder with non-hygroscopic in nature and insoluble in water but soluble in acidic soil solution.

In general, no rock phosphate, as such, is used as fertilizer. Adequate rainfall and long duration in the soil enhance the response.

2. MUSSOORIE PHOSPHATE

It is a black powder and considered to be a low grade phosphatic fertilizer. The other properties are same as rock phosphate.

3. SUPERPHOSPHATE

It is the most widely used phosphatic fertilizer and obtained by treating rock phosphate with sulphuric or phosphoric acid or both. It is distinctly acidic in reaction. In neutral or slightly acidic soils it remains in an immediately available condition for plant growth. But in acid soils where aluminium and iron content are high, reversion and fixation are rapid, superphosphate tends to become fixed in irreversible and irrecoverable forms resulting most of the applied phosphate becomes wholly ineffective. This fertilizer is manufactured in three grades: single superphosphate, dicalcium phosphate and triple superphosphate. In north-east India, it is applied as a basal dose to young tea as well as while planting. In case of single superphosphate the sulphur content is useful where it is required but the calcium can make problem. But in case of triple superphosphate contains high concentration of easily available water soluble phosphate and contains no sulphur as single superphosphate and having only about 1 % calcium. Dicalcium phosphate, since it is slightly alkaline its application should be restricted to suitable acidic (high) soils. In acidic soils having pH less than 5.5, liming is needed before application of single superphosphate to increase efficiency. Blending with biogas slurry, FYM etc can also increase its efficiency.

10.8.3 *Potassic fertilizers*

It should be applied only when soils are deficient in potash or to those which respond well to their application.

1. MURIATE OF POTASH

It consists of grey crystalline material and is highly soluble in water. But it is not lost from the soil, since it is absorbed on the colloidal surfaces. The unit cost of potassium in muriate of potash is cheaper than potassium sulphate.

2. POTASSIUM SULPHATE

It is with white crystals soluble in water. Since this fertilizer is imported, the price is very high. It is generally recommended for application in nursery mixture.

10.8.4 *Gypsum*

Gypsum (calcium sulphate) is a white, yellowish or occasionally brown opaque solid material ground to a fine powder form having particle size between 10 and 100 mesh. It is slightly soluble in water. Its hardness is low and it can be easily scratched with a finger nail. The commercial material contains varying amount of impurities including oxides of iron, calcium, aluminium and magnesium. Maximum reactivity of gypsum has been found at 30 mesh size. For best results, it should be mixed thoroughly with the surface soil so that it dissolves and saturates the soil solution relatively quickly. Use of green manure or FYM has been found to enhance the efficiency of gypsum. Its use should be preferred in non-calcarious, high pH soils.

10.8.5 *Iron pyrites*

It is a mineral containing iron and sulphur. The pyrites mineral occurs in various physical forms such as crystalline, massive and powdery varieties. It immediately reacts with air and water to form sulphuric acid and iron sulphate. The powdery variety is the most reactive one. It can be used efficiently on alkaline, moist and well-aerated soils.

10.9 Fertilizer recommendations for different tea growing regions

(A) NORTH-EAST INDIA

Tea seed and clonal nurseries

Now-a-days both seedlings and clonal plants are raised mostly in polythene sleeves. The clonal cuttings are either planted directly on to the sleeves or first planted on the rooting beds and then transferred to the sleeves when they have developed some roots. There is no necessity to apply manure on the rooting beds. 3 kg of single superphosphate may thoroughly be mixed per cubic metre of soil at the time of preparation of soil for filling the polythene sleeves.

Young tea manuring (YTD) containing NPK in the proportion of 2 : 1 : 2 can be used as follows :

Nutrient condition	Nutrient kg ha^{-1} $year^{-1}$	Remarks
N : P_2O_5 : K_2O	10 : 5 : 10	One part of fertilizer is mixed with nine parts of sand or powdered earth.
	or	
	10 : 5 : 15 where potash in soil is low	As the quantity of fertilizer applied is low, it is advised to increase its volume by addition of dry sand or dry powdered earth.

One tea spoonful of the mixture should be applied to each seedling or clonal plant in a ring little away from the collar. In all 6 to 8 fortnightly applications should be made avoiding contact on the leaves of the plant. Foliar spray, where necessary, zinc and / or magnesium sulphate @ 2 kg of either in 200 litre of water may also be applied. Manuring should start when seedlings or clonal plants have attained about two flush of growth.

At planting

4.5 kg of well rotted organic manure, or oil cake at the rate of 150-200 gram per pit or bone meal 100 kg ha-1 plus 30-40 gram single superphosphate per pit should be used. Concentrated organic manure should be thoroughly mixed with the excavated soil. A fresh stock of oil cake should be fermented by exposing to atmosphere with adequate watering to avoid injury to the root of the plant.

Young tea

10 : 5 : 10 mixture of sulphate of ammonia, superphosphate and rock phosphate in 1 : 1 ratio and muriate of potash is recommended during the formative stage but 10 : 5 : 15 is preferable in low K soil (below 100 ppm).

Frequent application in small quantities can be useful in promoting vigorous growth of plants. The manuring should be started as soon as plants are well established in the field (Table 10.19).

Table 10.19 : Nitrogen level for young tea (as 10:5:10 or 10:5:15 NPK mixture) during formative stage in north-east India.

Year from	N (kg ha^{-1}) planting	YTD mixture (kg ha^{-1})	Method of application
0	20 – 40	200 – 400	Ring in 4 splits
+ 1	80 – 100	800 – 1000	Ring in 4 splits
+ 2	100 – 120	1000 – 1200	Ring in 4 splits
+ 3	120 – 140	1200 – 1400	Ring in 4 splits
+ 4	140 – 150	1400 – 1500	Strip – 2 – applications
+ 5	140 – 150	1400 – 1500	Strip – 2 – applications

Source: Chakravartee and Sinha (1994)

Mature tea

North-east India is the largest tea growing tract in the world where fertilizer trials began as early as 1892. However, earliest statistically laid out field experiment on seedling tea was launched in 1930 followed by many others time to time. In 1973-74 number of NPK trials were undertaken with doses of N varying from 0-200 kg ha^{-1}, P from 0-100 kg ha^{-1} and K from 0-150 kg ha^{-1}. It was concluded that there was a positive response to high nitrogen and a significant interaction between N, P and K at 2 : 1 : 2 levels. But later it was observed that compared to the good response to high nitrogen in the past; nitrogen level beyond 165 kg ha^{-1} is not beneficial in the long run. Similarly positive response of phosphate upto 50 kg ha^{-1} was observed. In respect of potash it has been found that there was clearly a positive response upto 150 kg of K_2O per hectare when combined with appropriate levels of exchangeable K was in the range of 150 to 200 ppm. Potash requirement is to be, however, estimated based on existing status of potash in the soil. The current recommendation on nutrition of mature rea is given in Table 10.20.

The agroclimatic conditions of Darjeeling tea estates are different from those of plains estates. Due to low yield potential, slow decomposition process of organic matter and effect on overall quality of tea of the tea estates are in view, the highest level of N should not exceed 110-120 kg N ha^{-1}. Table 10.21 gives the recommendations for NPK nutrition of mature tea in Darjeeling tea estates. De (1997) reported that many tea gardens in Darjeeling instead of applying P_2O_5 every year as recommended the same

Table 10.20 : NPK manuring schedule for mature tea in plain districts of north-east India.

Yield range Processed tea (kg ha^{-1}) (Cycle average)	N (kg ha^{-1})	P_2O_5 (kg ha^{-1})	K_2O kg ha^{-1} at available K level		
			<60 ppm	60 – 100 ppm	>100 ppm
Up to 1500	up to 90	20	up to 90	up to 70	up to 50
1500 – 2000	90 – 110	20 – 30	90 – 110	70 – 80	50 – 70
2000 – 2500	110 – 140	30 – 50	110 – 140	80 – 120	70 – 100
2500 – 3000	140 – 165	50	140 – 165	120 – 140	100 – 120

Source: Tea Research Association – Tocklai Experimental Station, Jorhat, Assam (India)-Encyclopaedia on tea culture.

is applied only in the year of prune. He opined that regular application of phosphate is very much needed along with nitrogen and potash. Sarker (1989) opined that nitrogen may be increased further by 6 to 7 kg per 100 kg of made tea over 1400 kg ha^{-1}. If no extra yield is obtained after increasing the dose of fertilizer over a cycle, the other limiting factors should be looked into. If the yield is static or declining, the causes must be investigated and constraints removed before fertilizer levels are raised. This is most essential.

(B) NORTH WEST INDIA

Nursery

NPK mixture in the ratio of 2 : 2 : 1 by mixing ammonia sulphate (6 parts), super phosphate (7.5 parts) and muriate of potash (1 part) is recommended. The mixture should be mixed with sand (4 times). 250 gm

Table 10.21 : NPK manuring schedule for mature tea in Darjeeling hills.

Yield range Processed tea (kg ha^{-1}) (Cycle average)	N (kg ha^{-1})	P_2O_5 (kg ha^{-1})	K_2O kg ha^{-1} at available K level		
			<60 ppm	60 – 100 ppm	> 100 ppm
Up to 600	up to 60	20	up to 60	up to 50	up to 50
600 – 1000	60 - 90	20	60 – 90	50 – 90	35 – 50
1000 – 1400	90 – 120	20	90 – 120	70 – 100	50 – 70

Source: Tea Research Association – Tocklai Experimental Station, Jorhat, Assam (India) – Encyclopaedia on tea culture.

of the mixture to an area about 10 square metres when the plants are of 4-5 age should be broadcasted at fortnightly intervals. 8-12 applications are sufficient (Anonymous, 1986).

Mature tea

On the basis of marginal returns, through an analysis of yield data from 20 years of field experiments, a level of 90 : 90 : 90 :: N : P_2O_5 : K_2O was recommended (Jain, 1985). Sulphur @ 60 kg ha^{-1} has since been added for young tea. Highest yield of 3900 kg made tea ha^{-1} has been recorded with this fertilizer application. However, there is a tendency amongst tea growers of Kangra, Himachal Pradesh to increase nitrogen levels as an attempt to copy nutrition levels adopted elsewhere in the country.

(C) SOUTH INDIA

Nursery

After the tent / roof is removed, the plants may be sorted and restaked. 30 grams of the mixture of ammonium sulphate 8 parts by weight, ammonium phosphate 35 parts by weight, potassium sulphate 15 parts by weight, magnesium sulphate 15 parts by weight and zinc sulphate 3 parts by weight may be dissolved in 10 litres of water and applied at the rate of 10 litres of every 4 m^2 of bed at fortnightly intervals. The application can be done after watering with the help of a rose can or a hand sprayer.

At planting

No fertilizer or organic manure is applied except in areas where soil pH is above 5.0. Aluminium sulphate is applied at rate of 100 gram per pit, if the soil pH is between 5.5 and 6.0 and 200 gram per pit if the pH is above 6. Application of regular fertilizer begins two months later.

Young and mature tea

The fertilizer recommendation is given in Table 10.22.

(D) SRI LANKA

In Sri Lanka, the amount of N applied varies prorata with yield. P is applied as rock phosphate. K is used @ 60-80 kg ha^{-1} K_2O for yields of upto 1600 kg ha^{-1} or 90-180 kg ha^{-1} K_2O for higher yields. Zinc applied as zinc sulphate @ 22 kg ha^{-1} $year^{-1}$ as four foliar sprays annually. The fertilizer

Table 10.22 : Manuring schedule for young and mature tea in south India.

Stage	Nutrients	Nutrient (kg ha^{-1} *Year*$^{-1}$)	Remarks
1. Young tea			P in one blanket application placed at 15-25 cm depth N. K and Mg for soils with pH below 4.5 and N and K for soils with pH above 4.5 in 4 dressings in the first year and 4-6 dressings (broadcast) thereafter; foliar application of zinc sulphate at 6-10 kg/ha in 3-5 sprays as required.
- year 1	$N:P_2O_5:K_20:MgO$	90+180:270:30	
- year 2	$N:P_2O_5:K_20:MgO$	90+240:360:40	
- year 3	$N:P_2O_5:K_20:MgO$	90+300:450:50	
- year 4	$N:P_2O_5:K_20:MgO$	90+300:300:50	
2. Pruned fields			P in one blanket application placed at 15-25 cm depth. N and K in 4-6 dressing in lighter forms of pruning and 2-3 in harder forms of pruning. 3-5 foliar sprays of in total 6-10 kg/ha zinc sulphate.
- Rejuvenation pruning	$P_2O_5 + N:K_20$	90+180:360	
- Hard pruning	$P_2O_5 + N:K_20$	90+210:315	
- Medium, light, cut-across and skiffing	$P_2O_5 + N:K_20$	90+240:360	
3. Mature tea (low yielding less than 3000 kg/ha)		15 - 25 cm	P once in two years placed at depth. N and K broadcast in 4-6 dressings; 6 kg/ha zinc sulphate in 3 foliar sprays every year or 11.25 kg/ha in 5 foliar sprays in alternate years.
- 1000 kg/ha	$P_2O_5 + N:K_20$	90+120:100	
- 2000 kg/ha	$P_2O_5 + N:K_20$	90+120:170	
- 3000 kg/ha	$P_2O_5 + N:K_20$	90+250:190	
4. Mature tea high yielding (above 3000 kg/ha)			P once every year placed at 15-25 cm depth. N and K broadcast in 4-6 dressings. 6-10 kg/ha zinc sulphate every year in 3-5 foliar sprays. For fields yielding below 4000 kg/ha, 20-25 kg/ha/year Mg as magnesite and 75–125 kg/ha as magnesium silicate; for fields yielding above 4000 kg/ha 40-50 kg/ha Mg as magnesite and 125-150 kg/ha Mg as magnesium silicate.
- 3000 kg/ha	$P_2O_5 + N:K_20$	70+250:250	
- 4000 kg/ha	$P_2O_5 + N:K_20$	70+350:350	
- 5000 kg/ha	$P_2O_5 + N:K_20$	90+450:450	

- Liming to pH 4.5 – 5.0 recommended in year of pruning at rate based on soil pH. Yield level, precipitation and length of pruning cycle.

- N rates adjusted to yield of marketable tea. 10 kg N per 100 kg yield up to yield of 2000 kg/ha (minimum 120 kg/ha N), plus 5 kg N/100 kg additional yield over 2000 kg/ha up to 3000 kg/ha, and 10 kg N/100 kg additional yield thereafter.

Source: Ranganathan (1992)

recommendations for nursery and young tea are given in Table 10.23 and mature tea in Table 10.24.

Table 10.23 : Fertilizer recommendations for nursery and young tea in Sri Lanka.

Stage	Nutrient	Rate of application
Nursery	T65 mixture (Sulphate of Ammonia 15 parts, Di-ammonium phosphate 20 parts and Muriate of potash 15 parts)	50 kg mixture per 10000 plants/year
Young tea - 1st year	T260 mixture (SA 100 parts, rock phosphate 5 parts, MOP 25 parts and kieserite 25 parts)	1200 kg ha^{-1}
- 2nd year	-do-	-do-
- 3rd year	T750 mixture (SA 500 parts, RP 100 parts, MOP 100 parts and kieserite 50 parts)	1750 kg ha^{-1}
- 4th year	-do-	-do-

Source: Wickremasinghe and Krishnapillai (1986)

Table 10.24 : Fertilizer recommendations for mature tea in Sri Lanka.

Yield level Kg ha^{-1}	Kg ha^{-1} per 100 kg marketable tea Kg ha^{-1}		
	N*	P_2O_5	K_2O**
Below 1600	10	60	60 – 80
Below 1600	10	60	90- 120

* N rates adjusted to yield to marketable tea. 10 kg N per 100 kg yield up to yield of 2000 Kg ha^{-1} (minimum 120 kg ha^{-1} N), plus 5 kg N per 100 kg additional yield over 2000 kg ha^{-1} up to 3000 kg ha^{-1}, and 10 kg N per 100 kg additional yield thereafter.

**N : K_2O ratio of 2 : 1 recently introduced.

Source: Kemmler (1986)

(E) KENYA

A combination of N, P, K and S :: 25 : 5 : 5 : 5 is normally used for tea of all ages. The rate of application increase from 20 kg N ha^{-1} $year^{-1}$ in the first year to 57 kg ha^{-1} $year^{-1}$ in fourth year thence to between 100 and 250 kg N ha^{-1} $year^{-1}$ depending on yield of mature tea. P is applied as superphosphate in Africa. Aerial application of NPK is favoured. In Africa, the amount of N applied is based on the age of the bushes and on the region and response obtained in field trials. Zinc as oxide is applied mainly as foliar application in African countries. In Kenya, zinc oxide is applied twice a year at 3 kg ha^{-1} as a foliar spray. Zinc sulphate @ 10kg ha^{-1} as an alternative is also used (Otheino, 1989).

(F) MALAWI

Mature seedling tea

Table 10.25 shows the recommended rates for N, P_2O_5 and K_2O according to age and quantity per hectare of urea, triple superphosphate (TSP) and muriate of potash (MOP) for all seedling tea. Nitrogen can be applied between mid-September to mid-December and phosphate at any time. Potash can be applied best in October.

Table 10.25 : Fertilizer recommendations for mature seedling tea in Malawi.

Age years	Nutrient kg ha^{-1}			Fertilizer kg ha^{-1}		
	N	P_2O_5	K_2O	Urea	TSP	MOP
11 to 25	180	36	60	391	80	100
16 to 20	190	38	63	413	84	105
21 to 25	200	40	67	435	89	112
26 to 30	210	42	70	457	93	117
31 to 35	215	43	72	467	96	120
36 to 40	220	44	73	478	98	122
41 to 45	225	45	75	489	100	125
46 to 50	230	46	77	500	102	128
51 to 55	235	47	78	511	104	130
56 to 60	240	48	80	522	107	133
Over 60	245	49	82	533	109	137

Ratio of N : P_2O_5 : K_2O has been changed from 5 : 1 : 1 to 15 : 3 : 5.
TSP = Triple super phosphate.
Source: Grice et. al. (1988)

Clonal tea

The recommended rates N, P_2O_5, K_2O and S (sulphur) on the basis of age of clonal tea is shown in Table 10.26. Tea planted in virgin soil may require additional sulphur which are as follows :

1st year 57 kg S in 240 kg sulphate of ammonia

2nd year 87 kg S in 360 kg SO_4

3rd year 114 kg S in 480 kg SO_4

Application of four foliar sprays as zinc oxide @ 1.25 kg ha^{-1} in a year during February, March, August and November is recommended.

Table 10.26 : Fertilizer recommendations in Malawi for clonal tea planted into lands previously under tea.

Age – years	Nutrient kg ha^{-1}			
	N	P_2O_5	K_2O	S
+				
0 year at planting	-	50 – 80	-	32 – 50
3 – weeks after planting	55 – 90	-	-	65 – 100
1st year*	50	10	17	-
2nd year*	75	15	25	-
3rd year*	100	20	33	-
4th year	150	30	50	-
5th year	200	40	67	-
6th year	250	50	83	57
7th year to down prune	275 – 300	55 – 60	92 – 100	-
First two years after down prune	195 – 220	55 – 60	92 - 100	-

\+ Range is based on using 30 g/plant of SSP to supply P_2O_5 and S at planting and 30 g/plant of SOA 3–4 weeks later to supply N and additional S.

* Tea planted on virgin soil require additional S consequently SOA should be used instead of urea in the first three years to provide the required amount of N and Sulphur. Year 1 = 57 kg S in 240 kg SOA; Year 2 = 87 kg S in 360 kg SOA; Year 3 = 114 kg S in 480 Kg SOA.

(G) INDONESIA

In Indonesia the N : K_2O ratio varies with soil group and growth stage (young tea, pruned tea and mature tea). The current fertilizer recommendation for young and mature tea is given in Table 10.27.

(H) USSR (CIS)

Most widely followed recommendation (de Geus, 1973) is given below:

Nitrogen @ 200-250 kg ha^{-1} $year^{-1}$ split over two applications during March and June.

Phosphate is applied once every 2-4 years.

Potash is given as muriate of potash annually in April.

Mulching with peat moss at the intervals of four years is considered

Table 10.27: Fertilizer recommendations for young and mature tea in Indonesia.

Age of bushes/ yield level/ type of soil	Nutrients/ rate of application	Remarks
1. Young tea		
- year 1	8g N + 2g P_2O_5 + 2g K_2O/plant	MgO at 2 g/plant in podzols only
- year 2	12g N + 3g P_2O_5 + 3g K_2O/plant	
- year 3	12g N + 2g P_2O_5 + 4g K_2O/plant	
2. Mature tea in plucking		
- up to 1200 kg/ha	N at 12% pro rata on yield	N :P_2O_5 :K_2O ratio 4:1:2 in the year of pruning; from second
- 1200 to 2500 kg/ha	N at 10% pro rata on yield	N at 10% pro rata on yield year onwards 5:1:2 in wet months and 5:1:3 in dry months.
- above 2500 kg/ha	N at 8% pro rata on yield	
= andosols	120kg/ha N, 20 kg/ha P_2O_5, 40 kg/ha K_2O	
= podzols	120 kg/ha N, 30 kg/ha P_2O_5, 30 kg/ha K_2O, 30 kg/ha MgO	For every 1000 kg of made tea
= latosols	120 kg/ha N, 30 kg/ha P_2O_5, 24–72 kg/ha K_2O	

- N:P_2O_5 :K_2O = 6:1:2 in andosols; 5:1:1 – 5:1:3 with a mean of 5:1:2 in latosols.
- N:P_2O_5 :K_2O:MgO ratio = 5:1:1:1 – 6:1.5:1.5:1.5 in podzols.
- Placement of fertilizers at 10 cm depth between tea rows.
- Leaf tissue analysis is used to monitor the health of bushes under the recommended manuring schedules.

Source: Darmawijaya (1985)

to improve heavy soils. The annual application of fertilizer in the Krasnodar region are 300 kg N, 150 kg P_2O_5 and 100 kg K_2O per hectare. Despite this, additional phosphate @ 500 - 600 kg ha^{-1} is also given every four years.

(I) JAPAN

Very high levels of fertilizer usage are characteristics of Japanese tea industry in order to obtain a yield of 18 tonnes green leaves per hectare per year with high amino acid content. NPK fertilizer dressings of 800, 210, 360 kg ha^{-1} are applied annually in six splits i.e. early spring dressing, top-up fertilizer before first plucking, summer dressing after first crop, summer dressing after second crop, autumn dressing after third crop and late autumn dressing after a light autumn pruning (skiffing). Table 10.28 shows the fertilizer recommendations in Japan. Heavy fertilizer dose acidifies the soil and for compensation, application of lime magnesium (1- 2 tonnes ha^{-1}) in autumn is recommended (Ishigaki, 1978; Hoshina, 1985 and Takeo, 1992). Jain (1996) reported as fertilizer is sprayed in 10 cm wide bands between 2 meter spaced rows, its intensity in fertilized strips works out to 100000 kg ammonium sulphate per hectare. As a consequence of excessive feeding, clonal teas in Japan decline quickly with loss of absorbing root system, necessitating a rejuvenation pruning and root aeration at 15 years and uprooting at 30 years.

Organic fertilizers such as rapeseed oil cake and fish manure are abundantly used together with inorganic fertilizer (Okano, 1993). Application of 500 kg rapeseed meal or 1000 kg ha^{-1} chicken droppings is recommended for nursery soil.

Table 10.28 : Standard application rates for fertilizers in Japan.

Fertilizer	Nutrient (kg ha^{-1})		
	N	P_2O_5	K_2O
Spring dressing	160	70	120
Top-up 80			
Summer (1)	160		
(2)	160		
(3)	120	70	120
Autumn	120	70	120
Total	800	210	360

Source: Takeo (1992)

(J) VIETNAM

Recommendations for three regions of Vietnam are given in Table 10.29.

Table 10.29 Fertilizer recommendation for different regions of Vietnam.

Region	Nutrient (kg ha^{-1})		
	N	P_2O_5	K_2O
Tonkin Central	40	36	75
Haut Donnai	40	Nil	75
Pleiku	36	36	18

Source: Guinard (1953)

(K) TURKEY, IRAN AND TAIWAN

112 kg N and150 kg K_2O; 80 kg N accompanied by low application rates of P_2O_5 and K_2O; 150 kg N, 20-40 kg P_2O_5 and 40-60 K_2O ha^{-1} $year^{-1}$ are recommended for Turkey, Iran and Taiwan respectively (de Geus, 1973).

(L) MALAYSIA

Mature tea in the Cameron Highlands is given NPKMg @ 18 : 11 : 5: 2.5 at 153 to 270 kg N ha^{-1} $year^{-1}$ into two splits. In the low land, the same mixture is applied at the rate from 180 to 396 kg N ha^{-1} $year^{-1}$ (Sivaram, 1982).

(M) REPUBLIC OF CONGO (KIVU)

A combination of N, P, K, Mg was recommended except on drained swamps. The proportion varied in the range of 13-15, 4-6, 8-10, 3-5 for different soils. Swamps required only P and K. In the nurseries, combined fertilizer @ 50 g m^{-2} and 100 - 125 g m^{-2} for 6 to 8 months and 18 to 24 months old plants respectively was applied . For young teas, 250 kg, 350 to 450 kg and 500 to 600 kg ha^{-1} during 1st, 2nd and 3rd year respectively was applied. The rate of application was related to the yield for mature tea. A minimum 350 kg fertilizer per hectare for yield upto 500 kg ha^{-1} $year^{-1}$ was given. The fertilizer rate was increased by 125 kg ha^{-1} for every increase of yield of 250 kg ha^{-1} upto 1000 kg ha^{-1}, thereafter by 150 kg ha^{-1} (Flemal, 1960).

(N) BANGLADESH

Recommendations for young and mature tea are given in Table 10.30.

Table 10.30 : Manuring schedule in Bangladesh.

Age of bushes	Ratio of Nutrients ($N: P_2O_5 : K_2O$)	Rates kg ha^{-1} *$year^{-1}$*
Young tea	1 : 1 : 1	9, 18, 36, 54 and 72 kg ha^{-1} N in the first five Years from planting
Mature tea		
- Skiffing	1 : 0 : 0	90 kg ha^{-1} N plus foliar application of 1 – 2% urea spray at 15 days interval
- Light pruning	4 : 1 : 3	90 kg ha^{-1} N, 22 kg ha^{-1} $P_2O_{5,}$ 68 kg ha^{-1} K_2O
- Medium pruning	1 : 1 : 1	90 kg ha^{-1} N, 90 kg ha^{-1} $P_2O_{5,}$ 90 kg ha^{-1} K_2O

(O) QUEENSLAND, AUSTRALIA

When young plants are 10 cm high, NPK 10 : 3 : 3 @ 10 g m^{-2} (1 kg N ha^{-1}) should be applied. It should be repeated at the intervals of 2 months at a rate of which increases by 2 g m^{-2} at each application. NPKMg 33 : 7: 14 : 4 kg ha^{-1} at each of four applications applied for tea of 1-2 years old. For 2-3 years old tea, 40 kg N, 7 kg P_2O_5 and 14 kg K_2O at each four applications recommended. Three or four split applications of 182 kg N, 15 kg P_2O_5 and 38 kg K_2O ha^{-1} $year^{-1}$ is recommended for mature tea (Hobman, 1980).

10.10 Nutrient availability

It depends on the inherent fertility of the soil. Plants use nutrient in inorganic form. Therefore, organic sources of plant nutrient are first to be transformed to inorganic ones whereas inorganic sources are directly utilised. So the nutrients should be present in an available form before plant can use them. Since absorption of nutrients present in the soil solution takes place through the roots therefore absorbing roots must be well developed and conditions for absorption mainly aeration and moisture should be favourable. The availability of nutrients in the soil is also governed by soil properties and it differs from soil to soil. Tea grows well when the pH ranges between 4.5 and 5.5, a range at which availability of most nutrients is not favourable. Increasing soil acidity increases the amount of aluminium, iron and manganese in the soil solution and it is these elements, in ionic forms and especially Al_3^+ are less severe in presence of phosphate. But raising levels of P to ameliorate Al_3^+ effects can be costly although the placement, or banding, of P application can help. Availability of nutrients like N, Ca, Mg,

P, K and S decreases when pH goes down below 6. Despite this, cation exchange capacity of the root system also influences the uptake of nutrients. A deep root system has also a greater possibility in exploiting leached down nutrients.

Plant roots do not absorb fertilizer granules as they come out of the bag or dung particles as they are in a manure heap. Plant absorb nutrients in specific ionic forms (Table 10.31) which either a fertilizer furnishes when it dissolves in soil water or various chemicals and biological agents in the soil convert fertilizer currency into local currency acceptable to roots. K^+ and NH_4^+ ions are absorbed rapidly while Ca^{++} and Mg^{++} are absorbed at a much slower rate. The anions NO_3^- and Cl^- are absorbed faster than SO_4^{--} and $H_2PO_4^{--}$. This process of active nutrient absorption requires expenditure of energy by the plant. This energy is derived mainly from the oxidation of carbohydrates and other plant substances in respiration (Rahman, 1988).

Table 10.31 : Essential mineral nutrients, their forms absorbed by plants, concentration in dry matter and major functions.

Nutrient (symbol)	Forms absorbed	Content in dry matter	Major functions in the plant
Macronutrients			
Nitrogen (N)	NH_4^+, NO_3^-	1%-5%	Growth, protein production
Phosphorus (P)	$H_2PO^-_4$, HPO^-_4	0.1%-0.4%	Energy compounds, root development ripening
Potassium (K)	K^+	1%-5%	Crop quality, water regulation, stress tolerance
Sulphur (S)	SO_4^-	0.1%-0.4%	Protein and oil production
Calcium (Ca)	Ca^{++}	0.2%-1%	Cell structure and cell division
Magnesium (Mg)	Mg^{++}	0.1%-0.4%	Part of chlorophyll, energy transfer
Micronutrients			
Boron (B)	H_3BO_3	6–60 ppm	Cell wall development, seed setting
Chlorine (Cl)	Cl^-	0.2–2%	Tissue hydration, photosynthesis
Copper (Cu)	Cu^{++}	5–20 ppm	Enzyme activity, chlorophyll formation
Iron (Fe)	Fe^{++}	50–250 ppm	Respiration, chlorophyll formation
Manganese (Mn)	Mn^{++}	20–500 ppm	Enzyme activation, photosynthesis
Molybdenum (Mo)	MoO^-_4	<1 ppm	Nitrate reduction, N-fixation
Zinc (Zn)	Zn^{++}	25–150 ppm	Enzyme activity, protein synthesis

Source: Tandon (1994)

10.11 Time and method of application of fertilizers

The efficiency of fertilizer usage depends on the time and method of applications. Unless fertilizers are applied at the right time and in the right way, the fertilizers will largely be wasted. The choice of the time of application depends mainly on the moisture conditions of the soil. The principle to be followed is that there should be sufficient moisture in the soil at the time of application of the fertilizers and at the same time the chances of loss by leaching should be minimum. On dry soil, nutrients remain unavoidable to plants and volatilization loss of nitrogen as ammonia increases. Since the tea growing regions receive both south-west and north-east monsoons, application of fertilizer should logically be made during April-May (Pre-monsoon) and or during September-October (post-monsoon) periods. In areas where tea shoot grows throughout the year, fertilizers can be applied to unpruned tea at any time of the year provided the above mentioned conditions are fulfilled. The same can not be applicable in areas where growth is seasonal. A bush in the pruned year remaining virtually leafless for few months of the year and can not be expected to utilise as much nutrients as it does in the following years when it is fully foliated. In north-east India shoot production slows down from mid-October and shoot growth completely ceases by December and in Darjeeling shoot growth ceases by November. Under these conditions, fertilizers applied after June could not make the same impact on yield as that of earlier application although some residual effect of July-August applied fertilizer was observed in the following year. Application during April-May should invariably be made after the first few pre-monsoon showers and before the outbreak of the regular monsoon.

The method of application of fertilizers depends on the stage of growth of the plants and principle to be followed is that fertilizers should be applied in zones where the root of the tea plants are active. Uniform distribution of fertilizers over a large tea area, though it is difficult, should be the aim. However, to overcome the difficulty some inert materials like sand or powdered soil could be mixed with the fertilizers. To avoid wastage of fertilizers as well as to pay proper attention to the young plants in the early stage of growth, application of measured quantity of fertilizer in an annual strip around each plant is always advisable. Sometimes scorching of stem of young plants are observed when organic manures like ammonium sulphate and potassium chloride are applied too close to the collar of the plants. Fertilizers should be applied by the ring method, leaving a circle of

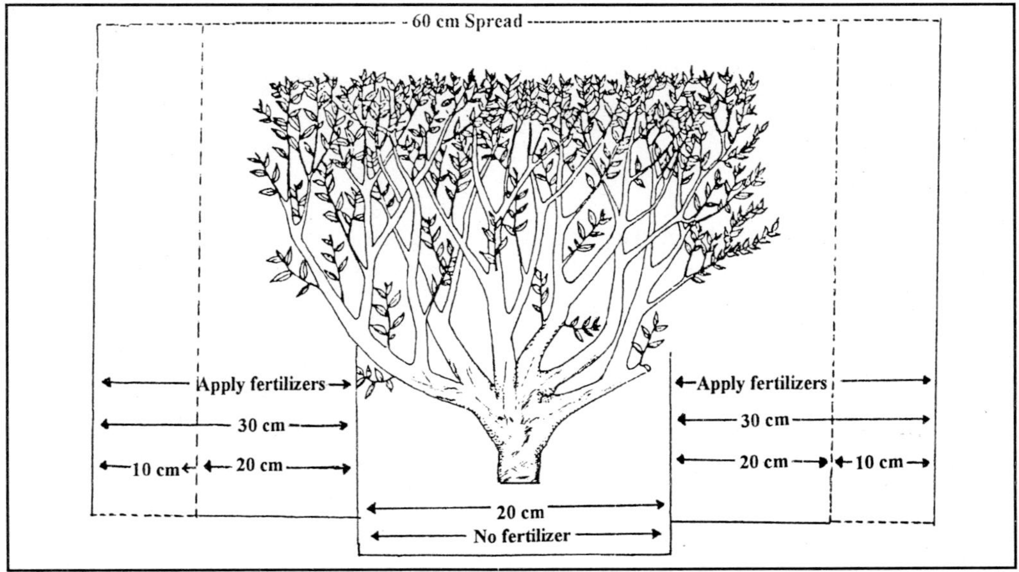

Fig. 10.3: Calculation of manure zone assuming the estimated spread of a tea plant to be 60 cm.

10 cm diameter around the collar unmanured. Since roots of the plant spread over a wider area from the top, the annual strip where fertilizers are evenly spread should be extended beyond the perimeter of the branches (Fig. 10.3). At each successive application, the width of the fertilizer applied band extends simultaneously with the spread of branch system of the plant and when one fertilizer strip making contact with the neighbouring strips then the fertilizer can be applied over the entire area as in the case of mature tea.

10.12 Symptoms of nutrient deficiency

When plants are grown under inadequate supply of any one or more of the essential nutrients can in no case give satisfactory yield unless the deficiency is corrected. The deficiency symptoms of various nutrients are describe below :

NITROGEN

The yellowing of leaves (chlorosis) and finally pale are the symptoms which are usually prominent in lower leaves of current season's growth.

Leaves become slender, small and fall off. The growth of plant slows down, internodes shorten and production of crop decline. Sometimes the leaves, petioles, veins and stems of the nitrogen deficient plants show purplish pigmentation due to synthesis of anthocyanin.

PHOSPHORUS

Deficient leaves become dull in appearance and lose the glossiness. Sometimes leaves are mottled with yellow spots showing bluish green with purplish tints. Premature leaf fall and stunted growth are also noticed.

POTASH

Mottled chlorosis followed by necrotic shrivelling of the margins of the leaves (Fig. 10.4), premature leaf fall and development of twiggy wood are the symptoms of deficiency. The bark of the stem becomes silvery white. Sometimes there will be profuse branching of shoots at the top and little or no branching from the lower older stems. The recovery from pruning is slow and irregular.

Fig. 10.4: Leaves showing potassium deficiency syndrome [Photograph from Bonheure and Willson, (1992)].

Fig.10.5: Leaves showing magnesium deficiency syndrome [Photograph from Arunachalam, (1995)].

MAGNESIUM

Yellowing of leaves near mid-rib spreads gradually between the main lateral veins (Fig. 10.5). The older leaves are affected earlier and fall off easily than the younger leaves.

SULPHUR

Deficiency of sulphur results in chlorosis. The yellowing first appears in young leaves. In some cases most of the leaves show chlorosis at the same time (Fig. 10.6). The internodal distance is reduced, leaves scorch and fall off and stems die back from the tips. The edge of the leaves curl up and tip turn brown under severe deficiency. The most striking symptom is the 'net veining' in the younger leaves where the leaf blade takes a striking yellow colour and the veins down to the finest branching stand out prominently dark green. At the advanced stage the bush is totally defoliated.

CALCIUM

Deficiency symptoms first appear in younger leaves and apices of

Fig. 10.6 : China-type bush with sulphur deficiency.

plants. They do not grow to normal size, distort and crack, turn yellow and curl backwards. The maintenance leaves are affected slowly but defoliation of leaves and simultaneous death of bush may occur eventually.

ZINC

Leaves of zinc deficient plants turn yellow to dull olive green in colour showing the characteristics chlorosis and twist to a sickle shape and may develop wavy edge. Stunted growth, bunchy appearance of shoots, reduction in the distance of the internodes and leaf size are also sometimes noticed.

COPPER

Copper of deficient plants have slightly darker leaves and very difficult to judge the difference in the field. Copper deficient leaves increase fermentation time than usual time in the factory and leaves become dark brown rather than bright orange.

BORON

Tip burning and die back from the top are characteristics symptoms. Leaves may become narrow, twisted, leathery, crinckled and dark green in

Fig. 10.7: Tea shoot with boron deficiency [Photograph from UPASI TRI Newsletter, 9(1): 1999].

colour (Fig. 10.7). There is cork formation on the petiole and on veins at the ventral surface of the leaves.

MANGANESE

The symptoms are common in young tea. Chlorotic and necrotic spots in interveinal regions of leaves have been seen in the deficient plants. Sometimes leaves develop a mottling of red-brown spots and number of green veins reduced.

IRON

Deficiency results in interveinal chlorosis in younger leaves and whose veins remain green. When the deficiency is severe, scorching of the leaf tips and margins may occur.

MOLYBDENUM

A bright green or pale orange colour of the leaves due to interveinal mottling, cupping of old leaves, necrosis at leaf tips and margins and drying out of leaves showing thin papery effect are the symptoms of deficiency.

10.13 Application of fertilizers and quality of tea

The applications of increasing levels of fertilizer led to increased yields. But the question arose as to whether the quality of processed tea changed. The studies undertaken so far are rather subjective and they show much variation in the quality of tea manured differentially. However, this controversy has not yet ended even today. The problem has been investigated by different scientists in various countries either by chemical analysis of volatile flavoury compounds or, subjectively, through organoleptic evaluation by the tea tasters. The effects of fertilizers reported until now are small and overshadowed by the effects of variety of plant, plucking standards and cultural operations (Bonheure and Willson, 1992). Quality of made tea broadly depends on the composition of leaves, which in turn is influenced by nutrition. The accumulation of non-essential elements entering the system depends on soil pH and impurities in agro-chemicals used. Changing manurial practices affect the quality such as tannins, caffeine and organoleptic compounds which have direct impact on marketability (Ranganathan and Natesan, 1987a).

Any increase above the optimum level of nitrogen causes an increase in fibre content (Cloughley, 1979) which affected the quality of the processed tea in Malawi (Hilton *et. al.*, 1973). It is evident from the accumulated findings that heavy application of nitrogenous fertilizer produces deterioration in quality (overall value) of processed tea though the terms like heavy or high etc do not clearly express any exact information. Harrison (1940) observed slight decrease in the value of tea from any kind of nitrogenous manuring if the same is sufficient enough to cause an increase in the yield of tea. The drop in value is proportional to yield. On the contrary, significant depression of theaflavins was observed by Hilton (1971) when the dose of nitrogen was raised from 45 to 224 kg ha^{-1}. Guseinov (1973) obtained better leaf quality at 90 kg than 180 kg or 360 kg ha^{-1} of nitrogen application. In former USSR, Utnelishvili (1974) reported application of nitrogen above 200 kg ha^{-1} in the soils had adverse effect on quality of leaf.

Rahman *et. al.* (1978) observed in north-east India that application of nitrogen at 200 kg ha^{-1} in three splits actually improved the strength of the liquor. In Darjeeling, the application of nitrogen at 120 kg ha^{-1} in two splits retain the Darjeeling quality. It was also observed that if any body likes to go for more splits then it will have less nitrogen per split and quality may not be affected. But single application of nitrogen more than 100 kg ha^{-1} may affect quality (S. K. Sarker, TRA, Jorhat, personal communication). In Darjeeling, Saha and Bisen (1995) observed that application of nitrogen @ 120 kg ha^{-1} with different doses of phosphatic (20 to 45 kg ha^{-1}) and potassic fertilizers (80 to 120 kg ha^{-1}) (basal) improved the strength of the liquor but deteriorated the quality. Malenga (1987) observed that the quality of processed tea from clones and hybrid seedlings in Malawi was highest at a nitrogen application level of 100 kg N ha^{-1}. Any increase of nitrogen above this level the quality will be deteriorated. He further observed that the application of nitrogen in splits than once in a year did not improve quality and value of processed tea (Malenga, 1996b). Whitehead and Temple (1990) in Malawi stated "it is now generally accepted that application of over 200 kg N/ha/year, which increase yields unfortunately have a deleterious effect on the quality of made tea". They observed that flushes of the clones PC1 and SFS 204 were pale in colour but with an increasing N the leaf appeared to darken. This suggested that chlorophyll could be making the factors responsible for quality. It was also observed that processed teas from clone MT12 which received more than 200 kg N/ha/annum had reportedly less TF and 30% more residual chlorophyll than tea obtained from unfertilized bushes. The processed tea having a high residual chlorophyll was reported to be responsible for 'dull' or 'muddy' colour. They also opined that some clones produce good teas despite high chlorophyll content in their green leaf, this was because of the green leaf cells contain an enzyme population which could degrade the chlorophyll during processing. In Kenya, Willson and Choudhury (1968) observed a reduction in quality when nitrogen level was increased and an improvement in quality with increase in phosphate application. But no change was observed with application of potash. However, the effect of nitrogen was comparatively lesser when ammonium nitrate fertilizer was used.

Different sources of nitrogen like ammonium chloride, urea and calcium ammonium nitrate did not have any adverse effect on liquor characters of tea, nor did they help in improving the general quality of tea. Comparison of forms of nitrogen showed a little improvement in CTC tea from calcium ammonium nitrate and foliar application of urea (Rahman *et al.*, 1978). Compared to urea, application of sulphate of ammonia resulted in better cup characters with higher valuation of tea (Anonymous, 1959). Tasters scores on colour of infused leaf, colour of liquor, quality, briskness and valuation were higher in teas from sulphate of ammonia treated than in urea treated plots. Owuor *et. al.* (1987a) reported that the application of higher rates nitrogenous fertilizers increases the overall levels of aroma compounds, but decreases the flavour index of black tea in Kenya. Owuor and Odhiambo (1994) further reported that high rates of nitrogen application reduced the theaflavin content and brightness of black tea liquor. The caffeine content of the black tea increased with increasing rates of nitrogen application. Phosphate fertilizers do not affect the composition of fatty acid in tea (Bajaj *et. al.*, 1984) are therefore unlikely to cause changes in the levels of aroma compounds produced from these acids (Robinson and Owuor, 1992). Foliar application of urea in quality period (first and second flush) is detrimental to Darjeeling quality but it can be applied in non-quality periods especially under adverse climatic condition (Tiwari, 1978). Aono *et. al.* (1982) found in Japan that foliar application of urea improved quality, chlorophyll and amino acids are also increased. Takeo (1979) reported that the application of high rates of sulphate of ammonia improved the quality of green tea in Japan. He also noticed the increment of amino acid content in green leaves. In Japan, Watanabe (1995) observed that the application of nitrogen fertilizer 5-3 weeks before first plucking did not affect the content of theanine (gamma-glutamyl ethylamide), one of the major constituents of quality of green tea, and thereafter the content decreased.

In Kenya, Owuor *et. al.* (1995) reported that theaflavins, thearubigins, flavour index and sensory evaluation decreased with nitrogen rates (80, 160, 240 and 320 kg ha^{-1} $year^{-1}$) when he experimented with clones viz. 6/8, 31/8 and S15/10. Owuor *et. al.* (1987b) reported the effects of two NPK fertilizers at various levels on the content of 32 volatile flavour compounds (VFC), theaflavins, thearubigins and caffeine in black tea.

In north-east India, Gogoi (1993) reported that the catechin, polyphenol oxidase activity, total oxygen uptake and caffeine content of the tea shoots were highest at 50 and / or 75 kg P_2O_5 ha^{-1}, with the quality of tea also being the best at these doses. These doses were also optimum for lipid biosynthesis in tea shoots, with 50 kg P_2O_5 ha^{-1} being optimal for phospholipids and 75 kg ha^{-1} being optimal for glycolipids. Phosphorus is essential for flavour and also offsets the adverse effects of nitrogen on flavour and quality (Subramanium, 1995). Owuor *et. al.* (1998) in western Kenya highlands found no quality variations arising from application of triple superphosphate or rock phosphate. They further observed that the increasing P-fertilizer levels improved the levels of thearubigins but reduced the levels of black tea brightness leading to a slight decline in sensory qualities. There are different opinions on the effect of potash on quality of tea. It seems potash may affect quality indirectly by increasing the fibre content of tea shoots (Barua, 1989). Potassium indeed partially compensates for the quality deterioration caused by very high rates of nitrogen (de Beaucrops, 1978). The application of high levels of potassium reduce the amounts of theaflavins and thearubigins in leaves which could lead to a reduction in quality (Devchoudhury and Bajaj, 1988). The effects of application of potash on the composition of aroma complex have not been established (Robinson and Owuor, 1992). Recently it has been observed that the application of increasing rates of potash in western Kenya highlands did not have any effect on plain black tea brightness levels leading to a slight decline in sensory qualities. But when applied together with K, increasing N fertilizer reduced the plain black tea quality and sensory evaluations; K rates had no significant effect. Where the application of K was staggered with the time of N fertilizer application or the time of pruning, increasing rates of K had no effect on quality, but staggering the application sporadically resulted in variations in theaflavins, total colour and sensory evaluations (Owuor *et. al.*, 1998).

Zyrin and Barabadze (1987) found that the levels of tannins and extractables fell as nitrogen doses increased. Potash and phosphate were unchanged throughout.

Sugianto (1985) reported that the quality was reduced when nitrogen and potassium applied separately and it was improved by phosphate. But all three nutrients when applied together the quality was reduced.

Krishnapillai and Ediriweera (1986) reported that increasing levels of nitrogen and potassium nutrients increased the content of the leaf chlorophyll which could increase the blackness rather quality of tea.

Ranganathan and Natesan (1987a) found that nitrogen impairs flavour, decreases tannins and water extractives and has varied influence on caffeine content, quality and valuation parameters. Phosphorus is essential for flavour and potassium has a positive influence on overall quality of tea. Quality of tea decreased when nitrogen alone is used at large quantities, but in combination with potassium at the recommended dose it has little influence in practical terms of quality of tea. Very high content of manganese in leaves leads to poor quality of processed tea. Aluminium has a positive effect on brightness and reduces of liquors of black teas; but it adversely affects the liquors and valuation of green teas. Zinc increases the crude fibre content and this is more pronounced in third leaf than in two leaves and a bud. Zinc along with copper, cobalt and molybdenum increases the contents of tannins. Chakravartee (1996) evaluated the response of quality of Assam CTC tea to sulphur application in a few tea estates of Assam and West Bengal. Two forms of sulphur viz. $MgSO_4$ and phosphogypsum were applied at 0, 20, 40 and 60 kg ha^{-1}. In these experiments, the basal NPK dose varied between 100-165 kg N ha^{-1}, 20-50 kg P_2O_5 ha^{-1} and 60-125 kg K_2O ha^{-1} depending on the yield and potash status of the soil. Quality parameters like theaflavin, thearubigins and flavonol glycosides increased by the application of sulphur at 20 to 40 kg ha^{-1}. Taster's scores in respect of strength and quality of liquor, valuations of processed tea were higher in teas receiving sulphur fertilization. Barbora (1995a) also observed that thearubigins and total colour was highest with 20 kg S ha^{-1} followed by 40 kg S ha^{-1}. Taster's scores on tea quality showed the significant effects of sulphur application. The quality of tea was significantly superior with the application of 20 kg S ha^{-1}. However, the increase in sulphur application to 40 to 60 kg ha-1 did not show any additional advantage in quality over control. HPLC analysis of black tea liquor revealed that sulphur application increased the theaflavin group of compounds, such as theaflavin, theaflavin-3-gallate, rutin and quercetin glycosides and flavonal glycosides, which may have contributed to improved tea quality (Barbora, 1995a). It was reported that the contents of various amino acids were increased by the addition of sulphur and the chromatographic analysis of aromatic compounds of processed tea indicated that sulphur treatments enhanced flavour components such as linalool, geraniol and B-lonone and fragrance quality of processed tea (Xun and Jianyun, 1994).

Zinc and other micronutrients do not normally affect quality of marketable tea if they are applied judiciously on the basis of plant requirement (Banerjee, 1993).

Ranganathan and Natesan (1987a) concluded that application of nutrients will have no effect on quality provided that yield and applied nutrients are so balanced that the amounts of the nutrient in leaves remain within critical limits. Owuor *et.al.* (1987a) have recommended the application of optimum fertilizer for most economic yield and best quality tea.

References

Ahsan, Q. 1994. Some physico-chemical properties of Bangladesh tea soils: their influence on tea plant. *Ph.D Thesis*, University of Dhaka, p.164.

Anonymous, 1959. *Annual Scientific Report 1959*, Tocklai Experimental Station, ITA, pp.104.

Anomymous, 1965. *Annual Report 1965*, Tea Research Institute of East Africa, Kericho, Kenya, pp.10.

Anonymous, 1966. *Annual Report 1966*, Tocklai Experimental Station, Tea Research Association, Jorhat, Assam, 45-66.

Anonymous, 1970. *Annual Report 1969-70*, Tocklai Experimental Station, Tea Research Association, Jorhat, Assam, 29-35.

Anonymous, 1980. *Annual Report 1979-80*, Tocklai Experimental Station, Tea ResearchAssociation, Jorhat, Assam, 25-36.

Anonymous, 1982. *Annual Report 1981-82*, Tocklai Experimental Station, Tea Research Association, Jorhat, Assam, pp.31.

Anonymous, 1984. *Annual Report 1983-84*, Tocklai Experimental Station, Tea Research Association, Jorhat, Assam, pp.30.

Anonymous, 1986. *Package and practice for tea cultivation in Himachal Pradesh*. Himachal Pradesh Krishi Vishva Vidyaalaya, Dept. of Pomology & Fruit Technology, Palampur, 1-14.

Anonymous, 1992. *Tocklai Experimental Station 1911 to 1991*. Tea Research Association, Jorhat, Assam, pp.232.

Anonymous, 1995. *Annual Scientific Report 1994-95*. Darjeeling Tea Research Centre, Tea Board, Kurseong, pp.25.

Anonymous, 1998. Monthly Report of Tea Research Association submitted to Tea Board, India, May 1998, 1-3.

Aoki, S. 1987. Varietal differences and effects of nitrogen fertilization on decline of photosynthetic rate in overwintered tea leaves. *Japanese J. Crop Sci.*, **56(2):** 252-256.

Aono, H., Tanaka, S. and Yanase, Y. 1982. Effect of foliar application of fertilizer on the growth and quality of new shoot of tea plant. *Study of Tea*, **3:** 23-32.

Bagyaraj, D. J. and Ravindra Padmavathi, T. 1993. Micorrhiza. In *Organics in Soil Health and Crop Production* (ed. P. K. Thampan), Peekay Tree Crops Development Foundation, Cochin, Kerela, India, 185-200.

Bajaj, K. L., Dev Choudhury, M. N. and Sukhija, P. S. 1984. Effect of phosphorus on fatty acids of tea leaves and on the quality of teas. *J. Food Sci., Tech.*, **21:** 392-394.

Banerjee, B. 1993. *Tea Production and processing*. Oxford & IBH publishing Co., Pvt. Ltd., New Delhi, p.336.

Banerjee, B. 1987. The role of elemental sulphur in crop biomass production in tea : an exploratory approach. *Proc. Int. Symp. Elemental sulphur in Agriculture*. **1:** 339-344.

Barbora, A. C. 1995a. Sulphur management for tea in north eastern India. *Sulphur in Agriculture*, 19: 9-15.

Barbora, A. C. 1995b. Organic matter management in tea estates. *Two and a Bud*, **42(2):** 8-12.

Barua, D. N. 1989. *Science and practice in tea culture.* Tea Research Association, Calcutta, p.493.

Barua, D. N. and Dutta, K. N. 1961. Root growth of cultivated tea in presence of shade trees and nitrogenous manure. *Emp.J. Expl. Agric.*, **29:** 287-298.

De Beaucorps, S. 1978. The potassium requirements of crops harvested green, with special reference to grassland. In *Potassium Research - Review and Trend. Proc. Cong. Int. Potash Inst.*, **11:** 325-342.

Bhattacharyay, N. G. and Dey, S. K. 1978. Phosphate in tea soils and changes on long-term manuring. *Two Bud*, **25:** 32-36.

Bhavanandan, V. P. and Manipura, W. B. 1969. Fertilizer responses of tea in up-country districts. *Tea Q.*, **40:** 135-140.

Bonheure, D. and Willson, K. C. 1992. Mineral nutrition and fertilizers. In *Tea Cultivation to Consumption* (eds. K. C. Willson and M. N. Clifford), Chapman & Hall, pp. 269-321.

Borpujari, N. 1982. Zinc status of tea soils and zinc nutrition of tea. *Ph.D. Thesis*, Assam Agricultural University, Assam, India.

Brady, N. C. 1984. *The nature and properties of soil.* Macmillan publishing Co., Inc. USA, p.750.

Brandram-Jones, R. E. 1963. Gypsum experiment. *Annual Report 1963.* Tea Research Institute of East Africa, Kericho, Kenya, 19-20.

Canvin, D. T. and Atkins, C. A. 1974. Nitrate, nitrite and ammonia assimilation by leaves. *Planta*, **116:** 207.

Chakravartee, J. 1996. Effect of sulphur on yield and quality of tea in N. E. India. *Two and a Bud*, **43(1):** 18-22.

Chakravartee, J. and Sinha, M. P. 1994. Soil management in tea for north east India. In *Field Management in Tea* (ed. J. Chakravartee), Tea Research Association, Tocklai Experimental Station, Jorhat, India, 34-40.

Chamuah, G. S. 1986. Calcium and potassium uptake by young plants in relation to root morphology. *Tropical Agriculture* (Trinidad), **64(3):** 177-180.

Chen, Zongmao, 1987. A perspective review on the tea science. *Proc. Int. Tea-Quality-Human-Health Symp.*, Nov. 4-9, 1987, Hangzhon, China, 3-6.

Chen, Zongmao, 1995. Tea in 21st century. *Proc. Int. Tea-Quality-Human-Health Symp.*, Nov. 7-10, 1995, Shanghai, China, 3-6

Chenery, E. M. 1955. A preliminary study of aluminium and the tea bush. *Plant and Soil*, 6(2): 174-200.

Cloughley, J. B. 1979. The effect of N-application on chemical composition, yields, quality and total values of clonal teas. *Annual Report 1978*, Tea Research Foundation of Central Africa, 117-126.

Clowes, M. St. J. and Mitini-Nkhoma, S. P. 1987. Copper deficiency on young clonal tea growing close to Mulanje mountain. *Tea Research Foundation (Central Africa) Quarterly Newsletter*, **88:** 13-15.

Dev Choudhury, M. N. and Bajaj, K. L. 1988. Effect of potassium on nitrogen and carbohydrate contents of the tea leaves and quality of made teas. *J. Food Sci. Technol.*, **25(2):** 105-107.

Dev Choudhury, M. N., Sarmah, T. C., Thakur, K. C. and Goswami, M. R. 1988. Nitrogen metabolism in tea plant: effect of different levels of nitrogen and potassium on some biochemical components. *Proc. 30th Tocklai Conference*, Tea Research Association, Jorhat, Assam, 47-56.

Dev Choudhury, M. N., Thakur, K. C., Goswami, M. R. and Rabindranath, S. D. 1983. Amides in tea. *Two and a Bud*, **30:** 7-9.

De, S. K. 1997. Limiting factors for productivity in Darjeeling. *Proc. Seminar Darjeeling Tea Industry*. Sept., 5., 1997, Calcutta, DPA & Tea Board, India, 6-18.

Dey, S. K. 1977. Role of potash on water economy by plant. *Proc. 28th Tocklai Conference*, Tea Research Association, Jorhat, Assam, 64-69.

Dharmawijiya, M. I. 1985. Tea manuring in Indonesia. *UPASI Tea Sci. Dept. Bull.*, **40:** 26-29.

Dong, D., Xie, Z., Du, Y., Liu, C. and Wang, S. 1999. Influence of soil pH on aluminum availability in the soil and aluminum in tea leaves. *Communications in Soil Science and Plant Analysis*, **30(5-6):** 873 - 883.

Dutta, S. K. and Sharma, K. N. 1966. Manurial value of pruning and shade tree litters in tea. *Two Bud*, **13(1):** 17-19.

Eden, T. 1944. Further studies on manurial response : Studies in the yield of tea V. *Emp. J. Expl. Agric.*, **12:** 177

Eden, T. 1952. The nutrition of a tropical crop as exemplified by tea. In *Report of 13th International Horticultural Congress*, 1952, 1138-1145.

Eden, T. 1965. *Tea*. Longmans, p.236.

Ellis, R. T., Grice, W. J. and Malenga, N. E. A. 1983. Response of young tea to nitrogen fertilizer. *Tea Research Foundation (Central Africa) Quarterly Newsletter*, **72:** 5-6.

Evans, H. J. and Sorger, G. J. 1966. Role of mineral elements with emphasis on the univalent cations. *Ann. Rev. Plant Physiol.*, **17:** 46-47.

Filippova, N. A. 1979. (Diagnosis of tea nutrition by means of soil and leaf analysis in mature plantation in the Krasnodar region). *Subtropicheskie Kul'tury*, **5:** 15-22.

Flemal, J. 1960. La fumure du theier au Kivu. *Bulletin Agricole du Congo*, 51(4): 807-832.

de Geus, J. G. 1973. Stimulants-Tea. In *Fertilizer Guide for Tropics and Subtropics, Centre* d'Etude de l'Azote, Zurich, pp.474-494.

Godziashvili, B. A. and Peterburgsky, A. V. 1986. Potassium and magnesium nutrition of tea on the red soils of Georgia. *Potash Review*, Subject 27, 114th Suite, 1-6.

Gogoi, A. K. 1993. Role of phosphorus in nitrogen metabolism in tea bush and quality of made tea. *Ph.D. Thesis*. Dibrugarh University, Assam, India.

Gohain, K. K., Bordoloi, P. K., Barooah, A. K. and Chakravartee, J. 1999. Modified soil sampling procedure. *Proc. JASC Seminar*, January 8 – 9, 1999, Tea Research Association, Tocklai Experimental Station, Jorhat, Assam, India, 12 – 18.

Gokhale, N. G. 1957. Effects of sulphate of ammonia treatment on soil acidity and calcium content. *Two and a Bud*, **42(2):** 4.

Goodchild, N. A. 1960. 4 x 2 x 2 Factorial experiment with N, P_2O_5 and K_2O. In *Annual Report 1960*, Tea Research Institute of East Africa, Kericho, Kenya, pp.21.

Grant, P. M. and Shaxson, T. F. 1970. The effect of ammonium sulphate on the sulphur content of tea garden soils in Malawi. *Tropical Agriculture* (Trinidad), **47(1):** 31-36.

Grice, W. J. 1984. Mulch, an essential prequisite for planting. *Tea Research Foundation (Central Africa) Quarterly Newsletter*, **75:** 16-18.

Grice, W. J. 1990. Collection of leaf samples for foliar analysis. *Tea Research Foundation (Central Africa) Quarterly Newsletter*, **97:** 15-16.

Grice, W. J., Clowes, M. St. J., Malenga, N. E. A. and Mkwaila, B. 1988. Update on fertilizer and foliar nutrient recommendations for tea grown in Malawi. *Tea Research Foundation (Central Africa), Quarterly Newsletter*, **89:** 4-9.

Grice, W. J. and Malenga, N. E. A. 1987. Are your prunings still being stolen? *Tea Resesarch Foundation (Central Africa) Quarterly Newsletter*, **86:** 11-12.

Guinard, A. 1953. La culture du the' en Indochine. *Arch. Rech. Agron. Past.*, Vietnam, 20.

Guseinov, R. K. 1973. The effect of long term fertilizer application on soil fertility and on tea metabolism and productivity (Russian). *Hort. Abs.* 1974, 44 no. 2937. Original not seen.

Hanson, J. B. and Koeppe, D. E. 1975. Mitochondria. In *Iron transport in plant cells and tissues* (eds. D. A. Baker and J. L. Hall), Elsevier, North Holland, Biotredical Press, Amsterdam, NewYork.

Harrison, C. J. 1940. Effect of plucking, pruning and manuring etc on quality of tea. *Proc. 4th Tocklai Conference*, Tea Research Association, Jorhat, Assam, 42-44.

Hasselo, H. N. 1965. The nitrogen, potassium, phosphorus, calcium, magnesium, sodium, manganese, Iron, copper, boron, zinc, molybdenum, and aluminium contents of tea leaves of increasing age. *Tea Q.*, **36:** 122-136.

Hayman, D. S. 1983. The physiology of vesicular-arbuscular endomycorrhizal symbiosis. *Can. J. Bot.*, **61:** 944-963.

Hilton, P. J. 1971. *Biochemistry Report 1970-71*, Tea Research Foundation (Central Africa), pp.103-104.

Hilton, P. J., Palmer-Jones, R. W. and Ellis, R. T. 1973. Effects of season and nitrogen fertilizer upon the flavanol composition and tea making quality of fresh shoots of tea (*Camellia sinensis* L) in central Africa. *J. Sci. Food Agric.*, **24:** 819-826.

Hobman, F. R. 1980. Tea in Queensland. *Queensland Agricultural Journal*, **106(5):** 386-407.

Hoshina, J. 1985. Studies on absorption and utilization of fertilizer nitrogen by tea plants. *Bulletin of the National Research Institute of Tea*, **20:** 1-90.

Hoshina, T., Kozai, S. and Honjo, Y. 1979. Turnover of storage nitrogen in tea plant and absorption of fertilized nitrogen following collar pruning. *Study of tea*, **56:** 66-69.

Huag, A. R. and Caldwell, C. R. 1985. Aluminium toxicity in plants. In *Frontiers of membrane research in agriculture* (ed. J. B. St. Jone), Owman & Allanheld, Totowa, 359-380.

Ikegaya, K. 1975. Studies on the method of application of phosphorus fertilizer to tea plant (Japanese). *Bull. National Inst. Tea*, Japan, **10:** 133-208.

Ishigaki, K. 1978. Studies on the nutritive characters of tea plants. *Bull. National Inst. Tea*, Japan, 40: 152.

Ishigaki, K. 1984. Influences of aluminium and boron on the growth and the content of mineral elements of tea plants grown on the sand culture method. *Study of Tea*, **66:** 33-40.

Jain, N. K. 1983. Future direction of tea research. *Tocklai Occasional Scientific paper*, **12:** 81-89.

Jain, N. K. 1985. Manuring of mature tea in Himachal Pradesh. *Information leaflet*, 1/85, CSIR Complex, Palampur, Himachal Pradesh, India.

Jain, N. K. 1996. Global advances in tea science and technology and the future of tea economy - an overview. *Paper presented to FAO Seminar*, July 8-10. 1996, Beijing, China,. 1-88.

de Jong, P. 1950. The role of nitrogen-phosphorus and potash in plant nutrition. *Planters' Chronicle*, **45:** 530-532.

Kalita, J. N. 1992. Some aspects of chemistry in production of tea in relation to mineral nutrition of boron and manganese under field and green house conditions. *Ph.D Thesis*, Dibrugarh University, Assam, India.

Kalita, J. N. and Mahanta, P. K. 1993. Analysis of mineral composition of some Assam and Darjeeling black teas. *J. Sci. Food Agric.*, **62:** 105-109.

Kanwar, J. S. 1976a. Calcium, magnesium and sulphur. In *Soil fertility theory and practice* (ed. J. S. Kanwar), ICAR, New Delhi, India, 202-228.

Kanwar, J. S. 1976b. Micronutrients. In *Soil fertility theory and practice* (ed. J. S. Kanwar), ICAR, New Delhi, India, 229-300.

Karasuyama, N., Yoneyama, T. and Kobayashi, H. 1985. ^{15}N study on the fate of foliarly applied urea nitrogen in tea plant. *Soil Sci. plant Nutr.*, **31(1):** 123 -131.

Kemmler, G. 1986. Nitrogen and potassium nitrification of tea in India, Sri Lanka and Central Africa. *Proc. Int. Conf. Management and fertilization of upland soils in tropics and subtropics*. Nanjing, People's Republic of China, 161-171.

Konishi, S. 1992. Promotive effects of aluminium on tea plant growth. *JARQ*, **26:** 26-33.

Konishi, S., Miyamoto, S. and Tazi, T. 1985. Stimulatory effects of aluminium on tea plants grown under low and high phosphorus supply. *Soil Sci. Plant Nutr.*, **31(3):** 361-368.

Koyamu, K. and Nambiar, E. P. 1978. Changes in the characteristics of virgin soil brought under plantation crops. *Proc. PLACROSYM 1*, March 20-23, 1978, Kottayam, Kerela, 1-16.

Krishnapillai, S. 1981. Effect of waste tea on growth of young tea plants. *Tea Q.*, **50(3):** 98-104.

Krishnapillai, S. and Ediriweera, V. L. 1986. Influence of levels of nitrogen and potassium fertilizers on chlorophyll content in mature clonal tea leaves. *Sri Lanka J. Tea Sci.*, **55(2):** 71-76.

Krishnapillai, S. and Pethiyagoda, U. 1980. Effects of calcium carbonate on ammonium and urea nutrition of young tea plants grown in sand culture. *Plant and Soil,* **55:** 455-463.

Kularatne, T. and Bhavnandan, V. P. 1971. Uptake and utilization of ammonium and nitrate ions by tea plants as determined by the use of ^{15}N labelled compounds. *Tea Q.*, **42:** 101-108.

Lauchli, A. and Pfluger, R. 1978. Potassium transport through plant cell membranes and metabolic role of potassium in plants. *Proc. 11th Cong. Int. Potash Inst.*, Berlin, Germany.

Ling, A. H., Harding, M. and Ranganathan, V. 1989. Phosphorus requirements and management of tea, Coffee and Cocoa. *Symp. Phosphorus Requirements for Sustainable Agriculture in Asia and Oceania.* March 6-10, IRRI, Manila, Philippines.

Magambo, M. J. S. and Othieno, C. O. 1977. Dry matter partitioning in the tea bush. *Tea in East Africa*, **17:** 15-17.

Malenga, N. E. A. 1979. The response of mature tea to aerial zinc application. *Tea Resesarch Foundation (Central Africa) Quarterly Newsletter*, **54:**12

Malenga, N. E. A. 1985. Response of young clonal tea to notrogen. *Tea Research Foundation (Central Africa) Quarterly Newsletter*, **79:** 7.

Malenga, N. E. A. 1987. The effect of different levels of nitrogen on the yield, quality and value of made tea from clones in agronomy trials. *Tea Research Foundation (Central Africa) Quarterly Newsletter*, **87:** 7-11.

Malenga, N. E. A. 1996a. Response of young unirrigated clonal tea to nitrogen. *Expl. Agric.*, **32:** 63-66.

Malenga, N. E. A. 1996b. Clonal tea response to time and rate of nitrogen application. *Tea Research Foundation (Central Africa) Quarterly Newsletter*, **124:** 15-18.

Mandal, S. C. 1975. Acid soils of India and liming. *Technical Bulletin, ICAR*, New Delhi, India, **51:** 1-126.

Manivel, L., Marimuthu, S. and Raj kumar, R. 1997. Physico chemical studies on nutrition of tea. *Scientific Publication Series*, 9. Tea Board, India, Calcutta, 1-13.

Mann, H. H. and Gokhale, N. G. 1960. Soils of tea growing tracts of India. *J. Indian Soc. Soil Sci.*, **8:** 191-200.

Marimuthu, S. and Manivel, L. 1992. Absorption of foliar applied zinc in tea tracer studies. *Newsletter*, UPASI Tea Research Institute, **2(1):** 3.

Marimuthu, S. and Raj kumar, R. 1998. Absorption efficiency of foliar applied urea nitrogen in tea. *Newsletter*, UPASI Tea Research Institute, **8(1):** 2-3.

Marwaha, B. C., Mehta, K. G. and Sharma, R. L. 1977. Response of tea to higher doses of nitrogen in relation to varying levels of precipitation on grey brown podzolic soils of Palampur. *Fertilizer Technology*, 14(3): 254-257.

Meskhidze, A. M. 1985. (The effects of forms of nitrogen fertilizers on the productivity of tea plants). *Subtropicheskie Kultury*, **5:** 55-58.

Mkwaila, B. 1992. Sulphur for tea. *Tea Research Foundation (Central Africa) Quarterly Newsletter*, **107(7):** 4-5.

Mondy, N. I. and Munshi, C. B. 1993. Effect of type of potassium fertilizer on enzymatic discoloration and phenolic, ascorbic acid and lipid contents of potatoes. *J. Agric. Food Chem.*, **41(6):** 849-852.

Nagarajah, S. 1979. The effect of potassium deficiency and cuticular resistance in tea (*Camellia sinensis*). *Physiol. Plant.*, **47:** 91-94.

Natesan, S. 1989. Physico-chemical studies on soils of the tea growing areas in South India. *Planters' Chronicle*, **84(2):** 41-49.

Natesan, S. and Ranganathan, V. 1990. Contents of various elements in different parts of the tea plant and in infusions of black tea from southern India. *J. Sci. Food Agric.*, **51:** 125-139.

Obatulu, C. R. 1985. Preliminary results on the comparative effects of two nitrogen sources on the growth of young tea cuttings. *Cafe, Cacao, Th'e*, **21(2):** 107-112.

Okano, K. 1993. Tea industry in Japan. *Proc. Int. Symp. Tea Science and Human Health*, Tea Research Association, January 11-14, 1993, Calcutta, 52-59.

Othieno, C. C. 1979. Estimates of removal of N, P and K by a clonal tea bush. *Tea in East Africa*, **19(2):** 11-13.

Othieno, C. O. 1988. Summary of recommendations and observations from TRFK. *Tea*, **9(2):** 50-65.

Othieno, C. O. 1992. Soils. In *Tea: Cultivation to Consumption* (eds. K. C. Willson and M. N. Clifford), Chapman & Hall, London, 136-172.

Othieno, C. O., Wanyoko, J. K., Owuor, P. O. and Lang'at, J.K. 1997. Effects of and rates of rock phosphate and triple superphosphate fertilizers on soil and leaf nutrient contents and yields of clonal replanted tea in Kericho, Kenya. *Tea*, **18(1):** 32-41.

Owuor, P. O. and Odhiambo, H. O. 1994. Response of some black tea quality parameters to nitrogen fertilizer rates and plucking frequencies. *J. Sci. Food Agric.*, **66:** 555-61.

Owuor, P. O., Othieno, C. O., Horita, H., Tsushida, T. and Murai, T.1987. Effects of nitrogenous fertilizer on the chemical composition of black tea. *Agric. Biol. Chem.*, **51:** 2665-2670.

Owuor, P. O. and Wanyoko, J. K. 1983. Fertilizer use advisory service. A reminder to farmers. *Tea*, **4(1):** 3-7.

Owuor, P. O , Wanyoko, J. K., Obanda, M. and Othieno, C. O. 1998. Potash and phosphorus fertilizers on black tea quality in the western Kenya highlands. *Tea*, **19(1);** 43 - 48.

Owuor, P. O., Wanyoko, J. K. and Othieno, C. O. 1987. Effects of nitrogenous fertilizers on leaf potassium contents of tea. *Tea*, **8(1):** 4-13.

Owuor, P. O., Wanyoko, J. K. and Othieno, C. O. 1988. Effects of fertilizers on tea: II Calcium contents in mature tea leaves. *Tea*, **9(1):** 10-20.

Owuor, P. O., Wanyoko, J. K. and Ruto, J. K. 1995. Effect of plant population and nitrogen on yield and quality of clonal black tea. *Proc. Int. Symp. Tea Quality-Human Health*, 7-10 Nov., 1995, Shanghai, China, 324-329.

Paul, K. K. 1992. Studies on nitrogen availability in soil and its use efficiency by tea (*Camellia sinensis*). *Ph.D. Thesis*, Guwahati University, Assam, India.

Paul, K. K. and Sen, A. 1984a. Consequences of high nitrogen fertilizer application on yield of tea (*Camellia sinensis*) with special reference to soil and plucked shoot nitrate nitrogen I. Under unshaded condition. *Bangladesh J. Soil Sci.*, **20:** 95-105.

Paul, K. K. and Sen, A. 1984b. Consequences of high nitrogen fertilizer application on yield of tea (*Camellia sinensis*) with special reference to soil and plucked shoot nitrate nitrogen II. Under shaded condition. *Bangladesh J. Soil Sci.*, **20:** 107-115.

Pethiyagoda, U. and Krishnapillai, S. 1971. Studies on mineral nutritions of tea in Ceylon, 4-Ammonium and nitrate as forms of N for tea plants in sand culture. *Tea Q.*, **42:** 109-117.

Phiri, G. S. N. 1985. Zinc oxide sprays in young clonal tea. *Tea Research Foundation (Central Africa) Quarterly Newsletter*, **79:** 4-5.

Pilani, N. and Verma, D. P. 1993. Magnesium fertilization in south Indian tea plantation.*Newsletter*, UPASI Tea Research Institute, **2(2):** 2.

Prasad, A. and Dey S. K. 1979. Studies on the uptake of manganese and boron by young tea plants. *Two and a Bud*, **26(1):** 9-13.

Rahman, F. 1988. Physiology of the tea bush. *Two and a Bud*, **35:** 1-14.

Rahman, F., Bhagavaty, H. N., Basu, R. P., Choudhury, R., Das, A. K., Dutta, S. K., Gilchrist, R. C. J. H., Sharma, K. N. and Trinick, J. M. 1978. Effect of different field management practices on crop characters and valuation of teas. *Two and a Bud*, **25:** 86-89.

Rahman, F., Sharma, P. B. and Nandi, N. C. 1978. Suitability of muriate of potash (KCl) and sulphate potash (K_2SO_4) for foliar spray in young and mature tea. *Two and a Bud*, **25(2):** 97-98.

Rahman, I. 1977. Nitrogen manuring of mature tea. *Tocklai Occasional Paper*, 11.

Rajagopal, B. and Ramarethinam, R. 1998. Interaction between vesicular arbuscular mycorrhiza. *Assam Rev. Tea News*, **86(12):** 30-34.

Raj, J., Bagyaraj, D. J. and Manjunath, A. 1981. Influence of soil inoculation with vesicular mycorrhiza and a phosphate dissolving bacterium on plant growth and ^{32}P uptake soil. *Biol. Biochem.*, **13:** 105-108.

Ramamorthy, B. and Velayuthan, M. 1976. Nitrogen, phosphorus and potassium in soil chemistry, forms and availability. In *Soil fertility theory and practice* (ed. J. S. Kanwar), ICAR, New Delhi, 156-201.

Ramaswamy , M. S. 1960. Copper in Ceylon teas. *Tea Q.*, **31:** 76-80.

Ramkrishnan, M. S. and Rangarajan, M. 1993. The limitations and potentials of green manuring. In *Organics in Soil Health and Crop Production* (ed. P. K. Thampan). Peekay Tree Crops Development Foundation, Cochin, Kerela, India, 43-68.

Ranganathan, V. 1969. Efficiency of nitrogenous fertilizers with special reference to urea. *UPASI Tea Sci. Dept. Bull.*, **27:** 16-19.

Ranganathan, V. 1971. Phosphorus nutrition of tea. *UPASI Tea Sci. Dept. Bull.*, **29:** 31- 45.

Ranganathan, V. 1973. Certain aspects of manuring of tea. *UPASI Tea Sci. Dept. Bull.*, **30:** 42-51.

Ranganathan, V. 1976a. Report of the chemistry division. *Annual Report 1976*, UPASI, Valparai, India.

Ranganathan, V. 1976b. Retention losses and utilization efficiencies of major nutrients in tea culture under south Indian conditions. *Planters' Chronicle*, **71:** 13-14.

Ranganathan, V. 1977. Nitrification inhibitors. *Annual Report 1977*, UPASI, Valparai, India, pp.46.

Ranganathan, V. 1979. Zinc deficiency and response to fertilizer. *UPASI Tea Sci. Dept. Bull.*, **36:** 27-29.

Ranganathan, V. 1981. Recent advances in agronomic practices-technology for increasing efficiency of fertilizer use. *UPASI Tea Sci. Dept. Bull.*, **37:** 92-103.

Ranganathan, V. 1992. Tea. In *IFA world fertilizer use manual* (eds. D. J. Halliday and M. E. Trenkel), International Fertilizer Industry Association, Paris, France. 532- 544.

Ranganathan, V., Ganesan, M. and Natesan, S. 1978. Methods of application of fertilizers in tea. *Proc. PLACROSYM 1*, March 20-23, 1978, Kottayam, Kerela, 249-260.

Ranganathan, V. and Narayanan, B. 1975. Root cation exchange capacity in tea and its practical application. *Planters' Chronicle*, **70:** 31-36.

Ranganathan, V. and Natesan, S. 1985. Potassium nutrition in tea. In *Potassium in Agriculture* (ed. Munson). American Society of Agronomy, Madison, Wisconsin, USA, 981-1022.

Ranganathan, V. and Natesan, S. 1987a. Nutrient elements and quality of tea. *Planters' Chronicle*, **82 (2):** 55-59.

Ranganathan, V. and Natesan, S. 1987b. Manuring of tea - recommendations for high yielding fields. *UPASI Hand book of Tea Culture*, Section 11, pp.16.

Ranganathan, V. and Natesan, S. 1988. Soil testing in plantation crops. *Planters' Chronicle*, **83(8):** 270-275.

Ranganathan, V., Natesan, S. and Palani, N. 1988. Response to tea to potassium application in the pruned year and effect of manuring on yield trend. *Planters' Chronicle*, **83(11):** 373-378.

Ranganathan, V., Venkata Ram, C. S. and Natesan, S. 1987. Superiority of AS and CAN as sources of nitrogen over urea to tea crop. *Planters' Chronicle*, **82:** 399-403.

Rikhter, M. A. and Lyashko, M. U. 1979. (The effect of high nitrogen rates on the yield and quality of irrigated tea). *Subtropicheskie Kul'tury*, **1:** 36-43.

Robinson, J. M. and Owuor, P. O. 1992. Tea aroma. In *Tea Cultivation and Consumption* (eds. K. C. Willson and M. N. Clifford), Chapman & Hall, London, 603-647.

Rojoa, H., Ramdaursing, K. and Ouradally, A. M. 1979. Fertilization of mature tea plants. *Revue Agricole et Sucriere de l' Ile. Maurice*, **58(3):** 142-152.

Saha, R. and Bisen, J. S. 1995. Phosphate and potash fertilizers in the optimisation of yield and quality of Darjeeling black teas. Tea Board, *Tech. Bull.*, **7:** 1-7.

Saha, R. and Bisen, J. S. 1996. Sulphur in Darjeeling tea soils its deficiencies and remedies. Tea Board, *Tech. Bull.*, **8:** 1-11.

Sarkar, S. K. 1989. Factors influencing productivity of tea in Darjeeling. *Proc. Joint Area Sci. Comm. Meeting*, Tea Research Association, Darjeeling, 9-11 Nov.,1989, 1-8.

Sarma, N. W., Mukherjee, S. and Kakoty, N. N. 1985. Can micronutrients cause increase in mite numbers? *Two and a Bud*, **32(1&2):** 29-31.

Sen, A. R. 1967. Response to mature tea to phosphate and potash in north east India. *Expl. Agric.*, **3(1):** 55-62.

Sharma, P. C., Jain, N. K. and Biswas, A. K. 1977. Response to nitrogen, phosphate and potash in mature tea. *Proc. 28th Tocklai Conf.*, Tea Research Association, Jorhat, India, 31-34.

Sharma, K. L. 1991. Response of dark leaf chinery hybrid tea (*Camellia sinensis* L.) to nitrogen, phosphorus and potassium in north west Himalayas. *J. Plantation Crops*, 18 (Supplement): 174-178.

Shuvalov, Yu. N. 1985. The role of horizontal and varietal displacement of ^{65}Zinc on introduction into tea plants. *Subtropischeskie Kul'tury*, **4:** 43-46.

Shuvalov, Yu. N. Kakauridze, V. V. and Kontridze, A. N. 1986. ^{32}P uptake from different layers on Krasnozem soil and determination or ^{65}Zn and ^{32}P uptake by tea leaves (Russian). *Subtropicheskie Kul'tury*, **1:** 83-86.

Singh, R. 1989. Address. *Proc. Joint Area Scientific Committee Meeting*. Darjeeling, Tea Research Association, pp.9-11.

Sinha, M. P., Thakur, K. C., Sarma, S. N., Sarma, J., Saikia, D. N. and Phukan, R. 1992. Effect of sulphur on yield of mature tea. *Proc. 31st Tocklai Conf.*, Tea Research Association, Jorhat, India, 20-21 Jan., 74-87.

Sivaram, D. 1982. Fertilizer application. In *Tea Production in Peninsular Malaysia*, Ministry of Agriculture, Malaysia, pp.34-41.

Sivasubramanium, S. and Talibudeen, O. 1971. Effect of aluminium on growth of tea (*Camellia sinensis*) and its uptake of potassium and phosphorus. *J. Sci. Food Agric.*, **22:** 32-39.

Subramanium, Bala, 1995. *Tea in India*. Publications & Information Directorate and Wiley Eastern Limited, New Delhi, p.123.

Sugianto, S. 1985. The effect of fertilizers on the quality of black tea. *Menara Perkebunan*, **53(2):** 42-46.

Swaminathan, P. 1992. Integrated nutrient management in tea. *Proc. 30th UPASI Sci. Conf.*, Sept. 14th, 1992, Coonor, **45:**13-28.

Takeo, T. 1979. Synthesis of amino acids induced by ammonium fertilizer application and seasonal fluctuation of the amino acid contents in tea. *Study of Tea*, **56:** 70-77.

Takeo, T. 1992. Green and semi-fermented teas. In *Tea: Cultivation to Consumption* (eds. K. C. Willson and M. N. Clifford), Chapman & Hall, London, 413-457.

Tandon, H. L. S. 1994. Fertilizers and their integration with organics and biofertilizers. In *Fertilizers, Organic Manures, Recyclable Wastes and Bbiofertilizers* (ed. H. L. S. Tandon), Fertilizer Development and Consultation Organisation, New Delhi, 12-35.

Tandon, H. L. S. 1995. Sources of sulphur. In *Sulphur Fertilizers for Indian Agriculture* (ed. H. L. S. Tandon), Fertilizer Development and Consultation Organisation, New Delhi, 17-18.

Tandon, H. L. S. and Ranganathan, V. 1993. Fertilizers and their management in plantation crops. In *Fertilizer Management in Plantation Crops A guide book*. (ed. H. L. S.

Tandon), Fertilizer Development and Consultation Organisation, New Delhi, 26-60.

Tiwari, R. S. 1978. Soil and tea nutrition. *Proc. Seminar Darjeeling tea "Operation 15 quintals"*. Area Sci. Comm. Darjeeling, Tea Research Association, Sept., 15-16, 1978, Darjeeling, TI1- TI5.

Tolhurst, J. A. H. 1960. Complexities in plant nutrition and their relation to manuring practice. *Tea Q.*, 31: 98-100.

Tolhurst, J. A. H. 1962. Zinc deficiency of tea in Ceylon. *Tea Q.*, **33:** 134-137.

Tolhurst, J. A. H. 1966. Field experiments and trials. In *Annual Report1966*. Tea Research Institute of Sri Lanka, Talawakelle, pp.26.

Utnelishvili, M. I. 1974. The effect of nitrogen fertilizers on tea productivity and quality. *Subtropicheskie Kul'tury*, **3:** 28-30.

Verma, D. P. 1993. Nutrient management of tea in south India. In *Tea Culture Processing and Marketing* (eds M. J. Mulky and V. S. Sharma), Oxford & IBH publishing Co., Pvt. Ltd., New Delhi, 55-69.

Verma, D. P. 1997. Sulphur nutrition in tea. *Proc. TSI/FAI/IFA Symp. on Sulphur in Balanced Fertilization*. Feb. 13-14, 1997. New Delhi, SII-2/1-2/6.

Verma, D. P., Marimuthu, and Rajkumar, R. 1997. Foliar application of nutrients and plant growth regulators in tea. *UPASI Tea Sci. Dept. Bull.*, **50:** 20-28.

Verma, D. P., Balasubramanium, K. and Udaya Bhanu, K. G. 1998a. Sulphur nutrition in tea. *UPASI Tea Sci Dept. Bull.*, **51:** 37-40.

Verma, D. P., Palani, N. and Udaya Bhanu, K. G. 1998b. Magnesium nutrition of tea. *UPASI Tea Sci. Dept. Bull.*, **51:** 12-14.

Venkataramani, K. S. 1987. Leaf analysis and foliar analysis. *Planters' Chronicle*, **82(2):** 49-52.

Wanyoko, J. K. 1979. Importance of potassium fertilizeron tea. In *Potassium Workshop,* Nairobi, Kenya. International Potash Institute, Berne, Switzerland, 116-124.

Watanabe, I. 1995. Effect of nitrogen fertilizer application at different stages on the quality of green tea (Japanese). *Soil Sci. Plant Nutr.*, **41(4):** 763-768.

Watanabe, I and Ikegaya, K. 1985. Influence of organic wastes on humic Fuji volcanic soil. *Study of Tea*, **67:** 33-36.

Whitehead, D. L. and Temple, C. M. 1990. Why does nitrogenous fertilizer affect tea quality. *Tea Research Foundation (Central Africa) Quarterly Newsletter*, **99:** 6 - 7.

Wickremasinghe, K. N. and Krishnapillai, S. 1986. Plant tissue analysis. In *Handbook on Tea* (eds. P. Sivapalan, S. Kulasegaram and A. Kathiravetpillai). Tea Research Instituteof Sri Lanka, Talawakelle, 65.

Willson, K. C. 1965. The long-term affects of fertilizers. *Tea*, **6(2):** 9-13.

Willson, K. C. 1967. Forms of nitrogen. *Tea*, **8(3):** 11-20

Willson, K. C. 1969. The mineral nutrition of tea. *Potash Rev*. Subject 27, 47th suite.1-17.

Willson, K. C. 1970. Foliar analysis of tea. *Expl. Agric.*, **6:** 263-265.

Willson, K. C. 1975a. Studies on the mineral nutrition of tea. II. Nitrogen. *Plant and Soil*, **43:** 501-516.

Willson, K. C. 1975b. Studies on the mineral nutrition of tea IV. Potassium. *Plant and Soil*, **43(2):** 279-293.

Willson, K. C. 1975c. Studies on the mineral nutrition of tea. VI. Magnesium. *Plant and Soil*, 43(2): 309-316.

Willson, K. C. and Choudhury, R. 1968. Fertilizer and tea quality. *Tea*, **9(3):** 17-19.

Willson, K. C., Hainsworth, E., Green, M. J. and O'shea, P. B. T. 1975. Studies on the mineral nutrition of tea. III. Phosphate. *Plant and Soil*, **43(2):** 259-278.

Wood, D. J., Bhatia, I. S., Chakravorty, S., Dev Choudhury, M. N., Deb, S. B., Ullah, M. R. and Roberts, E. A. H. 1964. The chemical basis of quality in tea. *J. Sci. Food Agric.*, **15:** 14-19.

Wyn Jones, R. G. 1975. Excised roots. In *Iron Transport in Plant Cells and Tissues* (eds. D. A. Baker and J. L. Hall), Elsevier North Holland, Biotredical Press, Amsterdam, New York.

Wyn Jones, R. G. and Pollard, A. 1983. Proteins, enzymes and inorganic ions. In *Encyclopaedia of Plant Physiology new ser. vol. 15B* (eds. A. Lauchli and R. L. Bieleski), Springer-Verlag, Berlin, Germany.

Xun, Wu and Jianyun Ruan, 1994. Effects of potassium and magnesium interaction in tea production and quality of green, oolong and black tea in China. *Proc. Intl. Seminar on Integrated Crop Management in Tea*. Colombo, Sri Lanka, April 26-27, 203-204.

Zaid, H. and Guerrier, G. 1983. Interactions calcium / potassium et calcium / sodium chez le theier. *Cafe, Cacao, The'*, **27(1):** 21-26.

Zyrin, N. G. and Barabadze, L. A. 1987. Effect of change in nutrient content of fertilizer on quality and leaf content of tea after long-term fertilization. *Subtropicheskie Kul'tury*, **1:** 57-62.

Chapter **11**

Weed management

11.1 Introduction

Weeds are plants, usually herbaceous, growing in an area where they are neither desired nor appreciated. They grow in the fields where they generate formidable competition with tea bushes for nutrients, space, air, light, moisture particularly during drought and when plants are small, thus reduce crop yields. Weeds harbour insects and diseases for which they serve as alternate host (such as *Melastoma malabathricum* Linn., *Urena lobata* Linn. are alternate hosts of red spider). These apart weeds restrict branching and development of frame in young tea and reduce plucking efficiency due to growth of creepers and thorny weeds (eg. *Mimosa pudica* Linn., *Amaranthus spinosus* Linn., *Solanum khasianum* Linn., *Rubus moluccanus* Linn, *Urtica dioica* Linn.). Weeds growing within tea bushes if plucked inadvertently with tea leaf (such as *Ageratum conyzoides* Linn., *Eupatorium odoratum* Linn., *Artimisia vulgaris* Linn.) may affect quality. It also acts as hosts of nematodes in the nursery. Weed growth block the drainage and impede the flow of water from the transport channels in the fields. It also protects the pests and prevents the pesticides from reaching them.

Weeds are not a major problem and regular weeding is not necessary in mature tea as its canopy cover the ground. But young tea plants may suffer a permanent setback in vigour if proper weed control measures are not taken at the appropriate time. In India, the critical weed competition

Fig. 11.1 : Infestation of dicot and monocot weeds in young tea field of Darjeeling hill.

period in young tea was found to be in between April and September (Fig. 11.1) and delay in weeding during this period adversely affects growth and yield (Sinha and Borthakur, 1992).

Weeds, by covering the ground, check oil erosion on sloping land like Darjeeling and other hilly terrains. It also helps in preserving soil structure and regulating fertility. Weeds are used for making compost in the plantation. A heavy growth of weeds would take up an appreciable proportion of the manure applied for the nourishment of the tea bushes, only some of which would be returned to the soil when the weeds buried or fork into it.

11.2 Crop loss

It was estimated that the losses of crop due to weeds below the potential maximum are 9, 6.6 and 6.8 percent in Asia, Africa and south America respectively (Cramer, 1967). Magnitude of losses for weed infestation in tea is quite high. In tea, broadleaved weeds reduce the yield to the extent of 12 percent whereas 21 percent losses are because of grasses in south India (Hudson *et. al.*, 1997). But the losses of tea crop due to weeds could be insignificant with high quality of management (Willson, 1999).

Plant architecture plays an important role in determining the losses caused by weeds. The China and China hybrid tea plants bear many stems arising from the collar level, which develop the soil pockets providing shelter to weeds. Such pockets are more evident in the gardens practicing the faulty application of inorganic fertilizers (mostly urea or ammonium sulphate) in the middle of the bush and eventually thinning the number of primaries at the centre of the bush frame. As such the weeds growing in these pockets escape various manual and chemical treatments for weed control. The problem is further aggravated in case of perennial grass as the same is difficult to uproot along with underground propagules. Further, the grass can not be dug out as it may damage stems of tea. Also application of herbicides over bush frame is a risky operation (Singh, 1999).

It has been reported that weeds in young tea in Assam accumulated 13090 kg dry matter/ha in the first year and 8400 kg in the second year and removed as much as 252 kg and 151 kg nitrogen/ha respectively. When weeds were controled during the critical period of weed infestation in between April and September, weeds dry matter between October and March was only 1890 kg and nitrogen removal was 34 kg/ha (Anonymous, 1978). In Japan, Hoshina *et. al.* (1980) reported nitrogen uptake by young plants was higher in weed free plots under herbicide treatments followed by manually weeded plots. It has been worked out that the weeds utilize 33% more of nitrogen, 50% more of P_2O_5 and 66% more of K_2O from the soil in comparison to tea bush.

11.3 Biology of weeds

For effective weed control, knowledge on their life cycles, growth habit, reproduction system is useful and essential. Weeds are classified on the basis of the duration of their life cycle into annuals, biennials and perennials. Annuals live and produce their seeds in a single growing season. Biennials need two growing seasons; in one season they pass through their vegetative growth followed by the reproductive stage in the next season. The propagation of both annual and the biennial are mainly through seeds. Perennials live indefinitely and are propagated through seeds and vegetatively. Morphologically major weeds belongs to either of the monocotyledons or dicotyledons. The monocotyledonous weeds have narrow leaves, are usually with parallel venation, clasping leaf sheaths, unbranched stem without central pith, vascular bundle scattered in a ground tissue, roots are fibrous and seed is with single cotyledon. The dicotyledonous weeds having broad leaves are usually with reticulate venation, stem is branched with central pith, vascular bundle arranged in a ring around the pith, seeds with two cotyledons, bark are easily separable and usually have tap root. In tea fields, major noxious weeds are dicotyledonous but some monocots such as *Imperata cylindrica* (L) Beauv., *Setaria palmifolia* (Koen). Stapf. are very much problematic. Despite monocots and dicots, sedges such as *Cyperus rigidifolius* in young tea was reported to be one of the predominant weeds in Kenya (Magambo and Kilavuka, 1975) and *C. rotundus* Linn. in Indonesia (Soedarsan *et. al.*, 1977). In the tea gardens having high humidity and limited sunny days throughout the year, mosses and ferns compete with tea. In Assam, it was reported that stirring the top soil every month increased the germination *Borreria hispida* (L) K. Schum. and *Cyperus pilosus* Vahl. indicating the wisdom of refraining from cultivation (Anonymous, 1978). In Sri Lanka, most vulnerable growth stage of *Pennisetum polystachion* (L.) Schult (a noxious grass in tea fields) was observed when the plants were taller than 45 cm and before flowering (Watson, 1986).

11.4 Reproduction of weeds

Weeds reproduce through seeds and regenerate through vegetative plant parts. Ferns and mosses reproduce by spores. The weeds are prolific seed producers, germinate seeds earlier, flower earlier and mature ahead of the crop they infest. They have the remarkable capacity to germinate under varied conditions. Seed germination is season bound and peak period of germination takes place in certain seasons in regular succession year after year. Weed seeds also can resist germination even under favourable conditions and they do not lose their viability for years even under adverse conditions. Seeds of *Portulaca oleracea* Linn. and *Cuscata reflexa* Roxb. germinate even after twenty years. The field should not be ploughed or hoed whilst weeds standing on the field bearing mature seeds. Weeds allowed to set seeds in one season will be menace for several years. In north-east India, the weed seeds germinate after onset of rains in late March and early April (spring) and continue to do so till September (autumn). The seedlings that emerged in April and May continue to grow actively till they mature and set seeds before winter sets in.

Weeds which reproduce vegetatively, multiply through their tubers, rhizomes, stolons, suckers, runners, bulbils, bulbs, nodes and also by a piece of stems. Weed seed could dispersed by water, wind birds, animals, man and through farm implements.

The absence of any winter in Sri Lanka enables weeds to grow vigorously all the year round except in a few of the drier districts, where their growth is comparatively dormant during the annual period of drought.

11.5 Weed control

Weeds are a problem chiefly in new clearings and pruned fields and the need for their control persists till the bushes cover the ground. Several species of grasses, broadleaved, ferns and sedge weeds infest the tea field under different agroclimatic conditions of the tea growing countries in the world. Some of these are:

Grasses and other monocotyledonous weeds — *Arundinella bengalensis* (Spreng). Druce., *Axonopus compressus* (Sw.) Beauv., *Commelina benghalensis* Linn., *Cynodon dactylon* (L) Pers., *Cyperus pilosus* Vahl., *C. rotundus* Linn., *Digitaria sanguinalis* (L.) Scope., *Eleusine indica* (L) Gaertn., *Eragrostis unioloides (Retz.) Nees., Imperata cylindrica* (L) Beauv., *Oplismenus compositus* (L.) Beauv., *O. Burmanni* Linn., *Paspalum conjugatum* Berg., *P. scrobiculatum* Linn., *P. distichum* Linn., *Panicum barbatum* Kunth., *P. repens* Linn., *Pennisetum clandestinum* Hochst. *Setaria palmifolia* (Koen). Stapf., *S. glauca* (L) Beauv. *and Saccharum spontaneum* Linn.,

Broad leaf or dicotyledonous weeds — *Achyranthes aspera* Linn., *Ageratum conyzoides* Linn., *Alternanthera sessilis* Br., *Amaranthus viridis* Linn., *A. spinosus* Linn., *Anaphalis contorta* Hook. f., *Artemisia vulgaris* Linn., *Biophytum sensitivum* (L.) DC., *Borreria hispida* (L) K. Schum., *B. articularis* (L.F.) Willd., *B. ocymoides DC., Calamintha umbrosa* Benth., *Calceolatia mexicana* Benth., *Capsella bursa-pastoris* Medic., *Cassia tora* Linn., *Conyza ambigua* DC., *Centella asiatica* (L.) Urban, *Crassocephalum crepidioides* (Benth.) Moore., *Digitaria longiflora* Pers., *Drymaria cordata*, Willd., *Drymaria diandra* Blume, *Dichrocephala integrifolia* L., *Eupatorium odoratum* Linn., *Euphorbia prunifolia* Jacq., *E. hirta* Linn., *Erigeron mucronatus* DC., *Erechthites valerianaefolia* DC., *Emilia sonchifolia* DC., *Gnaphaltium purpureum* Linn., *G. indicum* Linn., *Galinsoga parviflora* Cav., *Hydrocotyle himalaica* Linn., *Ipomoea digitata* Linn. *I. purpurea* Linn., *Lactuca runcinata* DC., *Melastoma malabathricum* Linn., *Mikania micrantha* HBK., *Mimosa invisa* Mart., *M. pudica* Linn., *Neonotis quadrilocularis* Lewis, *Oxalis acetosella* Linn., *O. corniculata* Linn., *Peperomia pellucida* HBK., *Polygonum chinense* Linn., *P. perfoliatum* Linn., *P. strigosum* BR., *P, runcinatum* Ham., *P. alatum* Spreng., *P. microcephalum* D. Don, *P. hamiltonii* Spreng., *Pollinia imberbis* Nees., *Portulaca oleracea* Linn., *Pouzolzia indica* Gaud., *Plantago major* L., *Rungia repens* Nees., *Richardia scabra* L., *Scoparia. dulcis* Linn., *Solidago sp., Sonchus brachyotus* DC., *Spilanthes acmella* Murr., *Sida rhombifolia Linn.*, *Sida acuta* Burm., *Spermacoce ocymoides* Burm., *Stephania japonica* (Thunb) Miers., *Synedrella nodiflora* L. Gaertn., *Tylophora tenuis* Blume, *Thunbergia alata* (Bojer), *Vernonia anthelmintica* L. Wild., *Xanthium strumarium* Linn., *Zehneria maysorensis* (L.) Wt.& Arn.

Ferns — *Pteridium aquilinum* Kuhn., *Dicranopteris linearis* (Burm. F.) Und., *Drymoglosum piloselloides* Presl., *Lygodium flexuosum* Sw., *Nephrodium* sp.

In south India, the grass weed problem is greatest in the clearings and in mature tea during first two years of the pruning cycle. In open situations, with poor or no shade grass weeds become prominent. Soon after the grass is killed it can be noticed that there is an invasion of broad leaf weed flora. Sedges are not serious weeds in tea fields but in Kenya, *Cyperus rigidifolius* (Magambo and Kilavuka, 1976) and in Indonesia *Cyperus rotundus* Linn. (Soedarson *et. al.*, 1977) were one of the predominant weeds. A thorough knowledge about weed flora, seed dormancy, critical period of weed growth are essential for proper weed management. Various methods of weed control are practiced.

11.5.1 *Manual or mechanical methods*

In tea, hand weeding, sickling, cheeling, forking and mulching are the convensional methods of weed control. Hand weeding is carried out in the nurseries (Fig. 7.16) and the newly planted tea areas where application of any chemical herbicide is hazardous for the plants. Cheeling or surface scrapping is an efficient means of suppressing weeds with a minimum of soil stirring. This operation is carried out mainly to remove shallow rooted weeds when their density is rather high and a large area to be weeded. Hand weeding is the most efficient method, but it is time-consuming and rather expensive. In southern Africa, Scarborough and Kayange (1977) reported that hand weeding in young tea was most effective and provided double the growth as compared to unweeded plots, or those received metribuzin (0.75–3.0 kg/ha), oxadiazon (0.48–1.92 kg a.i./ha), and asulam (3.4–8.8 kg a.i./ha). Cheeling should be restricted to cutting the weeds close to but above the level of the ground and best be done during dry season.

Sickling is generally done to control vigorous growth of tall weeds and it often precedes herbicide application. In the mature tea fields four to six rounds of sickling at four to six weeks intervals during rainy season is necessary to suppress weeds. In Darjeeling and other sloppy high rainfall areas, the concept of 100 % weed control is likely to increase the problem of soil erosion. Under such conditions the aim should be to control noxious weeds but leave a non-competitive ground cover of soft weeds.

Deep-rooted weeds growing around the collar of the bushes, probably not more frequently than once a year, are removed by 'forking'. This operation will prevent weeds from growing up into the bush and interfering with plucking. Weeds having suckers (eg. *Imperata cylindrica*) or bulbs (eg. *Oxalis acetosella*) or strong root stocks (eg. ferns) are removed by forking. In Sri Lanka, Sandanam and Jayasinghe (1977) observed that forking out rhizomes of *Imperata cylindrica* up to a depth of 45 cm gave long lasting control. Deep hoeing, scrapping and deep forking often damaged the feeder roots of tea remain in the top 10–15 cm of soil profile. Hence, utmost care should be taken during these operations. However, the best method of weed control is to smother the weeds by encouraging the tea bushes to cover the ground fully. Soil disturbing implements like fork and hoe better be avoided

Mulching very effectively suppresses weed growth. Thatching and mulching in the clearings play an important part in weed control in young tea until the tea plants themselves are well developed and cover the ground. The vacant areas in the field should be rehabilitated by planting Guatemala grasses (*Tripsacum laxum*) for one to three years (Fig. 11.2). This will help in keeping the weeds under control, protect the soil from erosion and rehabilitate the soil for infilling. In the nursery, planting of tea leaf cuttings by inserting through a black polythene mulch beneath a cover of hoop-supported polythene cloth has been reported to provide satisfactory weed control (Smale, 1991). Mulching with black plastic, tea factory residues, or peat also provide improved weed control in the nursery (Tabagari and

Fig. 11.2 : Vacant patches in the tea field of Darjeeling rehabilitated with Guatemala grass.

Kopaliani, 1984). In young and mature tea fields, mulching with black low density polyethylene (LDPE) sheet was also found very effective for control of weeds. LDPE mulch totally suppressed weeds in the inter-row spaces of young China hybrid tea planted on slopes of Himachal Pradesh (Singh *et. al.*, 1993). In pruned sections, mulching the inter-rows with tea prunings provided the best results in terms of weed management in the former USSR (Tabagari *et. al.*, 1988). It was also reported that rubber mulch checked weed growth and increased yield of mature tea by about 39%, while peat mulch was not found so effective (Zarnadze, 1972).

11.5.2 *Chemical methods*

Controlling weeds manually has become increasingly expensive and there is often a shortage of labour during peak cropping season. To cope up with the problems, herbicides were introduced in tea industry with satisfactory results. The benefits of chemical weed control in terms of tea yield were demonstrated by Rahman (1991). However, manual weeding should supplement chemical weed control and be a part of an integrated programme. For best results, a combination of herbicides, mechanical methods and mulching was recommended in southern Africa (Anonymous, 1975). Though chemical weed control initially is more expensive than mechanical or manual control but in the long run, rational application of herbicides can only optimize weed control. The success of a herbicide weed control depends on the identification of weeds, right choice of herbicide, dosage and time of application, and their use in suitable mixture. Weed management has to be such that it does not encourage soil erosion. In Darjeeling hills, chemical weed control was not recommended till 1960s apprehending soil erosion. Sarker (1972) suggested use of weedicides in Darjeeling tea area having 45% slopes or less and adoption of chemical weed control programme in combination with *in situ* mulching of pruning litter. However, herbicides should be used till June and then from September onwards to ensure that weeds do not compete for moisture in the cold weather. Herbicide can be given around the collar or in alternate strips in less sloppy areas even in the rains. Only sickling can be done in areas vulnerable to erosion (Rahman, 1977).

The herbicides used in tea plantations (Table 11.1) can be broadly classified into pre-emergent and post-emergent. Post-emergent herbicides further classified into contact and translocated.

Table 11.1 : Specificity, mode of action, approximate dilution of some common herbicides and spray fluid needed per hectare to cover uniform control.

Herbicide	Mode of action	Specificity	Dilution (amount per 200 litres of water)	Spray volume litres of water
Paraquat	Contact	Broad spectrum	500 gm	400-500
2,4-D	Translocated	Mainly used as broad leaf killer	500 gm	351-400
Diuron	Translocated	Broad spectrum	1.10 kg	400–450
Glyphosate	Translocated	Mainly used as Grass killer	0.50 litres	350–400
Simazine	Translocated	Broad spectrum	1.50 kg	400–450
Dalapon	Translocated	Specific to grassy Weeds	1.75 kg	350–400

(A) PRE-EMERGENT HERBICIDES

These group of herbicides have very little solubility and when applied they should stay in the top 1 to 5 cm of soil and kill any weeds as soon as they start emerging through soil layer. As the efficiency of these herbicides depends on their persistence in the soil, hence the factors viz. soil moisture, severity of the rainfall, soil type and temperature affect the efficiency of herbicides. The pre-emergent herbicides are absorbed by roots. Diuron and simazine used as pre-emergent herbicides in India are endowed with long persistency and generally do not loose their activity while remaining on the ground for the weeds to emerge. Diuron is applied in moist soil and get absorbed by clay particles. It is recommended for tea plants older than three years due to toxicity to young plants. The mobility of diuron in the soil is poor as they are generally bound in the top 2.5 cm layer. They are likely to be leached to the feeder root-zone only in light sandy soil. Diuron is capable of arresting weed growth for 6–8 weeks and therefore applied only once in a pruning cycle. Diuron is very effective on most of the broad leaf weeds and some annual grasses. It should not be applied in tea less than three years of age. Simazine (4 kg in 200 litres of water per hectare) is most effective on broad leaf weeds and can be very safely applied in young tea. Simazine can be conveniently used in nursery beds and sleeves 10–15 days before planting of leaf cuttings. In southern Africa, pre-emergence application of metribuzin at the rate of 1.5 kg /ha was found effective in controlling weeds as well as in enhancing growth of young tea plants (Scarborough and Kayange, 1976). Kayange (1980) concluded that metribuzin, oxadiazon (4–5 litres for Ronstar 26% EC and 2.5–3.25 litres/ ha for Ronstar 40% F) and metolachlor (1.5–2.5 litres/ha) could be applied as effective pre-emergence weedicides in southern Africa. In Mauritius, best pre-emergence weed control (100 and 85% at 1.5 and 3.5 months respectively after spraying), in 2 to 2.5 years old young tea was obtained with diuron at 4 lb/acre (about 4 kg/ha) (Lamusse *et. al.*, 1978). In Indonesia, simazine, atrazine and other triazines were reported to be promising herbicides (Sanusi, 1979).

In the former USSR, simazine was found to give 80–87% control of annual weeds in mature tea (Purtskhvanidze, 1973). Pre-emergence application of methazole at 5-6 kg/ha on cheeled (scraped) ground in Assam ensured satisfactory control of weeds for three months (Anonymous, 1972). Further, Barbora (1972) reported that linuron (50%) at 4 kg/ha applied before weed emergence in mature tea field of Assam was observed to control most broad leaved weeds and shallow rooted grasses. In south India, metribuzin and dichlormatem reported to hold promise for use as pre-emergence. Metribuzin is a herbicide with prolonged soil residual activity on a wide spectrum of broad leaved weeds and some annual grasses. It has potential for use in pre-monsoonal programmes and was found comparatively more effective at relatively lower doses (Anonymous, 1974). In Himachal Pradesh, pendimethalin (0.75 kg/ha), oxyfluorfen (0.44 kg/ha), simazine (1.25 kg/ha), atrazine (1.25 kg/ha) or simazine (1.88 kg/ha) followed by paraquat (0.3 kg/ha) were found to be more effective pre-emergence treatments suppressing seed-borne weeds (Anonymous, 1993). Subsequent weed growth in either of the cases could be controlled with spot treatment of 2,4-D and/or paraquat (Singh *et. al*., 1992).

Weed infestation becomes severe in the year following light pruning, deep skiffing and medium pruning due to exposure of the soil. Therefore, application of pre-emergence herbicide during March-April (normal season in north eastern region) becomes difficult as the pruning litter covers the ground which prevent the herbicide from reaching the soil uniformly after spray. This can be overcome by applying pre-emergence herbicides in autumn. Autumn application before pruning was found effective for better management of weeds in pruned sections.

(B) POST-EMERGENT HERBICIDES

These herbicides are applied from germination to maturity of weeds. In India, paraquat, 2,4-D, dalapon and glyphosate are used in tea fields as post-emergent herbicides for controlling weeds (Sinha, 1985). Glyphosate is a relatively recent addition to the available herbicides. Diuron can also be applied post-emergence in tea in combination with paraquat. In Sri Lankan tea plantations, paraquat was most commonly used herbicide followed by diuron (Somaratne, 1987).

i. Post-emergent contact herbicides

Paraquat (1,1-dimethyl 1-4, 4-bipyridium dichloride) popularly known as gramoxone is a broad spectrum contact herbicide and has no adverse effect on beneficial soil microbes (Chemical Industries Limited, 1989). The use of contact herbicides also ensures that there is minimum disturbance to the soil and no risk of building up of chemicals with residual effects in the soil. Paraquat is almost rain proof and absorbed by the foliage and green tissues and acts rapidly in light. Its movement is mainly through xylem. Paraquat forms H_2O_2 by oxidation process and series of chain reactions which kills the tissue. For maximum efficacy, contact herbicide should be sprayed at a concentration to kill the weed completely and also volume of fluid should be sufficient enough to wet the weed thoroughly. It kills the green portion of the weeds above the ground and root system remains intact. Very little of it is translocated to the plant. Paraquat was observed to be most effective against weeds reproducing from seed (Widjaja, 1977). Paraquat exhibited scorching effects on top of the weeds within four days and within two weeks regrowth from suckers of grasses like *Cynodon dactylon* and *Paspalum conjugatum* was visible (Kayange, 1978). It was reported that the paraquat increases the population of *Polygonum* weeds in Darjeeling (Kabir *et. al*., 1991).

ii. Post-emergent translocated herbicide

2,4-D (2,4-dichlorophenoxy acetic acid), dalapon (2,2-dichloropropionic acid) and glyphosate (N-(phosphomethyl) glycine) are examples. The requirement of volume of spray fluid of these herbicides is much less since wetting the weeds is not important in case of translocated herbicide. However, the object is to ensure maximum absorption and translocation of the herbicide by the weed. Translocated herbicides after spraying require four to six hours rain-free period, and with stickers they require at least two hours rain-free period.

2,4-D is available in sodium salt and dimethyl amine salt formulations. 2,4-D is readily absorbed by roots and leaves then moves down to the phloem along with carbohydrates manufactured by leaves. Spray of 2,4-D stops cell division and induces swelling of the cells. Rate of photosynthesis is reduced and absorption of water and soil nutrients and food transportation are disrupted. Due to varied rate of cell division and cell enlargement, twisting and curvature of leaves and stems occur. Dalapon is absorbed through foliage and roots and enters into the foliage through cuticle and stomata. It is a strong acid and protein precipitant and interferes with the synthesis of pantothenic acid, B vitamin used for growth. 2,4-D and dalapon are used only in mature tea fields as they are toxic to young tea. Kabir *et. al*. (1991) observed that individual spray of 2,4-D or dalapon was not very effective where mixed population of monocot and dicot weeds exist under the agroclimatic conditions of Darjeeling.

Most of the paraquat-resistant broad-leaved weeds could be controlled by 2,4-D at 1.12 kg/ha (Soedarsan *et. al.*, 1974; Sharma, 1977). Addition of paraquat @ 0.3 kg/ha made it more effective against grasses in Assam and Himachal Pradesh (Anonymous, 1978; Anonymous, 1989).

Glyphosate is a broad-spectrum herbicide readily absorbed by the foliage and translocated to the underground roots and storage organs in sufficient quantities to kill the plant. The efficacy of glyphosate is related to depletion of soluble sugar and starch content in the rhizome (Sarma *et. al.*, 1986). Glyphosate controlled perennial grasses (*Imperata cylindrica*, *Cynodon dactylon*, *Setaria palmifolia*) and deep rooted dicot weeds. Glyphosate acts slowly but it kills most of the weeds commonly found in tea field provided application is done under active growth. Normally it is used at 1.5 to 2.5 litre/ha. Kabir *et. al.* (1991) observed that glyphosate @ 0.92 kg or 1.23 kg/ha was most effective in controlling major weeds in Darjeeling tea plantations. Further, glyphosate applied @ 2 or 4 kg/ha was recorded to be highly effective, giving reduction of 86-100% in weed population, after 30 days of application, with no adverse effects on plant growth, yield and quality of tea (Ivakh *et. al.*, 1990). Doses of glyphosate should be selected considering the nature of weed infestation and its intensity.

(C) HERBICIDE MIXTURE

The effect of herbicides are often enhanced by simultaneous use of suitable mixture of more than one herbicide at a time. They also offer the possibility of reducing the total cost of herbicide and their application. Herbicide mixture perform better weed control due to synergistic effect. This apart, appearance and spread of resistant weed species are greatly reduced by use of suitable combination of herbicides. The selection of herbicidal combination is related to the weed flora, intensity of weed infestation and their compatibility. The widely used mixtures in tea are paraquat + diuron, paraquat + 2,4-D, dalapon + 2,4-D, paraquat + MSMA, paraquat + simazine, glyphosate + 2,4-D and glyphosate + oxyfluorfen. Paraquat + diuron mixture should be used over fresh growth. This mixture observed to be effective for long duration control of some of the annual grasses and broadleaved weeds. MSMA (monosodiummethanearsonate), is selective translocated chemical belonging to organoarsenical group. It is a post-emergence herbicide and best results are obtained when application is done in hot sunny conditions.

Paraquat + 2,4-D and dalapon + 2,4-D show better weed management in mature tea. Tank mixture of paraquat @ 0.3 kg + 2,4-D @ 1–1.5 kg/ha was recommended for mixed population of *Setaria barbata* and *Borreria* sp. and was found most economic and effective herbicide treatment (Rahman *et. al.*, 1973; Anonymous, 1989). In Darjeeling, the population of *Oplismenus compositus*, *O. Burmanni*, *Polygonum microcephalum*, *P. runcinatum*, *Ageratum conyzoides*, *Borreria articularis* were found very high in the plots treated with paraquat @ 0.36 kg a.i./ha and 2,4-D @ 1.2 kg a.i./ha in alternate rounds (Kabir and Ghosh Hajra, 1993).

In dalapon + 2,4-D mixture, dalapon controls perennial grasses, particularly *Imperata cylindrica* and 2,4-D eliminates broad leaf weeds (Chakravartee and Barbora, 1994).

Perennial grasses such as *Imperata cylindrica* could also be satisfactorily controlled by MSMA (3.75 l/ha) + dalapon (3.75 kg/ha) + 2,4-D sodium (0.7 kg/ha) (Sharma, 1975; Soedarsan *et. al.*, 1974). MSMA 48.6% (4 l/ha) + 2,4-D 72% (1.8 l/ha) + sodium chlorate (7.5 kg/ha) was reported to control *Borreria* sp. and *Paspalum conjugatum* but was non-economic against other species (Venkataramani *et. al.*, 1972). MSMA with paraquat, dalapon and 2,4-D give better and persistent weed control than any one of them used alone. Barbora (1972) reported that MSMA (34.8% + wetting agent) at 3 litres + paraquat (20%) 0.75 litres : (product) + wetting agent 0.75 litre with or without either 2,4-D @ 0.75 kg or dalapon @ 3.5-4 kg/ha provided very satisfactory control of most weeds in Assam.

The bio-efficacy of a tank mixture of glyphosate and sodium salt of 2,4-D was evaluated in south India. It was observed that glyphosate at 0.72 kg (1.75 litres of the formulation) and 2,4-D sodium salt at 1.12 kg (1.40 kg of the formulation) in 450 litres of water has been found to be effective against broad spectrum of weeds predominated by perennial hardy dicot weeds (Satyanarayana *et. al.*, 1992).

11.6 Adjuvants for herbicides

Use of adjuvants increases the efficacy of some herbicides. A sticker (tenac) or spreader (teepol) may help during rainy or sunny weather respectively at 1:500 parts of the spray fluid. A sticker helps in the retention of the spray fluid on the weed surface and the surface active agent favours uniform spreading of the spray fluid over the foliage. Murty and Sharma

(1990) have investigated the effect of different types of additives viz. non-ionic wetting agent (triton AE), a cationic wetting agent (ethokem), an oil based surfactant (crop oil) and a cheap native dust (kaolin) on the bioefficacy of glyphosate. They observed that addition of crop oil and kaolin enhances the efficacy of glyphosate even at lower rate; the efficacy is further increased when wetting agent triton or ethokem is added to kaolin–glyphosate mixture. In Assam, it was reported that addition of urea (1 kg in 200 litres) to 2,4-D, ammonium sulphate to glyphosate and triton AE (0.06%) to delapon (3.2 kg/ha) as well as 2,4-D (0.4–0.8 kg/ha) enhanced herbicidal efficacy (Anonymous, 1992). Kathiravetpillai and Punyasiri (1989) in Sri Lanka obtained effective control of *P. repens* with reduced rates of glyphosate (0.5%-1%) in combination with kaolin (3.6 kg/ha) or ethokem (1-2 l/ha). In Himachal Pradesh, DOWCO-453, a post-emergence @ 0.75 kg/ha, DOWCO-453 @ 0.50 kg/ha + Selwet 1% or glyphosate @ 0.5 kg/ha + uphar @ 0.6% were observed to provide effective weed control in mature tea (Anonymous, 1989).

11.7 Sprayers for weedicide application

For weedicide application, usually hand operated and pressure regulated Knapsack sprayers fitted with flood jet or flat fan nozzle is used since both the types give wider sweep of spray to cover larger areas (Fig. 11.3). The amount of spray fluid required to cover a particular area will vary with type of sprayer, intensity of weed growth and mode of application. For uniform spray, 450–500 l/ha of spray fluid are needed in case of contact herbicides, 350–400 l/ha for translocated post-emergence but for translocated pre-emergence, 400–500 l/ha are required. It is also better to use low volume nozzles discharging at around 40 l/ha for post-emergence and 60–70 l/ha for pre-emergence applications.

11.8 Herbicide toxicity

It can occur mostly due to overdosing, improper or non-targeted spray, surface run-off due to heavy rains, spray drift, health of the tea bush etc. There are reports that 2,4-D and diuron toxicities in tea cause rolling of young leaves followed by chlorosis and browning while in young tea toxicity of simazine and atrazine @ 12 kg/ha cause temporary chlorosis, defoliation and stunting of seedling (Lobzhanidze and Ivardava, 1972). Paraquat and glyphosate produce leaf scorching. Glyphosate even at rates of 0.18–0.90 kg/ha was reported to cause leaf narrowing and internode elongation

Fig. 11.3 : Herbicidal spray in young tea field in Darjeeling.

(Soedarson *et. al.*, 1973). In east Africa, Magambo and Kilavuka (1982) reported that the application of glyphosate @ 2 to 8 kg a.i./ha (300 l/ha) in young tea caused a reduction in internode length and after one week post-spraying the production of curled needle-like leaves. Symptoms of this type persisted for three months post-spraying. Sub-lethal effects are persisted and mainly involve shoot retardation, bud proliferation and leaf deformation. However, no ill effects were observed when applied on the frames of the tea bushes and bud break was found normal (Sharma *et. al.*, 1973). Thus, shield should be used while spraying herbicides in young tea. Yellowing of leaf margins followed by necrosis are the symptoms of dalapon injury. Oxyfluorfen sometimes produces leaf shrinkage in the next flush (Barbora *et. al.*, 1994).

11.9 Problem of resistant weeds

Introduction of new herbicide and timely change of herbicide programmes will certainly assist in overcoming the problem of tolerant and resistant weeds. In south India, Sharma (1975) observed that prolonged use of gramoxone in tea fields led to the invasion of newer weed species which are resistant to gramoxone such as *Achyranthus aspera*, *Borreria eradii*, *B. hispida*, *Crassocephalum crepidioides*, *Richardia scabra* and *Sida acuta*. Some perennial grasses like *Imperata cylindrica*, *Panicum repens* and *Paspalum conjugatum* are also found difficult to control with gramoxone. Sharma and Satyanarayana (1981) further reported where *Cynodon dactylon*, *Paspalum conjugatum* and other common grass species were eradicated by the continual use of dalapon and paraquat, *Panicum repens*, which is resistant to both the above herbicides has become a dominant weed which necessitated a change in weedicides. Kabir *et. al.* (1991) observed in Darjeeling that plots receiving paraquat, infestation of *Polygonum* weeds tended to increase. It has been reported earlier that *Borreria articularis* is resistant to paraquat (Basu, 1972). Rahman (1977) reported *Oxalis*, *Paspalum* and *Polygonum* weeds have become dominant because other fast growing species which used to keep them smothered were easily controlled by herbicides. As a result these weeds got opportunity to grow and multiply unhindered. In Indonesia, *Eupatorium riparium* found to become dominant in an area treated with asulum and infestation of *Murdannia nudiflora* Brenan. increased when mixed preparation of dalapon, MSMA and 2,4-D was used (Soedarsan *et. al.*, 1974).

Minimising the weeds by using sub-lethal dose of herbicide may become counter productive as this may produce resistant weeds in due course. Continuous cheeling sometimes shows a shift of weed population which alternates between *Borreria articularis* and *Paspalum conjugatum* (Mustafee, 1998).

Thus, regular monitoring on the shift in weed flora under different herbicide treatment is essential and the same should be kept in mind while deciding herbicide programmes (rotation) for specific weed problem. Sinha and Borthakur (1992) suggested following sequences in a rotational programme:

- A pre-emergence herbicide followed by a post-emergence herbicide.
- One post-emergence herbicide followed by a post emergence herbicide, and
- Repetitive application of the same or another pre-emergence or post-emergence herbicide.

11.10 Integrated Weed Management

The progressive modernization of agriculture involving intensive use of herbicides is gaining popularity in recent years due lower cost, easy and timely application and effectiveness in controlling weeds. Among the agrochemicals, the use of herbicides has been expanding more rapidly than that of other pesticides. The indiscriminate use of synthetic herbicides for controlling weeds may offer potential hazards to users, consumers, livestock, wild lives and the environment. All the herbicides mentioned in Table 11.1 are potentially hazardous. Integrated Weed Management approach is an appropriate way of minimizing the residue problems in plant, soil, air and water. This approach involves adoption of an ideal combination of different weed control methods. Greater emphasis is needed for quick growing and spreading cultivars, infilling of vacancies and close planting in order to give quick coverage of the inter-rows. Despite periodic slashing and hand weeding, resistant weeds should be controlled by deployment of plantation workers in an economical way. Hand weeding around the collar region of young tea bushes is always safe. Every care should be taken so that weeds do not flower and seed. The tea field remain weed free for several weeks if mulching is done immediately after chemical weeding. In general, all bio-degradable chemical herbicides are eco-friendly and use of these chemical herbicides will be suitable in controlling weeds without affecting the agro-ecosystem.

(A) SYNTHETIC HERBICIDES

Sulfonylurea herbicides require very less amount for controlling weeds and hence suitable for integrated weed management. The important sulfonylurea herbicides are chlorimuron ethyl, metsulfuron methyl, tribenuron methyl, bensulfuron methyl, sulfosulfuron, ethoxysulfuron, clodinofop etc; these herbicides are being used for selective control of weeds in soybean, wheat, rice etc (Mishra and Raghuwanshi, 1999). A large number of other herbicides containing heterocyclic moiety have been developed and found effective and have got bio-degradable property (Roy and Dureja, 1998). Imidazolinone herbicides, Triazolopyrimidine herbicides, Pyrimidylether herbicides and Flumichlorac pentyl herbicides are effective on susceptible broadleaf weeds and have good selectivity between crops and weeds.

(B) BIO-HERBICIDES

Bio-herbicides refer to microbial plant pathogens which are applied as sprays that uniformly kill or suppress the growth of weeds. Fungi, bacteria and viruses offer great promise for use as bio-herbicides in modern weed management. Spray of alke-strain of Tobacco Mosaic Virus on herbicide resistant *Solanum carolinense* effectively controlled the weed (Izhevskii *et al*., 1981). However, only fungal pathogen have been exploited commercially as micoherbicides. This apart, many secondary plant metabolites (allelochemicals) and microbial toxins are reported to have good herbicidal properties. The recent development in herbicide technology is the use of herbicide 'safeners' in order to protect crop plants from possible damage by a herbicide and also permit higher rates of herbicide application for effective weed control (Mishra and Raghuwanshi, 1999). Serious attempts are to be made for evaluation of the above mentioned chemicals and introduction of suitable bio-agents in the integrated weed management system on tea. Further, attention for exploring the feasibility of using allelopathic properties of certain potential weed species in tea weed management is also needed.

References

Anonymous, 1972. *Annual Scientific Report 1971 – 72*. Tea Research Association, Jorhat, Assam, India, pp. 27.

Anonymous, 1974. *Annual Report 1972*. United Planters' Association of South India, Valparai, Coimbatore, India, pp.107.

Anonymous, 1975. Herbicides for young tea. *Tea Research Foundation (Central Africa) Quarterly Newsletter*, **39:** 10-12.

Anonymous, 1978. *Annual Scientific Report 1977–78*. Tea Research Association, Jorhat, Assam, India, pp.95.

Anonymous, 1989. *Annual Report 1988–89*. CSIR Complex Palampur, Himachal Pradesh, India, pp. 11.

Anonymous, 1992. *Tocklai Experimental Station 1911 to 1991*. Tea Research Association, Jorhat, Assam, India, pp. 78

Anonymous, 1993. *Annual Report 1991–92*. CSIR Complex Palampur, Himachal Pradesh, India, pp.1,4-5.

Barbora, A. C., Baishya, J. K. and Bhattacharya, K. G. 1994. A study of toxicity symptoms of herbicides in tea. *Pestology*, **18(9):** 13–18.

Barbora, B. C. 1972. Some new weedicides. *Two and a Bud*, **19(1):** 29–30.

Basu, S. 1972. The need of chemical weed control. *Two and a Bud*, **19(2):** 65–67.

Chakravartee, J. and Barbora, A. C. 1994. Weed control in tea – choice of weedicide and efficient weed control schedules. *In : Field management in tea* (ed. J. Chakravartee). Tocklai Experimental Station, Jorhat, Assam, India, pp. 109–114.

Cramer, H. H. 1967. Plant protection and world crop production (translated by J. H. Edwards). *'Bayer' Pflanzenswchutz,* **20(1):** 381.

Hoshina, T., Kozai, S. and Honjo, Y. 1980. Influence of weed control on the absorption of fertilizer nitrogen N^{15} by tea plants. *Study of Tea*, **59:** 27–32.

Hudson, J. B., Durairaj, J., Muraleedharan, N. and Dhanakumar, V. G. 1997.*Guidelines on tea culture in south India. UPASI – KVK Publication*, **30**, p.212.

Ivakh, M. I., Sanikidze, A. D., Toidze, M. G., Spiridonov, Yu. Ya, Nikitin, N. V., Raskin, M. S., Nazarova, T. A., Makeev, A. M. and Andguladze, L. Z. 1990. Application of the herbicide Utal in control of Japanese *spiraea* in tea plantations. *Subtropicheskie Kul'tury*, **1:** 135–140.

Izhevskii, S. S., Livshits, A. E., Murusidze, G. E. and Gogoladze, G. G. 1981. Prospects of using Alke strain in integrated control of Solanum carolinense in tea plantations. *Subtropicheskie Kul'tury*, **4:** 60–65.

Kabir, S. E., Chaudhuri, T. C. and Ghosh Hajra, N. 1991. Evaluation of herbicides for weed control in Darjeeling tea. *Indian Agric.*, **35(3):** 179–185.

Kabir, S. E. and Ghosh Hajra, N. 1993. Chemical weed control in Kurseong tea plantation. *J. Hill Res.*, **6(1):** 37–40.

Kathiravetpillai, A. and Punyasiri, P. A. N. 1989. Effect of additives on the phytotoxicity of glyphosate to *Panicum repens*. *Sri Lanka J. Tea Sci.*, **58(2):** 118–123.

Kayange, C. W. 1978. Herbicides for tea. *Tea Research Foundation (Central Africa) Quarterly Newsletter*, **51:** 11-12

Kayange, C. W. 1980. Herbicides for young tea. *Tea Research Foundation (Central Africa) Quarterly Newsletter*, **57:** 7

Lamusse, M., Ramlogun, H. G. S. and Forget, J. 1978. Weeds in tea fields of Mauritius and their control by chemical weedkillers in newly established plantations. *Revue Agricole et Sucriere de l'Ile Maurice*, **57(3):** 115–139.

Lobzhanidze, A. and Ivardava, Sh. 1972. The efficacy of using herbicides. *Subtropicheskie Kul'tury*, **1:** 135.

Magambo, M. J. S. and Kilavuka, C. I. 1975. Preliminary observation on the effect of glyphosate on *Cyperus* weeds in young tea. *Tea in East Africa*, **15(1):** 11–14.

Magambo, M. J. S. and Kilavuka, C. I. 1976. Preliminary observation on the effect of glyphosate on *Cyperus* weeds in young tea. *Tea in East Africa*, **15(2):** 17–19.

Magambo, M. J. S. and Kilavuka, C. I. 1982. Effect of glyphosate on shoot growth of tea. *Tropical Pest Management*, **28(3):** 315–316.

Mishra, J. S. and M. S. Raghuwanshi. 1999. Eco-friendly herbicides for integrated weed management. *Pestology*, **23(3):** 17–18.

Murty, R. S. R. and Sharma, V. S. 1990. Effect of certain additives on the bio-efficacy of glyphosate. *Planters' Chronicle*, **85:** 267–271.

Mustafee, T. P. 1998. Integrated approach towards efficient weed management in tea (*Camellia sinensis* (L.) O. Kuntze.). *Assam Rev. & Tea News*, **87(5):** 6–12.

Purtskhvanidze, S. 1973. The effectiveness of simazine and mulching on a fully bearing tea plantation in Kvemo Imeretiya. *Subtropicheskie Kul'tury*, **4:** 24–26.

Rahman, F. 1977. Chemical weed control in tea. *Two and a Bud*, **24(2):** 33–35.

Rahman, F. 1991. Effect of different forms of cultivation and chemical weed control on yield of tea. *Two and a Bud*, **38(1-2):** 28–30.

Rahman, F. 1997. Suggested remedial measures for augmenting production in Darjeeling.

Proc. Seminar on Darjeeling Tea Industry, September 5, 1997, Calcutta, DPA and Tea Board of India, 24–32.

Rahman, F., Dutta, A. K. and Saikia, M. C. 1973. Biology and control of *Oxalis acetosella*. *Two and a Bud*, **20(2):** 46–47

Roy, N. K. and Dureja, P. 1998. New eco-friendly pesticides for integrated pest management. *Pesticides World*, April – May, 16–19.

Sandanam, S. and Jayasinghe, 1977.Manual and chemical control of *Imperata cylindrica* on tea land in Sri Lanka. PANS, **23(4):** 421–426.

Sansui, M. 1979. Problems and control of weeds on young tea in Indonesia. *Proc Sixth Asia–Pacific Weed Science Society Conference*, February 2, 1979, Jakarta, Indonesia, 427–432.

Sarma, S. N., Singh, M. P. and Thakur, K. C. 1986. Effect of glyphosate on thatch grass (*Imperata cylindrica* (L.) Beaur) control. *Two and a Bud*, **33(1/2):** 53–55.

Sarkar, S. K. 1972. Chemical weed control in Darjeeling. *Two and a Bud*, **19(1):** 22–24.

Satyanarayana, N., Sreedhar, C and Sharma, V. S. 1992. New recommendations on the use of glyphosate. *UPASI Handbook of Tea Culture*, Section 7: 2.

Scarborough, I. P. and Kayange, C. W. 1976. Herbicides in young tea. *Tea Research Foundation (Central Africa) Quarterly Newsletter*, **44:** 11–13.

Scarborough, I. P. and Kayange, C. W. 1977. Herbicides for young tea. *Tea Research Foundation (Central Africa) Quarterly Newsletter,* **48:** 10–11.

Sharma, V. S. 1975. Gramoxone-resistant weeds and their control. *Planters' Chronicle*, **70(3):** 59.

Sharma, V. S. 1977.Chemical weed control in tea fields. *Planters' Chronicle*, **72(2):** 79–80.

Sharma, V. S., Haridas, P. and Venkataramani, K. S. 1973. Roundup – a promising broad spectrum weed killer. *Planters' Chronicle*, **68(14):** 287–288.

Sharma, V. S. and Satyanarayana, N. 1981. Changing pattern of weed flora in tea fields in south India. *Bulletin of the Botanical Survey of India*, **23:** 121–125.

Singh, R. D. 1999. Weed control. *In : Global Advances in Tea Science* (ed. N. K. Jain). Aravali Books International, New Delhi, India, 659–674.

Singh, R. D., Shanker, A.and Gulati, A. 1992. Management strategies of weeds, insects and diseases in Kangra tea plantations. *Kangra Tea Festival Souvenir*, June 5–7, 1992, Palampur, India, 13–14.

Singh, R. D., Sohani, S. K., Singh, Brajindar and Chakrabarty, D. N. 1993. Influence of long-term mulching on yield of young tea and weeds. *Proc. Intl. Symp.*, November 18 20, 1993, Indian Society of Weed Science, Hissar, India, **3**(S): 34–36.

Sinha, M. P. 1985. A perspective of weed control in tea. *Two and a Bud*, **32(1-2):** 35–39.

Sinha, M. P. and Borthakur, B. 1992. Important aspects of weed control in tea. *In: Field Management in Tea*, Tocklai Experimental Station, Jorhat, Assam, India, pp. 134–139.

Smale, P. E. 1991. New Zealand has its own green tea industry. *Horticulture in New Zealand*, **2(2):** 6–9.

Soedarsan, A., Amri, J. and Santika, H. 1973. Glyphosate. I. Some notes on its effectson tea, coffee and rubber. *Menara Perkebunan*, **41(4):** 175–177.

Soedarsan, A., Kuntohartono, T. and Mangoensoekarjo, S. 1977. Weed control in plantation crops in Indonesia. *Menara Perkebunan*, **45(4):** 183–188.

Soedarsan, A., Mubijanto, B. O., Suhendar, E. and Santika, H. 1974. Screening of herbicides in productive tea, assessed from the view point of the weed situation. *Menara Perkebunan*, **42(2/3):** 121–140.

Somaratne, A. 1987. The use of herbicides in tea in Sri Lanka. *Tea Bulletin*, **7(1):** 10–18.

Tabagari, L. G. and Kopaliani, R. Sh. 1984. Preliminary data on agrotechnical measures in the interrows of young tea plantations, cv. Kolkhida, on podzolic soil. *Subtropicheskie Kul'tury*, **3:** 77–79.

Tabagiri, L. G., Mikeladze, A. D. and Cheishvili, I. D. 1988. Results of preliminary studies on some agricultural measures for restoring neglected and unproductive tea plantations. *Subtropicheskie Kul'tury*, **5:** 67–75.

Venkataramani, K. S., Padmanaban, R. and Haridas, P. 1972. Recommendations on the use of Ansar 529 in the control of weeds in mature tea fields. *Planters' Chronicle*, **67(13):** 311–312.

Watson, M. 1986. Notes on the control of foxtail grass (*Pennisetum polystachion* (L.) Schult)

Widjaja, S. 1977. The effect of some herbicides on weed population in unproductive tea plantations. *BIOTROP Newsletter*, **19 :**10.

Willson, K. C. 1999. *Coffee, Cocoa and Tea*. CABI Publishing, Wallingford, UK, p. 300.

Zarnadze, D. N. 1972. The results of interrow mulching in young tea plantation in the Kolkhida lowland. *Subtropicheskie Kul'tury*, **1:** 121–123.

Chapter 12

Insect pests and management

12. 1 Introduction

Tea plant is an evergreen woody perennial and grown mostly as a monoculture, provides a suitable condition for the pest to feed on and breed. It supplies plenty of food on one hand and on the other the problem of finding a suitable host for breeding purposes is much reduced. About 300 species of insects, mites and eelworms are active in tea areas but all of them are not economically important. The monographs by Green (1890) and Watt and Mann (1903) are the earliest contribution on tea pests which in fact form the basis of present day knowledge of tea pests occurring in Asia. Besides, information on tea pest biology are available from India (north-east—Hainsworth, 1952; Das, 1965; Banerjee, 1993; south India—Rao, 1970; Muraleedharan, 1991a), Sri Lanka (Cranham, 1966a), Indonesia (Kalshoven, 1950), Malaysia (Dammermann, 1929), China (Anonymous, 1974), Taiwan (Sonan, 1925), Japan (Minamikawa and Osakabe, 1979; Kodomari and Horikawa, 1989), Turkey (Alkan, 1957), former USSR (Dzhashi, 1970) and Africa (Benjamin, 1968, 1968b; Rattan, 1992a). Various aspects of tea pest control in Asia were again discussed by Banerjee (1993) and Muraleedharan (1992).

It was recently estimated that 1034 species of arthropods and 82 species of nematodes infest tea plant (Chen and Chen, 1989a, 1989b). Among the arthropod pests, Lepidoptera is the largest order containing 32% of the pest species followed by Hemiptera with 27%. The dynamic

Table 12.1 : Number of insects and mite pests recorded on different parts of the tea plant.

Plant part	Pests (no.)	As % of total number of pests
Foliage	734	58.9
Stem	300	24.1
Root	160	12.8
Flower	30	2.4
Fruit	9	0.7
Seedling	14	1.1

Source: Chen and Chen (1989a).

adaptations of insects have enabled them to attack every part of the tea plant and the maximum number of pests occur on foliage (Table 12.1). Each geographic region has its own distinctive pest fauna though many species have been recorded from more than one region. The number of insects and mites associated with tea depends on the length of time for which the crop is cultivated in that region. The area under tea became important only after allowing for 'age effect' and the latitude had no influence on species richness (Banerjee, 1981). The north Asian countries with the longest history of tea cultivation have the largest number of pests, while the South American countries where tea cultivation was rather introduced recently have few pests (Table 12.2). However, pest damage in tea often leads to a significant impact on productivity, in both short and long term, the latter leading to almost total debilitation and thus eroding into the capital asset itself.

12.2 Crop loss

Since the magnitude of pest infestation varies seasonally depending on altitude, climate and cultural practices therefore an exact assessment of crop loss is very difficult. However, Cramer (1967) estimated 8 % crop loss in tea in Asia due to the ravages of pests. In north-east India, Glover *et al.* (1961) assessed a loss of 20 million kg of tea which accounted for 13 % of the production whereas another estimate places this figure anywhere between 6 and 14 % (Banerjee, 1976a). In China, 10% decrease in crop due to pests and diseases was reported by Chen (1986). The intensity of

Table 12.2: Mean number of tea pests occurring in different regions of the world.

Global Section	Countries	Insect pests
North Asia	China, Japan, Korea	219
South Asia	Bangladesh, Burma, India, Indonesia, Malaysia, Philippines, Sri Lanka, Thailand, Vietnam	136
East Africa	Congo, Kenya, Malawi, Nigeria, Rawanda, Tanzania, Uganda	39
South America	Argentina, Brazil, Peru	4

Source: Chen and Chen (1989a).

losses may be higher today in view of the increased production and productivity engineered by improved agro-technology (Muraleedharan, 1992).

The loss of crop caused by individual pest species have also been assessed several times. Damage due to *Helopeltis theivora* Waterhouse, in south Indian tea estates resulted in 11 to 100 % crop depression during the months of highest pest incidence (Rao and Murthy, 1976) and 20-25 % crop loss reported in north-east India (Prasad, 1992). The annual loss caused by *Helopeltis schoutedeni* Reuter, in Malawi reported upto 55% and short term loss can be considerably higher (Rattan, 1984a). A highly significant negative correlation has been found between loss in crop and percentage of damaged shoot harvested (Rattan, 1984b).

In south India, chemical control of *Scirtothrips bispinosus* (Bagnall), resulted 13 to 56 % increase in crop (Rao and Murthy, 1976) while in Malawi it was 18 to 83 % increase in crop by controlling *Scirtothrips aurantii* Faure (Rattan, 1992b).

The loss due to red spider mite *Oligonychus coffeae* Nietner, varies between 6 and 11 % in the Dooars, 5 to 7 % in Assam (Sen and Chakraborty, 1964) and 6 to 20 % in certain parts of north-east India (Awasthy and Venkatakrishnan, 1977). Efficient control of eriophyid mites resulted in 8-17 % increase in crop in south India (Rao and Subramanian, 1968) though around 5 % loss due to eriophyids in Anamallais was reported by Muraleedharan and Radhakrishnan (1989a). Subramaniam (1995) reported losses due to pests is estimated at 7 to 10 % of the total tea crop; of this three-quarters is due to the red spider. It can at times be 15-25 % in individual gardens or in localized areas. In Sri Lanka, *O. coffeae* and *Brevipalpus spp.* (scarlet mite) could cause 0.56 to 4.99 % reduction in yield (Danthanarayana and Ranaweera, 1970) while in Indonesia about 13 % crop reduction by scarlet mites was observed (Oomen, 1981). In Kenya, infestation of *B. phoenicis* (Geijskes) and *B. obovatus* (Donnadieu), responsible upto 12 % crop depression (Rattan, 1992a). The control of yellow mites *Polyphagotarsonemus latus* (Banks), in an experimental trial resulted 80 % increase in crop (Cranham, 1966a).

Around 20 % loss of crop in severely affected areas by bunch caterpillar *Andraca bipunctata* Walker, was reported by Banerjee (1982). A highly significant negative correlation exists between the intensity of infestation by the shot-hole borer *Euwallacea* (= *Xyleborus) fornicatus* (Eichoff), and yield in the preceding months (Sivapalan, 1977). Muraleedharan and Radhakrishnan (1989b) observed when 10 % bushes were infested by borer, the loss in crop in the field was around 3 %. In north-east India, termite infestation accounts for 10 to 15 % reduction in yield (Sen and Chakraborty, 1964). Plant parasitic nematodes or eelworms limit yield of tea resulting crop reduction ranging from 4 to 40 % depending on the severity of infection (Gnanapragasam, 1989). It was estimated that approximately 5 % infestation by lesion nematode, *Pratylenchus loosi* Loaf, in the high altitude areas of Sri Lanka can cause a loss of 225-350 kg processed tea ha^{-1} $year^{1}$ (Visser, 1958).

12.3 Pest infestation and quality

Despite crop loss, pest infestation also adversely affects the quality of processed tea. Damage by sucking pests like thrips and mites results in dull appearance of tea. Teas made from flushworm infested shoots have low levels of extractable solids and high crude fibre content (Murthy and Chandrasekaran, 1979). Liquors obtained from such teas are 'flat' and the presence excreta of the larvae perhaps responsible for the deterioration in quality (Das, 1957). The severe infestation of flushworm, pink mite, thrips and tea mosquito bug adversely affected flavour and decrease in polyphenolic contents (Murthy and Chandrasekaran, 1979). But tea manufactured from cicadellid *Empoasca flavescens* Fabr. infested leaves had higher amount of pigments. An increase in polyphenol oxidase activity and total polyphenols was associated with high pigment formation in these teas (Mahanta *et. al.*, 1992).

12.4 Leaf and shoot pests

A. MITES

Various species of phytophagous mites infest tea. Among the leaf feeders, the red spider mite (Eriophyidae) *Oligonychus coffeae* Nietner, is widely distributed in India, Bangladesh and Sri Lanka. In Africa, it is also a serious pest in Burundi, Kenya, Malawi, Uganda and Zimbabwe. In Japan and Taiwan the common spider mite is *Tetranychus kanzawai*. Despite red spider mite, the pink or orange mite *Acaphylla theae* (Watt), purple mite *Calacarus carinatus* (Green) and scarlet mite *Brevipalpus phoenicis* (Geijskes) are also economically important pests in most of Asia.

O. coffeae breeds throughout the year and may be seen any time on the bushes. It is the largest among tea mites and can be seen by the naked eye. The population of red spider and other mites is found throughout the year in varying intensities. Tea bushes near roadside, labour lines especially those covered with dust are usually harbour more number of red spider. Infestation starts along midrib and vein and gradually spreads to the upper surface of mature leaves (Fig. 12.1) but when severe it attacks young leaves and spread to the undersurface. They often form a thin web of silken threads and remain protected against adverse weather conditions. The larvae, nymphs and adult mites suck cell sap and the infested leaves turns reddish or coppery and finally brown in colour and defoliation may occur leading to huge loss in crop. A severe attack of red spider is usually followed by the brown and the grey blight. Further, build up of mite population following hail infestation was also observed.

The scarlet mite *Brevipalpus phoenicis* (Geijskes), is widely prevalent species in Africa, China, Bangladesh and Indonesia. In India, it was originally restricted to the Dooars and Darjeeling but it has now spread to Assam and south India (Muraleedharan, 1983). The mites feed on the underside of the leaf particularly around the petiole and sometimes prevents the flow of nutrients into the leaf lamina (Banerjee, 1976b). Its feeding leads to yellowing of leaves and brownish discolouration of the midrib. Severe infestation results premature defoliation.

Fig. 12.1: Leaves showing damage symptoms of red spider mite.

The pink or orange mite *Acaphylla theae* (Watt) Keifer, occur in China, Bangladesh, Indonesia, Malaysia, Taiwan and USSR (CIS). In India, the pink mite is more prevalent in north-east India and widely spread in all the tea districts of south India. It attacks both surfaces of mature and young tender leaves. Petioles and tender stems are also invaded. Affected leaves turn pale yellowish, leathery texture and curl upwards. It could affect young tea during cold weather. The rate of development of *A. theae* varied seasonally. Its life cycle in north-east India during July-August is completed in 6 days (Das and Sengupta, 1958). In Taiwan and China 28 to 30 generations of *A. theae* are produced in a year, the development being completed in 6.2 to 9.5 days (Yeh and Huang, 1977).

The purple mite *Calacarus carinatus* (Green), is a active pest both in south and north-east India. The affected symptom being characterised by a copperish discolouration of leaves particularly at the margin of the upper surface of the mature leaves. In north-east India, the minimum development period of *C. carinatus* was 6 days (Das and Sengupta, 1962). It requires 20 days to complete a generation at 28^{0}C or 50 days at 16.5^{0}C (Shiao, 1975). In Kenya, it has been observed that bushes receiving no or inadequate fertilizer are comparatively more susceptible to this pest (Rattan, 1992a).

The yellow mite *Polyphagotarsonemus latus* (Banks), sporadically attacks the flush and the young leaves and causes damage in isolated patches in north-east and south India. But it is a serious pest in the nursery in east and south Africa, Mauritius and Malwi. Young leaves of mature tea also infested in Mauritius and southern Africa (Rattan, 1992a). Although they occur on both surfaces of leaves but is more abundant on the abaxial surface. It prefers cool, humid sheltered conditions and disappears with onset of dry and hot conditions. The affected leaves become more dark green and lose their natural gloss. The internodes become very shortened with many leaves arising out almost at the same point.

The outbreaks of red spider, scarlet mite and purple mite in Sri Lanka have been related to monthly precepitation patterns (Danthanarayana and Ranaweera, 1971). The adverse effect of persistent rainfall has been documented on the abundance of *O. coffeae* in Malawi (Mkwaila, 1985), *B. phoenicis* in Indonesia (Oomen, 1982), *C. carinatus* in Taiwan (Shiao, 1975), and *A. theae* in India (Muraleedharan and Chandrasekaran, 1981) and China (Kuang and Zhao, 1988). Since rain physically dislodges the mites from the adaxial surface of the unshaded tea leaves, therefore Banerjee (1993) opined its effect is perhaps more of mechanical nature than physiological. As scarlet and pink mites are inhabit the abaxial surface of leaves, they are least affected. China hybrids in general are more susceptible to the attack of red spider mites than Assam jats while Assam varieties are more prone to the attack of pink and purple mites. Rao (1978) reported in south India that the clones B/6/129 (UPASI-12) and B/5/149 (UPASI-149) harboured more number of pink mites while on leaves of B/6/62 (UPASI- 10, Pandian) population density of these mites was very low. Its rate of development is dependent on an optimum range of temperature and humidity. However, it fails to develop at any temperature beyond 35^{0}C (Das and Das, 1967). Further, light penetration within the canopy regulates the distribution of red spider mites. The highest concentration of mites found in the middle zone of the bushes where optimum combination of temperature and light intensity exists (Banerjee, 1971a). Pruning method, time of pruning and degree of cleaning out during pruning influences considerably the attack of mites. In north-east India, it has been observed that prunned tea was considerably less attacked by red spider mite in comparison with unpruned or skiffed areas (Das, 1965). In south India, it was seen that population density of pink mite in fields was higher in 2nd and 3rd year in comparison to first year fields (Radhakrishnan *et. al*., 1991). The number of *Brevipalpus phoenics* increased from the first year to the fourth year in a pruning cycle (Widayat and Oomen, 1981). In Anamallais (south India), an analysis of the distribution of pink and purple mites on tea bushes showed that the top portion of the bushes harboured 70.82 % of pink mite, 25.18 % was present in the middle portion, whilst the bottom portion has the minimum at 4 %. The occurrence of purple mites was more or less equal in top (45.85 %) and middle (42.4 %) portions. The population of purple mites in the bottom portion of the bushes was also found to be slightly higher (11.75 %) than that of pink mite at the same level (Muraleedharan and Chandrasekaran, 1981). The abundance of *O. coffeae* on mature leaves is correlated to the rhodoxanthin content which probably act as a feeding and reproductive stimulant (Fernando, 1967).

Control: Dicofol, ethion, thiometon, dimethoate at 1:200 dilution (low volume sprayer) and 1:400 (high volume sprayer). Sulphur formulations are also used to control mites but it is not advisable to spray in fields under plucking as sulphur taints the tea. Wetting of all leaves on both surfaces with spray fluid is recommended.

B. THRIPS

Several species of thrips (Thripidae) infest tea and cause heavy damage by feeding on the buds, tender leaves and older tea under plucking. Pruned tea is usually the worst affected. Tea plantations in south India, *Scirtothrips bispinosus* (Bagnal), *Heliothrips haemorrhoidalis* (Bouche) and *Haplothrips tenuipennis* Bagnal, are the three species of thrips attacking tea. *Taeniothrips setiventris* Bagnall, is a common thrip and cause considerable damage to tea in Darjeeling hills when new shoots start coming up. *Scirtothrips dorsalis* Hood, is a major pest in north-east India, Japan, China and Taiwan. *Scirtothrips aurantii* Faure, infests tea in Malawi, Zimbabwe, and South Africa. *Scirtothrips kenyesis* Mound, occurs in Kenya throughout the year. Black tea thrips *Heliothrips haemorrhoidalis* (Bouche), is also serious pest in some areas of Indonesia, Kenya, Tanzania and Uganda. In India, thrips cause considerable loss in tipping fields recovering from pruning. Continuous feeding of young leaves by adults and nymphs causes lacerations of the plant tissues. It sucks cell sap which oozes out from the damage leaf. They make small spots in scattered patches and/or in continuous lines on the leaf surface. Two or more corky lines often called sand papery lines appear on either side of mid-rib. Affected leaf become rough and deformed and may curl up. The life cycle of *S. dorsalis* is completed in about 12 to 15 days depending on environmental conditions (Dev, 1964). Developmental period of *S. bispinosus* varied from 5 to 18 days but it was prolonged upto 25 days during cold (Ananthakrishnan, 1963). Population of thrips gradually declines with the onset of monsoon. Prolonged droughts are associated with outbreaks of black tea thrips, *Heliothrips haemorrhoidalis* (Bouche), in Kenya (Rattan, 1992a).

Control: Tea thrips can be controlled by spraying endosulfan, phosalone, monocrotophos, quinalphos, dimethoate or chlorpyriphos at 1:200 (low volume) and 1:400 (high volume). Results obtained from the field experiments show that the effective control of thrips is possible if the control measures are taken up during early or middle of February when the initial attack is noticed. During peak period of incidence (May-June) two sprays at fortnightly intervals should be undertaken immediately after hard plucking (Das, 1983). Application of insecticides on foliage, frame and soil around the collar during the early part of the season gives better control of thrips

C. JASSID

The Jassid (Cicadellidae) *Empoasca flavescens* Fabr. commonly known as green fly, is another cell sap feeder occasionally causes serious damage to young and mature tea bushes in north-east India, Bangladesh, Japan, China and Vietnam. *E. onuki* and *E. formosana* occur in Japan and Taiwan respectively. Jassids are not common pest and rarely seen on tea in south India. Infestation of Jassid causes the margins of the young leaves to turn brown. The affected leaves curl downwards with recurved margins and growth hampered severely (Fig. 12.2). This characteristics symptoms is known as rim blight. The attack is restricted to the abaxial surface of the young leaves. In China, Chen *et. al*. (1978) observed that around 48 % of its population are present in the upper part of the plant, while middle and lower parts harbour the rest of the population. Prune tea in the plains is prone to Jassids particularly during March-April and reaches the peak in May-June. Developmental period completed in 8-10 days but the period may prolonged upto 15 days during cold.

Control: Regular plucking removes a large part of the eggs, nymphs and keeps the pest under control. Application of Monocrotophos,

Fig. 12.2: Leaves showing typical symptoms of damage by Jassid.

quinalphos, endosulfan, malathion, dimethoate, chlorpyriphos etc at fortnight intervals at 1:200 (low volume) or 1:400 (high volume) control the pest. Similar measures on the peak incidence of this pest should be undertaken as mentioned in case of thrips but while spraying care should be taken that the lower surface of the leaves are also properly sprayed.

D. TEA MOSQUITO BUG

The tea mosquito bug (Miridae) *Helopeltis theivora* Waterhouse, popularly known as 'tea mosquitoes' is one of the major pest causing extensive damage (Fig. 12.3) in north-east, south India, Darjeeling despite

Bangladesh and Indonesia. In Africa, *H. schoutedeni* Reuter, is a serious pest in east Africa, Malawi and Cameroon. Another species *H. orophila* Ghesq. occurs as a minor pest in Uganda, Burundi and Zaire. *H. bradyi* and *H. cinchona* Mann, infest tea in Malaysia while *H. clavifer* Walker and *H. antonii* Signoret, attack tea in New Guinea and Sri Lanka respectively. Assam clones are more resistant to *H. theivora* than Cambod or China clones (Gope, 1991). The nymphs and adults of *Helopeltis* are the active suckers of sap from buds, young leaves, tender stems and shoots. They feed at early morning, late evening and night hours. On sunny and warm day, *Helopeltis* takes shelter in lower layer of bush canopy. The insect inject toxic saliva through their needle like rostrum which causes necrosis of the tissues and turn brown

Box 12.1 : A belief

The incidence of Jassid used to welcome the tea planters of Darjeeling. There is a consensus of opinion that the green fly (Jassid) damage improves the quality of tea by increasing the flavour. But there is no definite evidence on the belief. However, if there is even some improvement in flavour, it is doubtful how far the loss of crop is compensated by the improvement of quality of processed tea.

Source: Anonymous (1988)

Fig. 12.3 : Leaves showing damage symptoms by mosquito bug.

first then black and they dry up subsequently. A female may lay about 220 white sausage-shaped eggs in 36 days during March-April, 4-10 eggs are laid per day (Chakravartee, 1995). Eggs are laid in tissue of tender stem, mid-rib and petiole of leaves. Das (1984) reported that a single first instar nymph of *H. theivora* could make as many as eighty feeding lesions in 24 hours. Infestation of *Helopeltis* may become destructive during the cropping season.

A total of at least 3–4 rounds of rehabilatory spray (zinc sulphate 500 g, boric acid 200 g, urea 500 g in 200 litres of water) can be done in between the pesticidal applications or at weekly intervals.

Control: Endosulfan, chlorpyrifos, fenitrothion, quinalphos, monocrotophos, at 1:200 (low volume) and 1:400 (high volume) can be sprayed during early mornings and late afternoon when these bugs are active. The ideal choice of insecticides would be to opt for a cycle: contact then systemic again contact. The powerful contact insecticide is used to knock down the mobile fraction of the population at a shortest possible time and the systemic one to take of the emerging populations within the bush and unfolding leaves. Before spraying the whole section, about 6 to 10 rows of the bushes covering the outer periphery of the section would be sprayed first and then the operators should move from the periphery towards the section of the centre systematically. Barrier spraying helps to trap the migratory and mobile stages of the population of tea mosquitoes. Spraying should be done after removing the infested leaves by skiffing from the table or by black plucking. The sections prone to *Helopeltis* attack every year may be pruned or deep skiffed if possible instead of leaving them either unpruned or under lighter form of skiff. Pruning / skiffing in the pest prune areas should be done from the periphery towards the centre where few bushes may be left unpruned. Spray on untouched bushes may be done after three days and then only pruning / skiffing of the bushes can be done. Abundant growth of weeds should not be allowed in *Helopeltis* prune areas.

E. APHID

The tea aphis (Aphididae) *Taxoptera aurantii* Boyer de Fonscolombe, is a common pest and widely distributed almost all the tea growing countries. The adults and nymphs of this insect continuously suck sap from lower surface of tender leaves along the midrib, tender stem, bud and petioles resulting crinkle leaves curl inwards and growth of the shoot retarded. The field recovering from pruning suffer worst attack of aphids from February onwards. The life history is completed in 5-10 days depending on season. Reproduction takes place parthenogenetically. Males are not found during any part of the year.

Control: Endosulfan, phosalone, dimethoate, quinalphos can be sprayed when infestation is severe.

F. SCALE INSECTS AND MEALY BUGS

Coccoids are sucking pests which attack the foliage, stems and roots of the tea bushes (Fig. 12.4). Their increasing trend have been found

Fig. 12.4: Branches of tea plant infested by the mealy bugs.

throughout north-east India and Darjeeling. Over 30 species of scale insects and mealy bugs (Coccoidea) have been recorded from tea; out of which 12 are serious pests infesting leaves, tender stem (Das and Ganguli, 1961). In Darjeeling, *Eriochiton theae* (Green), *Pinnaspis theae* (Maskell), *Phenacaspis manni* Green, *Ceroplastes rubens* (Maskell), *C. cerifera* (Anderson), *Hemiberlesia rapax* (Comstock), *Andaspis dasi* Williams, *Asteroleucanium pustulans* (Ckll.) have been recorded as major pests of mature tea. *A. dasi* caused severe damage to branches of mature tea under cover of mosses. Cultural practices in Darjeeling gardens to keep the bushes unpruned and also the weather conditions are perhaps the main contributing factors for this pest. In south India, two coccids viz. *Saissetia coffeae* Walker, *Coccus viridis* (Green) are commonly found in tea plantations. Other species reported from south Indian tea are *Aspidiotus destructor* (Signoret), *Pinnaspis theae*, *Chrysomphalus dictyospermii* (Morgan) and *Ceroplastes floridensis* Comstoc. In south India, very recently *Nipaecoccus viridis* (Newstead), the shoot mealy bug, has started appearing in many estates in Anamallais (Muraleedharan, 1993). Infestation by *N. viridis* on tea is localised and honey dew excreted by the insect favours the development of sooty mould on tea leaves and this affects normal photosynthetic activity (Muraleedharan and Radhakrishnan, 1990). Most of these coccoids also occur either throughout the year or occasionally in Dooars and Assam. The mulberry scale, *Pseudaulacaspis pentagona* Trag. and Toss has been a serious pest and it occurred in 40 % of tea areas in Japan (Tatara, 1999). The coccid and scale infested shoots become unproductive leading to drying up of leaf and fall. When attack lasted for long irregular swelling noticed was on the branches due to growth of callus inside.

Control: Quinalphos, dimethoate, monocrotophos, fenthion can effectively controll these pests. In the plucking season two rounds of drenching spray with appropriate chemical at 15 days interval would generally be sufficient. Addition of spreaders/stickers will be helpful in achieving better control.

G. CATERPILLARS

The looper caterpillar (Geometridae) *Buzura suppressaria* Guenee, is a well known pest of tea in India and Indonesia. It is responsible for considerable damage in upper Assam, north bank and Dooars. But it is a minor pest in south India. In China, *Ectropis obliqua* Warren and in Sri Lanka, Indonesia, *E. bhurmitra* Walker, are the main geometrid caterpillar

Fig. 12.5 : A mature larva of looper caterpillar.

pest. The young early instars feed on the margin of the tender leaves while the late instars prefer older leaves (Fig. 12.5). In severe cases of attack caterpillars can completely strip a bush of its leaves and infestation becomes devastating within a short period. The young looper is dark brown with pale greenish white lines on the back and sides and has a brownish head. The body colour gradually changes to green to brownish grey or dark brown. The life cycle is completed in about 60-72 days. Young larva of looper is highly mobile and can therefore, be carried away to long distances by wind when suspended on silken thread.

The bunch caterpillar (Bombycidae) *Andraca bipunctata* Walker, is widely distributed tea pest and found in north-east India and occasionally observed in Darjeeling gardens. The destructiveness of caterpillars increases as they grow in age (Fig. 12.6). They remain in clusters during day hours and become active in the night and start feeding ravenously. The branches on which they form clusters are completely stripped of their leaves thereafter they migrate to adjacent bushes. They lay oval shaped yellow eggs in rows on the abaxial surface of the leaves. The life cycle is completed in 55-65 days in March-June.

Fig. 12.6 : Tea bush defoliated by bunch caterpillar.

The black or Sungma caterpillar *Lymantria albulunata* Mre. found to cause extensive damage to young and mature teas during early 1980s in Sungma tea estate of Darjeeling hills. This is an endemic pest restricted to some particular gardens only. It is normally found at an elevation of about 1300 metres and highest population found in May. The female lays eggs in cluster on the abaxial surface of the leaf.

The Darjeeling black hairy caterpillar *Euproctis latifascia* Walker, is observed in many gardens of Darjeeling hills with limited damage caused by them. It was also reported to cause damage in estates of north bank in Assam. It attacks old leaves of young and mature tea. They make small and big perforations on the leaves and occasionally whole leaf is eaten up. Sometimes the young plants in the nursery are defoliated and death occurs. The hairy caterpillars remain concealed underneath the stone, fallen bamboos and lathframes or clod of earth during the day and the damage is caused during night. The body colour of the full grown caterpillar is blackish brown.

Other major caterpillar pests are the red slug caterpillar (Zygaenidae) *Eterusia magnifica* Butler, a serious pest in north-east India; *E. virescens*

Butler, occur in tea in south India, while *E. singala* Moore, infests tea in Sri Lanka. Usually the caterpillar attacks the leaves but in severe outbreak, the bark is also eaten away resulting in die back of the branches. Leaf folding caterpillar such as flushworm (Tortricidae) *Cydia leucostoma* Meyrick, a main tortricid pest in India, Indonesia and Bangladesh. In south India, flush worm is abundantly seen in the mid elevation tea areas of Anamallais and Vandiperiyar during July to December (Muraleedharan, 1993). It has become a serious pest of young tea in Darjeeling hills (Bisen and Ghosh Hajra, 1997). The larvae immediately after hatching tie up the margins of two or three tender leaves thus forming a protective pocket where the caterpillars live and feed (Fig. 12.7). The affected leaves become rough, thick and somewhat brittle in texture with crinkled appearance. Some tortricids like *Homona magnanima* Diakonoff and *Adoxophyes* sp. cause extensive damage to tea in Japan (Kawai, 1997) and Taiwan while *Homona coffearia* Nietner, is an important pest in Sri Lanka, Mauritius, Bangladesh, China and India. These caterpillars make nests by webbing the leaves, often one above the other, using silken threads and feed from inside. It causes considerable damage in the nurseries. The tea leaf roller (Gracillariidae) *Caloptilia theivora* (Walsingham), is a minor lepidopterous pest in India

Fig. 12.7 : Shoot showing typical damage by flush worm larvae.

and major pest in Japan (Kawai, 1997). This species is also known to infest tea in Sri Lanka, Bangladesh, Indonesia and Malaysia. This caterpillar, in general, attacks young leaves and is found in fields recovering from pruning. The limacodids such as *Thosea cervina* Moore, *T. recta* Hampson, *T. sinensis* Walker, *T. cana* Walker, *Latoia lepida* Cramer, *Parasa pastoralis* Butler, *Belippa lalaena* Moore, *Phocederma velutinum* Koll. are the common occasional defoliators of the different tea growing areas. The psychidae group of caterpillars are very primitive and live inside silken bag covered with bark, small twigs and leaves. Female moths are devoid of wings. Some of the well known psychids are large fagot worms *Eumeta crameri* (Westwood), *Clania cramerii* Westw. and small fagot worm *C. destructor* Dudg., bag worm *C. sikkima* Moore, busket worm *Mahasena theivora* Dudge and conical psychid *Manatha assamica* Watt. These caterpillars as geometrids and limacodids feed mainly on the maintenance foliage. The lobster caterpillar (cochllidae) *Stauropus alternus* Walker, eats away all the leaves of the branch. Their attack in general is confined to a few scattered bushes. The highly polyphagus noctuid, the cut worm *Spodoptera litura* (Fabricius), enjoying wide distribution in Asia and Australia at times feeds on tea leaves. This caterpillar first invades the weeds and shade trees and in absence of their natural host migrate to the maintenance foliage of the tea plant. The young caterpillars feed on leaves by scraping the epidermal cells. Feeding by mature larvae is characterised by the irregular holes made on the leaves.

Control: Endosulfan, quinalphos, monocrotophos, phosalone are effective. Synthetic pyrethroids are very efficient to control bunch, Sungma and hairy caterpillars. The application rate should be 1 : 2000 and 1 : 4000 with low volume sprayer only for full grown caterpillars and early instars respectively. The infestation of flushworm can be controlled by hand picking of infested shoots at the time of plucking. Manual removal of caterpillars, pupae will be an easy method in controlling them.

H. WEEVILS AND BEETLES

Various kinds of weevils attack tea in India, Sri Lanka, China and Africa. *Systates smeii* and *Entypotrachelus meyeri* infest tea in Malawi and Kenya respectively. These weevils devour young and old leaves and if they attack in the nursery or in newly planted field the damage will be serious. *Myllocerus curvicornis* Fab. attacks tea in Sri Lanka. In south

India, *Myllocerus sp.* occasionally eats tea leaves. In Anamallais, south India, *Geotragus sp.* was found attacking tea leaves. The large green weevil *Astycus chrysochlorus* Wield. and small green weevil *A. lateralis* Fabr., often caused damage to maintenance leaves of tea in north-east India. *Geotragus subfaciatus*, attacks tea leaves in Darjeeling hills. The beetles such as *Mimela xanthorrhina* Hope, *Callistethus gemmula* Arrow and *Serica assamensis* Brenske, have been reported to feed on leaves leaving the midrib and veins.

Control: Endosulfan, chlorpyrifos and quinalphos are effective against *Myllocerus* and *Geotragus sp*. Deltamethrin, cypermethrin at 1:2000 dilution using low volume sprayers are effective against large and small green weevils.

12.5 Stem and branch pests

A. STEM BORER

The cerambycid *Haplothrix griseatus* Gah., often bored the stem of mature tea. As the larva grows it tunnels through the branches and the affected branches become unproductive and they die-back.

The red borer *Zeuzera coffeae* Nietner, often damage young plants in Sri Lanka, Malaysia, Taiwan, Indonesia, India and it causes considerable damage in some parts of Assam. Another species *Z. leuconotum* causes damage in Japan. The larva tunnels through the branches and reach the main stem (Fig. 12.8). They even sometimes reach upto root of the plants which result in death. The red borer is generally pest of new tea clearings and appears in patches. The adult beetles lay eggs in strings and reddish brown colour larvae emerge out of the eggs, settle on the stems to bore tunnels.

In upper Assam and in China, the oecophorid stem borer *Casmara patrona* Meyrick, occasionally damage tea stem.

The hepialid borer *Sahyadrassus malabaricus* Moore, endemic to south India, is an occasional pest. In the temperate regions of China and Georgia, *Parametriotes theae* Kusnetsov, damages tender stems. The eggs are broadcasted by female moth in flight and the larvae affect the stem of the bush. The full grown larvae is around 8-9 cm long, pencil thick in size and pale yellow in colour. The larva makes long cylindrical tunnels opening

Fig. 12.8 : The larva of red borer showing damage.

into extensive cankers. Wood of the stem and callus tissue are used by the larva as food. The entrance hole of the tunnel is covered by frassy mat of chewed wood and silk.

The scolytid shot-hole borer *Euwallacea (Xyleborus) fornicatus* Eichoff, found to damage tea in India, Indonesia and Taiwan. This is a well known pest of tea in the middle and low elevation areas of southern India and occasional pest in north-east India. It is a major pest in Sri Lanka (Cranham, 1966b). Females of this beetle are black in colour and capable of flying enter tea stem by boring a hole through the bark and construct galleries within the branches in which eggs are laid. These beetles have symbiotic association with certain groups of fungi called 'ambrosia fungi'.

In Sri Lanka, the symbiotic fungus of the shot-hole borer was recorded as *Monacrosporium ambrosium* and in south India it is *Fusarium bugnicourtii* Brayford (Parthiban and Muraleedharan, 1992). The larvae of the beetle feed on the mycelia of the fungus growing on the walls of the gallery; pupation takes place within the gallery and the newly emerged adults mate within the gallery (Banerjee, 1993).The new branches formed after pruning are more susceptible to this borer attack. Infestation by the beetle observed more in the 3rd and 4th years of the pruning cycle and the population exhibit peaks in April/May, July, October and December in south India (Muraleedharan, 1991b).

Despite boring into wood, bark eating borer *Indarbela theivora* Hamps and *I. quadrinotata* Walker, also feed on the bark of the branches of tea. A spiral web round the stem made up of bark and excreta spin by the larvae by their silky secretion. The larvae feed on the bark under cover of this web.

Control: Endosulfan, quinalphos, deltamethrin, cypermethrin, fenvalerate are effective.

B. CARPENTER MOTH

Carpenter moth *Teregra quadrangula* is a potentially serious pest, especially of young tea in central Africa. Caterpillars feed on bark and ring bark branches resulting in calluses and knots; ring barking of young plants may, however, result in death of plants. Most of the branches of mature tea remain productive though some may stop flushing or die.The presence of this pest has a strong bearing on infilling and replanting programmes (Anonymous, 1995).

Control: Carbaryl is effective.

C. BEETLES

The stem eating beetles in small numbers are found on tea bushes in north-east India as well as in Darjeeling causing damage to tender stem during March to October. It includes orange beetle *Diapromorpha melanopus* Lecord and metallic green beetle *Chrysolampra flavipes* Jacoby.

Control: Spraying with Deltamethrin, cypermethrin, fenvalerate at 1:2000 (low volume) or quinalphos at 1:400 (high volume) give effective control.

D. TERMITES

Termites causes considerable damage usually in dry period to young and mature tea including cuttings in the nursery beds and shade trees in north-east India. Cachar and Darrang districts of Assam are the most affected areas. They are active during cold weather (Banerjee, 1975). However, these are not serious pests in south India and considered as secondary pests but they pose a serious threat to tea in Sri Lanka, Bangladesh and Cachar in north-east India. Termites sometimes girdle branches and attack the frame of mature plants. In Africa, *Macrotermes natalensi*, *Pseudoacanthotermes militaris* etc often sever or ring bark young plant resulted wilting and in severe cases dies. Killing of bushes in Mauritius due to building of galleries inside the main stem, branches and roots by *Coptotermes havilandi* was observed (Rattan, 1992a). A survey of termites in Kawalazi and estates in southern Malawi showed *Ancistrotermes* to be particularly prevalent in Kawalazi where it was causing economic damage (Anonymous, 1995). *Glyptotermes dilatatus* Bugnion and Popoff, *Postelectrotermes militaris* Desneux and *Neotermes greeni* Desneux, attack tea in Sri Lanka (Muraleedharan, 1991a). *Microtermes* sp., *M. pakistanicus* Ahmed, *Microcerotermes* sp., are live wood eating termites in north-east India and Bangladesh while *Odontotermes assamensis* Holmgren, *O. feae* Wasmann, *Neotermes buxensis* Roonwall and Sen Sharma, are scavenging termites damaging tea in north-east India (Anonymous, 1994).

Control: Drenching spray of lower frame and soil near collar with endosulfan 35 EC or chlorpyriphos 20 EC at a dilution of 1:300 is effective in preventing the damage. Application of Phorate 10G, 5 to 10 g depending on the age of the plant is also effective. Further, soil should be forked before application of this granular insecticide and after application slight watering of soil is essential.

12.6 Root pests

A. NEMATODES OR EELWORMS

Plant parasitic nematodes of tea which are found in the soil affect the root system of the plant. There are 40 species belonging 20 genera of plant parasitic nematodes are associated with tea in different tea growing countries. But most of them have no positive evidence of pathogenicity

(Sivapalan, 1972). However, infestation of root-knot nematodes *Meloidogyne incognita* Kofoid & White, *M. hapla* Chitwood and *M. javanica* Treub, are very destructive to young plants in the nursery as well as fields in north-east and south India. *M. javanica* is commonly associated with tea in India, Sri Lanka, China, Indonesia, East and Central Africa and Queensland of Australia. *M. brevicauda* Loos, parasitises the roots of mature tea in the high altitude areas of Sri Lanka and in isolated pockets of south India and China. The root lesion nematodes *Pratylenchus lossi* Loaf, is a serious pest in Sri Lankan tea plantations located in the elevation between 900 and 1800 m and to a lesser extent in the plantations of China and Japan (Muraleedharan, 1991a). *Pratylenchus curvitatus* Van der Linde, is reported from the feeder roots of tea in north-east India, Japan and Sri Lanka. *Helicotylenchus dihystera* Cobb and *H. erythrina* Zimm. are predominant in Sri Lanka and Japan respectively. The invasion of roots by eelworms stimulates the root tissues to abnormal growth with the result that galls or root knots are produced and normal differentiation of cells into vascular tissues is virtually blocked and transport of water, solutes are impaired. Therefore, nutrient imbalance is created, photosynthetic rate is reduced and symptom of nutrient deficiency is expressed causing a general decline in the vigour of the plant which ultimately becomes more susceptible to various kinds of stresses. Premature flowering and fruiting some times noticed in case of severely infested plants. The resultant symptoms in the aerial parts are stunted growth, pallor of the leaves and later wilting. Tea plantation infested with nematodes are characterised by patches of unthrifty teas having weak frames. In the nursery, though clonal plants and seedlings are equally affected by root knot nematodes but clonal plants can withstand nematode attack comparitively better than seedlings since they possess greater amount of roots.

Control: No specific method ensuring complete eradication of nematodes in the nursery is available but some precautionary measures can be taken to reduce the intensity of attack to seedlings and cuttings. However, nursery soils may be treated with systemic granular formulations such as carbofuran 3G, phorate 10G @ 1 g / sleeve / 1 kg of soil. Second application at the same rate after one month of first application should be given. Watering should follow after each application for quick absorbtion of chemical. Phytotoxicity was noticed when applied at higher dosages of granular nematicides. Basamid G, a broad spectrum soil sterilant, @ 20, 30 and 40 gram / sq. m soil was also found very effective against eelworm.

B. WHITE GRUBS

Holotrichia sp. normally causes damage to young tea around the collar region which is debarked in the form of a ring or in patches. In such cases the mulch materials during late February or early March should be kept in between rows of tea plants instead of keeping them around the collar, so that the adult cockchafer do not lay eggs near the collar region in the mulch. The early instar grubs feed on the tender roots of grasses and on dry decomposed leaves. The later instars feed on the young tea plants in the collar region. Cockchafer beetles emerge between March and June in north-eastern region of India. They mostly affect plants in new clearings. But damage is not uncommon to three to four year old plants. In Sri Lanka, adult *H. disparilis* Arrow, lay eggs in April-May and the grubs emerge from June onwards. The pupation takes place in December-January and the life cycle completed in about a year (Thirugnanasuntharan *et. al.*, 1988). In the plains of north east India, the duration of life cycle of *H. impressa* Burm. is about a year but the life cycle lasts for two years at an altitude above 1600 m (Das, 1965). This insect is abundantly found in Darjeeling hills.

Control: In the affected area, the surface soil around the collar of each plant may be loosened and treated with endosulfan / chlorpyriphos solution at a rate of 1:300. After 15 to 21 days of first application, second application at the same rate may be given. If the proposed new clearing area is infested, application of 10 g dust formulation of aldrin, chlordane, heptachlor or endosulfan per pit give effective control.

The root mealy bugs *Pseudococcus theaecola* (Green), *Crisicoccus* sp., damage roots of the tea bushes in Darjeeling (Das, 1965). It can be controlled by soil drenching with endosulfan / chlorpyriphos at 1:300 dilution after soil forking of the collar region.

Rats (*Rattus* sp.) and bandicoots (*Bandicota* sp.) at times make burrows in the fields and eat away the roots of the tea bushes. Grass infested fields are more prone to rat damage than clean fields. Baiting with zinc phosphide and anticoagulants viz. bromadiolone, coumafuryl, warfarin have been found useful for rodent control.

Fig. 12.9 : The tea seed bug – fifth instar nymph.

12.7 Flower and seed pests

There are several species of thrips such as yellow thrips *Taeniothrips lefroyi* Bagnall, black thrips *Haplothrips tenuipennis* Bagnall infesting tea flowers. The Pentatomid, *Poecilocoris latus* Dall., medium sized coloured bug is the most destructive pest in tea seed gardens in India (Fig. 12.9). The bug sucks the juice of the seed cotyledons and prematurely fall off the young seeds taken place. In case of mature seeds which can not be punctured through the shell, the feeding is restricted to the pericarp only. A weevil *Curculio styracis*, causes injury to tea seeds in China. Sometimes the grubs growing inside the tea seeds feed the contents (Muraleedharan, 1992).

Control: Endosulfan, quinalphos, monocrotophos, thiometron or dimethoate at 1: 400 dilution may be sprayed during July-August. If necessary, second and third spray may be done during February-March and May-June respectively.

12.8 Seasonal abundance of pests

Since insects are poikilothermic (cold blooded), their body temperature varies with the surrounding temperature and so they are especially strongly influenced by climate and weather. Temperature, precipitation, humidity, wind speed and other climatic variables can directly influence pest by affecting their rate of development, reproduction, distribution, migration and adaptation. In addition, indirect effects can occur through the influence of climate on the insects' host plants, natural enemies and interspecific interaction with other insects (Porter *et. al.*, 1991).

The red spider mite *Oligonychus coffeae*, persists on tea bushes all the year round with varying numbers. Its population increases as temperature rises in March-April. In Assam and Dooars it causes severe damage to tea in May and June until monsoon sets in. During rain the active forms are washed down resulting decrease in population from July onwards. At the end of rainy season sometimes a second peak in the populaton occurs but occasionally it assumes serious proportion. The population of red spider is reduced during winter and extremely low during December and January even on unpruned tea (Anonymous, 1988). Its incidence in Africa also increases in dry periods. Intermittant rains in the hot dry period favour a rapid build-up. However, the population decrease markedly with the onset of rains (Rattan, 1992a). The population of purple mite *Calacarus carinatus*

Box 12.2 : Effects of climatic change on agricultural insect pests

Climate and weather can substantially influence the development and distribution of insects. Anthropogenically induced climatic change arising from increasing levels of atmospheric greenhouse gases would, therefore, be likely to have significant effect on agricultural insect pests. Current best estimates of changes in climate indicate an increase in global mean annual temperatures of 1°C by 2025 and 3°C by the end of the next century. Such increases in temperature have a number of implications for temperature-dependent insect pests in mid-latitude regions. Changes in climate may result changes in geographical distribution, increased overwintering, changes in population growth rates, increase in the number of generations, extension of the development season, changes in crop-pest synchrony, changes in interspecific interactions and increased risk of invasion by migrant pests. Some of these effects can be illustrated using an example of the European corn borer (Ostrinia nubilalis) (ECB). Lepidopteran ECB is probably native to southern Europe, but is now widely distributed in the northern hemisphere between latitudes 10 and 58° in Europe, Asia, northern Africa and North America. The ECB can adapt to a wide range of climates and it is likely that it will eventually cause damage wherever maize is grown.

Source: Porter et. al. (1991)

Green increases from October in north-east India and severe attack can be seen even before winter. As the ambient temperature goes higher from February or early March the multiplicity of the adults increases resulting severe damage of the leaves of young and mature bushes by this mite.

The scarlet mite *Brevipalpus phoenicis*, of all stages can be found on tea bushes in Assam round the year but in general its population increases from March reaching a maximum in May-June. A part of population washed away during rains but a majority continues to multiply till January. A temperature ranged between 25 to 30^{0}C and humidity of 70 to 80 % were found to be optimum for growth of scarlet mite population (Anonymous, 1992).

The comparative abundance and seasonal distribution of mites vary in the different tea districts of south India. In the Anamallais, the populations of pink and purple mites exhibited a gradual increase from November, reached their peaks in January-February and started declining from March onwards (Muraleedharan *et. al*., 1988). The weather factors have shown great influence on the seasonal incidence of pink mite *Acaphylla thea* Watt and purple mite in Anamallais. Rainfall adversely affect the growth of both the species. Minimum temperature significantly negatively correlated to populations of *A. thea* (Muraleedharan and Chandrasekaran, 1981). In Central Travancore, purple mite infestation starts early in August-September, reaching peak during November-December and continues upto May-June. The scarlet mites cause severe defoliation of maintenance leaves during the peak dry weather in April-May (Joseph, 1992). Severe red spider attack observed in certain areas of Vandiperiyar during July-August. In Wanaad and Nilgiri-Wynaad, pink mite is comparatively more commonly observed than purple mite. The infestation is observed mainly in the flush from August-September onwards and continuing to December-January. The mite infestation in Nilgiri-Wynaad is recorded to be very low in comparison with Anamallais, High Range and the Nilgiris (Joseph, 1991). The pink mite infestation in Wynaad and Nilgiri-Wynaad increases after scanty summer showers in March-April. In the Nilgiris, heavy infestation of mite is observed during May. In the High Range, incidence of of purple mite in general is more severe than that of pink mite and its severity starts from January and continues upto April with a peak attack in March-April (Surendra Mohan, 1989). In Indonesia, the infestation of different mites are generally seen during the dry months.

Although a temporary decline in mite population was observed in the beginning of the monsoon in May but the insect sap feeders viz. thrips, Jassid, tea mosquito and aphis become active. They attack mostly young leaves and buds.

The thrips *Taeniothrips setiventris* Bagnall and *Scirtothrips dorsalis* Hood, depending on pruning cycles are seen more or less throughout the year on tea bushes in Assam. The population starts building up in March-April when prolonged droughty conditions prevail and its peak incidence observed during May to July and gradually declined with the onset of monsoon. During cold very few thrips are found in the pruned sections. In south India, thrips are known to attack during April-May and August to October in Nilgiris, the Anamallais and Munnar; in Peermade and Vandiperiyar these are seen throughtout the year with peak infestation from August to Novenber/December and February to May/June (Joseph, 1992). In Africa, highest populations coincides with the dry hot period but the number reduced abruptly in the monsoon (Rattan, 1992a).

In the foothill regions of Darjeeling the incidence of aphids, thrips and red spider mites was observed to be correlated with maximum-minimum temperature, humidity and rainfall. In general, the month of February and March appeared to be most favourable for incidence of aphids *Toxoptera aurantii* Boyer de Fonscolombe, followed by a greater incidence in July. The pest population showed insignificant correlation with aforestated weather factors. The population of red spider mites mainly occurred in late winter (December to March) and summer months (May to July). The mite population was found to be very much weather dependent (Sannigrahi and Mukhopadhyay, 1993). In the Anamallais, south India, infestation of aphids was observed throughout the year. The field recovering from pruning suffer worst attack of aphids from February onwards. Their peak period of abundance was from January to March and again from August to October. It was observed that aphid populations are adversely affected by rainfall, low temperature and high relative humidity and the population density is found high under shade (Radhakrishnan, 1989).

The population of Jassid found on tea bushes throughout the year declined during cold weather but with the rise of ambient temperature during March-April its population increases. The population reaches peak in May-June in plains and June-July in Darjeeling.

In plains of north-east India, the tea mosquito *Helopeltis theivora* Waterhouse, breeds all the year round but its number fluctuates. The population during cold dwindles down and reappears in February-March from the residual population. Serious infestation seen during June to September when bushes virtually cease to form shoots and thereby may be total loss of crop. The infestation subsides in November and the adults are rarely found in December. The smallest population of *H. schoutedeni* Reuter occurs in Africa in the hot dry period (Rattan, 1992a).

The peak period of incidence of red borer *Zeuzera coffeae* Nietner, in Assam was found to be in March to May and again in August to October (Anonymous, 1992).

The tea leaf hopper *Empoasca formosana* reached high population density during the summer months in Taiwan (Chen *et. al.*, 1978).

In the tea field of Anamallais, south India, the infestation of tortrix *Homona coffearia* Nietner, gradually increased from September-October reaching a peak in December-January and declined by March-April (Selvasundaram and Muraleedharan, 1989).

In general, sap feeders cause the maximum damage during main flushing period because different pests almost at the same time infesting various parts of the plant. In north-east India, heavy precipitation during July-August cause decrease in the abundance of sap feeders but this does not deter an abundance of defoliaters which damage mostly maintenance foliage viz. looper caterpillar *Buzura suppressaria* Guen., bunch caterpillar *Andraca bipunctata* Walker, lobster caterpillar *Stauropus alternus* Walker, lymantriid caterpillar *Euproctis? subnotata* Walker, Darjeeling black hairy caterpillar *Euprocfis latifascia*. Looper attack was worst during autumn in Assam. The young leaf eating species that are active during heavy rains include cockchafer beetle *Serica assamensis* Brenske, cockchafer grubs *Holotrichia seticollis* (Moser) and *Sophrops plagiatula* (Brenske) (Banerjee, 1993). When rain started declining in September onwards the activity of the defoliaters gets slow but scarlet and purple mites become active on mature lower leaves.

The population of shot-hole borer *Xyleborus fornicatus*, in tea growing areas of south India found mainly correlated with the temperature and elevation. In the higher elevated areas in Nilgiris (2500 m) and high ranges (1900 m) where the temperature regimes are low, infestation of borer found negligible but severe infestation was noticed in Central Travancore areas where the elevation is low (1400 m) and the temperature regimes are higher (Kandasamy and Muraleedharan, 1983). In Anamallais of south India had four peaks in a year; the periods of maximum abundance were April or May, June or July, October and December or January (Muraleedharan, 1985).

The root lesion nematode *Pratylenchus* sp. and root knot nematode *Meloidogyne* sp. occur round the year in the soil in odd patches. Their infestation was noticed very severe between the months of June and August and declined during cold in north-east India (Anonymous, 1994).

12.9 Pest management

The use of conventional insecticides and acaricides are the most common method known to bring about immediate reduction in pest populatoins, increased yield and relible economic returns. In fact, chemical pesticides have virtually become *sin qua non* for cultivation of tea and are often an index of progressiveness of the tea industry. But they have significant limitations. These are:

Development of insecticide resistant strains of insects

Virtually all insect species against which a particular insecticide has been used over a period of time have evolved strains which are considerably less susceptible to the insecticide than the original population. It is believed that generally such resistant insects are already present in the original population but their proportion increases as the insecticides selectively eliminate susceptible individuals. The resistant strains may also develop by mutation and interbreeding of less susceptible individuals in a pest population during the course of the use of an insecticide (Pathak and Dhaliwal, 1993). The problems of pest resurgence, pest resistance and secondary pest outbreaks were reported in the past (Das, 1959). It was shown that resurgence of tetranychid mite, aphid and scale insects following application of chlorinated hydrocarbon such as DDT, was due to the destruction of their natural enemies. It was also inferred that poor control of *Helopeltis* with chlorinated hydrocarbon (DDT) might be due to the outbreak of resistant strains of *Helopeltis*. Pest outbreaks in the past had been noticed when insecticides like heptachlor and dieldrin were used against tea tortrix and the shot hole borer (Baptist, 1956; Ananthakrishnan, 1964). In Japan, the problem of insecticide resistance has recently become important for *Tetranychus kanzawai* Kishida, *Homona magnanima* Diakonoff, *Adoxophyles* sp., *Scritothrips dorsalis* Hood and *Empoasca onukii* Matsuda. It is observed recently that *Amblyseius womersleyi*, which controls effectively the population density of *T. kanzawai* in Japan, has become resistant to many kinds of insecticides (Kawai, 1997). In Malawi, *S. aurantii* seems to have developed resistance to many of the pyrethroids resulting in the withdrawal of the recommendation on their use against thrips (Rattan, 1992c).

Effects on non-target organisms

An insect pest in a crop is only a part of a complex ecosystem comprising other pest, parasites, predators and pollinators. The insecticides are usually also toxic to the natural enemies of the pest which affects a fast resurgence of the pest population and also encourages the build up of other pest species against which the insecticide applied was not effective. Such changes have been recorded on many crops (Pathak and Dhaliwal, 1993). Therefore, there is growing apprehension that if proper precautions are not taken similar situation may develop on tea as well.

Influence on environment

The wide use.of insecticides have had far reaching side effects on the environment. Insecticides have been detected from soils, water and air, and by getting in the food-chains they have contaminated all forms of life including birds, fish and mammals including mother's milk. The wild birds like crow, ergot, kite and vulture were reported to show more HCH residues in blood plasma and brain when compared to DDT and its metabolites (Kaphalia *et al*., 1981). The residual levels of DDT, endosulfan, HCH and Heptachlor were higher enough in the eggs of light species of colonial water birds breeding at Keoladeo, National Park, Bharatpur, India to raise concern (Muralidharan, *et al*., 1992). HCH was predominantly a pollutant in certain wild species Tamil Nadu (Regupathy and Kuttalam, 1992). Fish mortality was observed in lakes and canals lying between paddy fields sprayed with DDT (Wadhwani and Lal, 1972). High concentration of dimethoate 20–200 ppm was detected in Hooghly river in West Bengal (Dhaliwal and Arora, 1996). Market samples of various fish species were contaminated with DDT and HCH residues (Mahapatro, 1999). Fish samples at Bombay, India contaminated upto 34 ppm of DDT. Such residues ultimately passed on to humans (Singh and Dhaliwal, 1994). Higher concentration of DDT, HCH and Carbaryl in the soil inhibited the nitrogen fixing capacity of Rhizobium, Azotobacter and blue green algae (Dhaliwal and Arora, 1996). Nonjudicious use of insecticides causes environmental pollution in different ways.

Chandrasekaran (1992) generalised the dynamics of pesticide as below:

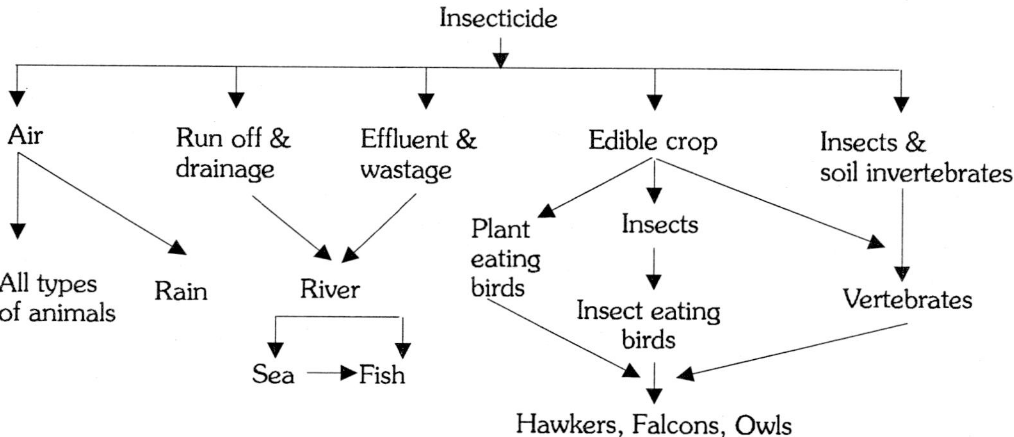

Although these contaminated quantities have been generally low. There is the danger of a built up of their amounts with time and of course there is no way of accurately knowing the long range effects of these minute quantities on the health and survival of human beings and other animals. Despite this, there are large number of cases of insecticide poisoning either by direct contact or through food. These have aroused concern about the safety on the application of insecticides.

Cost of insecticides

The increasing cost of insecticides enforces the small tea growers to reduce the use of insecticides in their tea gardens particularly in developing countries.

In spite of these limitations, till date insecticides play major role in controlling insect pests problems in tea. Therefore, ways and means need to be devised to use ecofriendly insecticides with higher efficacy.

12.10 Pesticide residue in tea

Box 12.3 : Pesticides in soil

A. Factors influence the behaviour and fate of pesticides in soil
   1. *Adsorption to clay and organic matter*
   2. *Leaching with the downward percolation of water*
   3. *Volatilization to the atmosphere*
   4. *Uptake by soil organisms or plants*
   5. *Movement with runoff water or eroded soil*
   6. *Microbial degradation*
   7. *Chemical degradation*
   8. *Photolysis*

B. Effects of insecticides on living organisms in soil
   1. *They may be directly toxic to animals and/or plant life in the soil.*
   2. *They may affect these organisms genetically to produce populations resistant to the pesticides.*
   3. *They may have sublethal effects that result in alterations in behaviour or changes in metabolic or reproductive activity.*
   4. *They may be taken into the bodies of soil flora and/or fauna and passed on to other organisms.*

Source: Edwards (1973)

Recent focus on pesticide residues on processed tea has led issuance of Maximum Residue Limits (MRLs) on a group of pesticides by different producing and consuming countries and various International agencies (Table 12.3). A time frame has been prescribed to achieve these limits. It is evident from Table 12.3 that the MRLs of different pesticides adopted by various countries are not uniform and some of the standards are very stringent. On the otherhand, some countries go on tightening the standards, making it extremely difficult by producers to achieve such standards. Since the surface area per unit weight of tea leaf is large and the interval between pesticide application and harvest is short, the levels of pesticide residues on tea at the same application rate will usually be higher than in other crops (Chen and Wan, 1988). However, a pesticide after spray on the canopy of the tea plant, most of the pesticide is retained on the leaf surface and constitute the initial deposit which penetrates gradually with time into the cuticle of the leaf. Some systemic pesticides may also be translocated to the other part of the plant along with transpirative flows (Chen *et. al.*, 1987). Certain synthetic pyrethroids are highly lipophilic and so remain attach on to the waxy layers of the leaf cuticle, whereas many of the organophosphate compounds penetrate into the leaf surface rather quickly. The application of acaricides and other insecticide on tea are usually carried out soon after plucking. These chemicals consequently are deposited on the mature maintenance foliage and also on the shoots which are in

Table 12.3 : Maximum Residue Limits (MRL) of pesticides used in tea plantations prescribed by various agencies/ countries.

	MRLs (mg kg^{-1})									
	1	2	3	4	5	6	7	8	9	10
Bromopropylate	5.0	-	5.0*	5.0	3.0	1.5	-	5.0	-	-
Captan	-	-	15.0*	-	15.0	15.0	-	-	-	-
Cartap	2.0	-	-	20.0	0.1	-	-	20.0	20.0	-
Chlorbenzilate	-	-	1.5	-	-	1.5	-	-	-	-
Chlorfenvinphos	-	-	1.0	0.2	0.5	0.3	-	-	-	-
Chlopyrifos methyl	0.1	-	0.1*	0.1	0.2	0.1	-	0.1	0.1	-
Cyhexatin	2.0	-	2.0*	2.0	-	1.5	-	2.0	-	-
Cypermethrin	-	-	-	-	-	-	3.0	20.0	0.5	20.0
Deltamethrin	-	-	-	-	-	-	2.0	10.0	5.0	10.0
Diazinon	-	-	0.3	0.1	0.5	0.5	-	-	-	-
Dichlorvos	-	-	2.0	-	2.0	0.3	0.1	-	0.1	-
Dicofol	8.0	45.0	2.0*	5.0	2.0	2.0	-	5.0	0.1	-
Dimethote	-	-	1.0	0.05	1.0	0.5	0.1	-	0.20	-
Dinabuton	-	24.0	1.0*	-	-	-	-	-	-	-
Endosulfan	30.0	10.0	30.0*	30.0	3.0	0.1	-	30.0	-	-
Ethion	5.0	-	2.0*	-	0.1	1.0	-	7.0	2.0	-
Fenitrothion	0.5	-	0.5	0.5	0.5	0.2	0.3	0.5	0.1	-
Fenvalerate	-	-	-	-	-	-	2.0	-	-	-
Flucythrinate	-	-	-	-	-	-	-	20.0	-	-
Heptachlor	-	-	0.10*	-	-	-	-	-	-	-
Hexachlorobenzene	-	-	-	-	-	-	-	-	0.01**	-
Methidathion	0.1	-	2.0*	0.1	0.2	0.5	-	0.1	0.1	-
Malathion	-	3.0	-	8.0	0.5	1.0	-	0.1	-	-
Monocrotophos	-	-	5.0	-	0.05	0.2	-	-	0.1	-
Permethrin	-	-	-	-	-	-	3.0	20.0	2.0	20.0
Parathion	-	-	0.5	0.1	0.5	0.5	-	-	-	-
Parathion methyl	0.2	-	1.0	0.2	0.2	0.5	-	0.2	-	0.2
Phosalone	-	8.0	2.0	-	1.0	2.0	-	-	-	-
Phoxim	-	-	-	-	-	-	0.5	-	0.1	-
Piperoxyl butoxide	-	-	3.0*	-	3.0	3.0	-	-	-	-
Propargite	-	-	5.0*	-	2.0	1.5	-	10.0	0.1	-
Quinalphos	-	-	0.20*	-	-	-	0.2	-	0.1	-

1 = CODEX (Commission appointed by WHO and FAO to determine pesticide residue level), 2 = Environmental Protection Agency, USA, 3 = Germany (*Specific limits for tea), 4 = The Netherlands, 5 = Italy, 6 = Switzerland, 7 = Peoples Republic of China, 8 = FAO / WHO, 9 = Council of European Communities (**Indicates lower limit of analytical determination), 10 = Japan.

Box 12. 4 : A profile on ill-impacts of pesticides

The World Health Organisation estimates that there are at least three million acute, severe cases of pesticide poisonings and 20,000 unintentional deaths each year, mostly in South (WHO, 1990). This estimate excludes cases of chronic pesticide poisoning. A study estimated that 25 million agricultural workers in the South are acutely poisoned every year (Jeyaratnam, 1990).

Atwal and Dhaliwal (1997) estimated that more than one million poisoning cases occur in the world annually with about 15000 deaths, and India accounts for one third of the total pesticidal poisoning cases.

The International Agency for Research on Cancer of the WHO, on the basis of limited evidence recently found that "spraying and application of non-arsenical insecticides entail exposures that are probably carcinogenic to humans. This has profound implications for agricultural workers and particularly for plantation workers who apply pesticides on a regular basis" (Dinham, 1993).

The heavy use of chemical pesticides has taken a serious toll in Thailand on Hmong farmers health (Ekachai, 1992).

The first major incident of pesticide poisoning in India on a large scale is that of Kerala in 1958, where more than 100 people died consuming wheat flour and sugar contaminated with parathion leakage during shipment from Bombay to Cochin. In the worst ever tragedy in Bhopal in 1984, more than 3000 died, 30000 people disabled in varying degrees. The surviving population is still expressing teratogenic, mutagenic, carcinogenic and other consequences involving vital body organs (Mahapatro, 1999). HCH poisoning resulting in 207 deaths due to consumption of flour mixed with BHC witnessed in Lakhimpur Kheri of Uttar Pradesh (Nag, 1983).

Chronic health effects in almost every part of the body are associated with prolonged pesticide exposure (Pingali et. al., 1992). The hypothesized health effects of chronic exposure to pesticides are as follows :

System	*Effects*	*Critical signs and symptoms/ laboratory findings*	*Source*
Eye	*Pterygium*	*Vascular membrane on eye, decreased visual acuity*	
Skin	*Eczema*	*Lichenification and fissuring*	*Morgan (1982) WHO (1990)*
	Nail destruction		*FPA (1988)*
Respiratory			*Hock (1987)*
(pulmonary tract)	*Bronchial asthma*	*Wheezing cough, bronchospasm dyspnea*	*Morgan (1982)*
Cardiovascular	*Electrocardiograph (EKG) changes, high blood pressure*		*Morgan (1982)*
Gastrointestinal tract	*Chronic gastritis*	*Epigastric pai, nausea, vomiting*	*WHO (1986)*
Kidneys	*Asymptomatic urinary abnomalities*	*Albuminuria, hematuria, elevated blood urea nitrogen (BUN), creatinine*	*WHO (1986) Hock (1987)*

different stages of growth. The weight of these immature shoots increases over a period of 6-14 days, depending on the plucking interval. Therefore, the pesticide applied on such shoots undergo a growth dilution. In other words, the weight of a bud increases during the plucking interval and this helps in bringing down the residues of pesticides if any on a weight by weight basis (mg/kg). The immature buds by the time attain the size of pluckable shoots (2/3 leaves and a bud), the residue of applied pesticides on them will be either nil or very low (Agnihothrudu and Muraleedharan, 1990). In China, it was observed that if the residue levels declines by 90 % in 8 days after spray, 45% may have been caused by growth dilution (Chen and Wan, 1988). In Darjeeling, the residue level of malathion, quinalphos and fenvalerate declined more than 90 % in 7 days after spraying and the percentage caused by growth dilution may be little less than 45 % (Bisen and Ghosh Hajra, 2000). Further, between application of a pesticide and the consumption of the tea, a considerable amount of the pesticide present in green leaves is lost at the time of processing at high temperature. Generally, 30-60 % of pesticide residues are lost in the manufacturing process (Table 12.4). Degradation at this stage is primarily due to evaporation and thermal decomposition. It has been reported that the higher the vapour pressure, the greater the degradation during processing (Chen and Wan, 1988). Chen *et. al.* (1987) observed that rainfall, evaporation, biodegradation and photolysis are some of the ways in which pesticide residues are lost. Sensitivity of pesticides to light is an important factor which determines their residues on tea. The mechanism for most

Table 12.4 : Degradation of certain pesticides during processing of tea.

Pesticides	Degradation (%)
Cypermethrin	19.8
Deltamethrin	18.2
Dimethoate	67.0
Ethion	25.2
Fenitrothion	66.9
Fenvalerate	13.0
Quinalphos	65.6

Source: Chen and Wan (1988).

photochemical reactions involves the homolytic cleavage of covalent bonds. Since these reactions are based on functional groups, it makes no difference whether a pesticide is an insecticide, fungicide or herbicide (Agnihothrudu and Muraleedharan, 1990). In an experimental trial in Darjeeling, no residue of monocrotophos after 4 hours of its application was detected during dry and wet seasons. The residues of malathion, fenvalerate, dimethoate on 5th day after application in dry season were found below the MRLs established by various organisations and countries. Further, in case of quinalphos it was on 7th day after application. While in case of dicofol the residue on 7th day during wet season was found below MRLs but it was higher in dry season. Bhattacharya *et. al*. (1995) also observed in Darjeeling that the residue of organophosphate and chlorpyriphos in processed tea on 5th day after application at the recommended dose (1:400) are well below the prescribed MRLs. They further observed there is no necessity to discard first plucking when cypermethrin is applied at the recommended dose (1:4000 dilution). Residue of dicofol were 5.5 ppm on 7th day after spraying and 3.75 ppm on 14th day after application were detected in south India (Singh and Agnihothri, 1984). But Raju Kannu *et. al*. (1981) reported that the residue of dicofol was only 2.3 ppm on the 10th day after application. The observed difference in the result of dicofol on tea may be on account of variations in agroclimatic conditions prevailing on the different experimental sites, jat of plant and also due to different doses. Loss of pesticide residue also takes place in storing and with the passage of time between processing of tea and use of the same. When tea is infused with boiling water, pesticide residues contained in tea are partly degraded by high temperature and partly by dissolving in the liquor which will be consumed. Generally, only those pesticides with high water solubility may potentially be transferred to the tea cup in significant amounts. If the applied chemical is not easily soluble, more residues may remain on the spent tea and less in the infusion (Muraleedharan, 1992). It has been observed that teas with 17.01 ppm dicofol (practically insoluble in water) when infused with water leaves only 0.005 ppm residues in the brew (Muraleedharan, 1994).

Among the various pesticides used in tea plantations, ethion and dicofol deserve special mention in view of their extensive use for mite control. A few busket surveys have been conducted during 1993 in the tea estates of south India for determining the residues of these acaricides. The results in Nilgiri survey showed that residue of ethion was below 1.0 ppm in 92.86% of the sample and exceeded the upper limit of 2.0 ppm in 7.14 % of the samples. The residues of dicofol revealed that it was below 1.0 ppm in all the samples (Muraleedharan, 1994). The results of the similar kind of survey conducted in the tea estates of Kerala and Tamil Nadu in south India on the residues of dicofol, ethion and endosulfan revealed that the residues of these pesticides were below the MRL stipulated by Germany (Muraleedharan, 1994). A wet season trial conducted in UPASI experimental farm to determine the residue of endosulfan applied @ 350 g a.i./ha. Analytical data indicated that the residue level of endosulfan and its isomers was only 11.75 ppm on teas manufactured on the first day after application which is below the levels stipulated by Germany (Anonymous, 1997). A high incidence of HCH in Japanese tea observed by Niimura *et. al.* (1989). Similarly residues of 30 different organophosphorus pesticides including fenitrothion, phosalone and chlorpyrifos in Japanese tea reported by Nagayama *et.al.* (1989). Residues of organochlorine pesticides were reported to be high in Sri Lankan teas (De Silva and Thiemann, 1991). Residues of chlorinated hydrocarbon in black tea in Germany (Buchholz, 1988) and in Spain (Fernandez *et. al.* 1993) was detected below the levels stipulated by EC.

The 'half life' of insecticides on tea varies greatly. Chinese workers found that organophosphate compounds like dimethoate, fenitrothion and quinalphos have half life below 2 days, those of ethion and dicofol were 3.4 and 3.9 days respectively which are rather high. The synthetic pyrethroids have also comparatively long half lives, but these insecticides are characterised by very low mammalian toxicity (Table 12.5). In Darjeeling, pyrethroid pesticides had very similar half life in dry and wet seasons while the half life of organophosphate vary from 13 hrs to 26 hrs in dry season and 6 to 21 hrs in wet season. The half life of quinalphos in wet season was recorded high (2.73 days) while in dicofol it was 0.4 and 1.2 days in dry and wet seasons respectively . No pesticides on tea shoots, will have a half life longer than eight days (Chen and Wan, 1988).

The consciousness about the pesticide residues in tea calls for immediate attention of the tea growers regarding very restricted and judicious application of insecticides, herbicides, fungicides etc in the plantation. Further, judicious use of broad spectrum chemical pesticides and integrated planning are required for several reasons: lowering production cost, reducing environmental contamination and effect on non-target organisms, delaying the development of resistant pest bio-types and above all minimising residues in produce (Ghosh Hajra *et. al.*, 1994b).

Table 12.5 : Average half-lives ($T_{1/2}$) of some pesticides on tea.

Pesticides	$T_{1/2}$ (day)
Dichlorvos	0.20
Phoxim	0.27
Phoxim (under shade)	0.87
Malathion	0.50
Dimethote	0.96
Dimethote + Omethoate	1.20
Phosmet	1.20
Quinalphos	1.20
Fenitrothion	1.30
Bifenthrin	2.50
Cypermethrin	2.50
Permethrin	3.20
Fenvalerate	3.20
Deltamethrin	3.20
Ethion	3.40
Chlordimeform	3.60
Dicofol	3.90
Demeton	8.00

Source: Chen and Wan (1988)

12.11 Alternative to targeted pesticides

The use of chemical pesticides can be minimised by integrating it with other available methods of control.

A. BIOLOGICAL CONTROL

Numerous biocontrol agents are active in tea fields exerting a natural regulation of several pests. A considerable number of natural enemies of pests of tea have been identified (Table 12.6) and some of them have been proved capable under ideal conditions of keeping the pest populations below economic threshold levels. Since the mid 1980s considerable progress has been made in the study of indigenous natural enemies. There are two methods:

i. Identification and augmentation and conservation of natural enemies.

ii. Mass rearing of natural enemies in laboratories and releasing against pest at frequent intervals.

Table 12.6 : Important natural enemies of some tea pests.

Pest	Natural enemy
Red spider mite Oligonychus coffeae (Nietner)	Verania vincta Garh., Stethorus qilvifrons Mulsant, Scymnus sp., Jaurevia quadrinotata Kapur, J. soror, Verticillium lecanii (fungus), Agistemus sp., Chrysopa sp., Phytoseilus persimilis (Exotic sp.), Chrysoperia carnea Stephens, Exothoris caudata, Pronematus sp., Cunaxa sp., Amblysieus herbicolus (Chant), A. ovalis, A. mcmurtry, Staphilinid beetle.
Pink and purple mites, Acaphylla theae (Watt), Calacarus carinatus (Green)	Amblysieus herbicolus (Chant), Lestodiplosis sp., Scolothrips rhagebianus Priesner
Bunch caterpillar Andraca bipunctata Walker	Cylindromya sp.(Techinid fly), Cantheconidea furcellata (Wolff), Pencilium and Rhizopus (Fungi), Bacteria, Dipterous fly
Red slug caterpillar Eturesia magnifica Butler	Apanteles sp., Exorista heterusiae, Bacteria
Looper caterpillar Buzura suppressaria Guenee	Sarcophagus sp., Apanteles taprobanae Cameron, Aphanogmus monilae, Bacteria
Flush worm Cydia leucostoma (Meyrrick)	Apanteles aristaeus Nixon, Calleida nilgirensis Straneo, Asympiesella indis
Leaf roller Caloptilia theivora (Walsingham)	Sympiesis dolichogaster Ashmead, Mestocharella javensis
Tea tortrix (Walker) Homona coffearia Nietner	Phytodietus spinipes (Cameron), Palexorista solennis
Cut worm Spodoptera litura (Fabricius)	Peribaea orbata (Wiedemann)

Aphid Toxoptera aurantii, (Boyer de Fons.)	Syrphus balteatus, Asarcina aegrota, Scymnus pyrocheilus (Mulsant), Verania vincta, Cryptogonus bimaculata, Cryptogonus or Cryptogonus biculus (Gyllenhal), Coccinella repanda, Coccinella septempunctatus, Lemnia bissellata (Mulsant), Pseudaspidimerus circumflexus (Mots.), Aphelinus sp., Menochilus sexmaculatus (Fab.), Jaurevia pubescens (Fab.), Aphidius colemani Vierick, Lipolexis scutellaris Mackauer, Episyrphus balteatus De Geer, Paragus tibialis (Fallen), Leis dimidiata var quindecimmaculata, Trioxys indicus, Chrysoperla carnea Stephens
Jassid Empoasca flavescens Fabr.	Cephalosporium sp. (Fungus), Drynid wasp,
Helopeltis Helopeltis theivora Waterhouse	Oxyopes sp., Melamphus sp., Preying mantis, Chrysoperla carnea Stephens, Reduviid bug, Spider, Mermethrid Nematode, Ichneumonoids
Scale insects	
Eriochiton theae (Green)	Coccinella septempunctata, Spalgis epius, Blastothrix sp.
Chrysomphalus aonidium Ashmead	Chilocorus circumdatus, C. nigritus, Pharoscymnus horni, Comperiella bifasciata, Prospeltella sp., Aphytis chrysomphali, Aspidiotiphagus citrinus
Fiorinia theae Green	Jauravia quadrinotata, Scuymnus sp., Aphytis theae, Prospeltella sp., Aspidiotiphaus sp.
Aonidiella aurantii (Maskell)	Aphytis chrysomphali, C. bifasciata
Saissetia formicarii (Green)	Coccophagus acanthosceles
Saissetia coffea Walker	Encyrtus infelix Embleton , Coccophagus cowperi Girault
Shoot mealy bug Nipaecoccus viridis (Newstead)	Aprostocetus purpureus (Cameron), Cacoxenus perspicax (Knab).
Cockchafer grub Holotrichia sp.	Gliocladium roseum (Fungus).
Termite	Metarrhizium anisopliae (nematode).

A single ladybird can devour an entire colony of thrips or mites in a day. Collection and subsequent release of ladybirds in badly infected areas has given excellent result. The areas have shown almost complete recovery within 45 days followed by prolonged mite free period (Ghosh Hajra and Kabir, 1993). There are 20 species of predatory mites belonging to phytoseiidae and stigmaeidae feed on *Brevipalpus phoenicis* in Indonesia (Oomen, 1982). Similarly a spider can kill five hoppers in a day. The aphid, *Toxoptera aurantii* is naturally controlled by syrphid, coccinellid and hemerobiid predators and aphidiid parasitoids (Radhakrishnan, 1989). The braconid, *Apanteles aristacus* Nixon, is the principal parasitoid of *Cydia leucostoma*. This parasitoid, reported from India and Indonesia appears to be host specific and prefers third instar caterpillars for oviposition. Adults and larvae of a cerabid, *Calleida nilgrirensis* Straneo feed on late instar caterpillars, the beetle consuming more than 140 flushworms during its life span (Muraleedharan *et. al.*,1991). In India, *Sympiesis dolichogaster* Ashmead, is the major natural enemy of leaf roller, *Caloptilia theivora*. In Japan, *Amblyseius womersleyi* controls effectively the population density of *Tetranychus kanzawai* (Kawai, 1997). Biological methods of control are not available at present for all pests affecting tea. However, they are in the process of evolution and refinement.

B. COVER CROP AND LIVING MULCHES

It may provide habitat for natural enemies of insects, mites and nematodes. Pests may be suppressed through antagonistic or allelopathic effect, or by preferential attraction to the cover crop over the desired crop.

C. MICROBIAL AGENTS

It includes bacteria, fungi, viruses, protozoa, and nematodes that are pathogenic or antagonistic to pest species, including insects, plant pathogens and weeds. Pathogens can cause epizootics in pest populations, especially caterpillars. Many adults and nymphs of Jassids are found to be attacked by a whitish growth of a species of the fungus *Cephalosporium* in north-east India. The dead insect usually remains attached to the under surface of the leaf (Barua, 1983). Encouraging results were obtained with introduction of *Bacillus thuringiensis* (Bt) against *Homona magnanima*, *Caloptilia theivora* and *Adoxophyes* sp. (Kariya, 1977). Its efficacy has also proved against *Andraca bipunctata* and *Buzura suppresaria* (Reddy *et. al.*, 1990; Borthakur, 1986) and *Cydia leucostoma* (Bisen and Ghosh Hajra, 1997). Bt is safe for human beings but its crystal protein produces a fatal endotoxin in the gut of the target caterpillar pests. To avoid insects developing resistance, several strains of Bt should be combined. Spore load of Bt is reduced quickly in the field after application. Liposome microlipid encapsulation has been used to increase the persistence of Bt in the field. Infection of cocoons of bunch caterpillar, *A. bipunctata*, by fungi like *Fusarium* sp. has also been reported (Banerjee, 1979). Many fungal species of different genera are known to parasitize nematodes. Of these, species of the genera *Arthrobotrys* and *Dactylaria* sp. were recorded in Tocklai, Assam, soils (Agnihothrudu, 1961). The fungus, *Verticillium catenulatum* (Kamyshko ex Barron and Onions) W. Gams was isolated from egg masses of root-knot nematodes (Barua, 1983). In Sri Lanka *Pasturia* sp is employed to kill parasitic nematodes. Nuclear Polyhedrosis Virus (NPV) is used for control of looper caterpillar in Sri Lanka and China. In field-scale trials, most of the 4th and 5th instar caterpillars were killed in 7-9 days and pest population was reduced from 42 pupae to only 0.03 per plant. A fungus *Paecilomyces tenuipes* (Peck) Samson was recorded in tea estates of Darjeeling parasitising pupal stage of various Psychid pests (Basket worm) of tea during early June to late October (Devnath, 1986). The fungal biocides like *Metarhizum* and *Beauveria bassiana* are used against live wood-eating termites in Sri Lanka. *B. bassiana*, an entomopathogenic fungus isolated from membracid bug, a shade tree pest, was multiplied and tested against *Helopeltis*. Trials conducted by TRA in Assam and Darjeeling provided 27% and 49% control respectively by spraying 20% suspension for two rounds at 15 days intervals. It was more effective in cool climate. Coccoidea is the largest group attacked by diverse fungi species of *Aschersonia*, *Septobasidium*, *Fusarium* and *Podonectria* (Barua, 1983). Different families of viruses cause disease in insects. The major limitation for their use is the concern for human safety. High mortalities of the tortricids, *Adoxophyes* sp. (50-88%) and *Homona magnanima* (50-79%) were reported when mixture of two granulosis viruses (GV) isolated from *Adoxophyes orana fasciata*, summer fruit tortrix and *H. magnanima* were applied in the tea field of Japan. The efficacy of the virus application appeared in the succeeding two generations by suppressing larval density (Kodomari, 1993). It was also reported that when GV are sprayed in spring, they are able to control effectively the abovementioned two species of tortrix moths until autumn, but the viruses do not persist in winter (Kawai, 1997). GV is also used for control of nettlegrub in Sri Lanka.

D. PEST RESISTANT CLONES

It would provide inexpensive, environmentally sound, and highly effective way to avoid pest problems. Even low levels of resistance are important since the need for other control methods can be reduced. Different cultivars of tea with varying growth habits differ in their resistance, tolerance and susceptibility to pests. In Sri Lanka and Kenya, number of tea clones against *O. caffeae* have been screened (Thirugnansuntharan and Amarasinghe, 1990; Sudoi, 1990). The preference of Assam clones by pink mite, *Acaphylla theae* was observed by Loganathan (1992). Incidence of *Helopeltis theivora* is found more in China varieties and Assam clone TV1 (Muraleedharan and Chen, 1997). The clone UPASI-17, cambod in nature, is susceptible to attack of flush worm (Muraleedharan, 1991c). Several tea selections in Indonesia are found susceptible to the attack of flushworms (Dharmadi and Sudjatmoko, 1981). Clones with high content of alpha spinasterol are found susceptible to attack by the shot-hole borer, *Euwallacea fornicatus.* Saponins could bind sterols and act as determinants in the plant's resistance to the attack of shot-hole borer (Wickramasinghe *et. al.*, 1978). In Sri Lanka, considerable work had been done on the evaluation of clones against shot-hole borer, white root grub, *Holotrichia disparilis* etc which lead to the identification of several susceptible clones (Calnaido and Kanapathipillai, 1967; Thirugnansuntharan and Calnaido, 1968; Thirugnansuntharan and Jayachandran, 1989 and Thirugnansuntharan *et. al.*, 1988).

E. BOTANICAL INSECTICIDES

These are derived from extractions, pressing, infusions, pulverizing, etc and they come from any plant part. The fact they are plant derived, however, does not distinguish them from other pesticides by virtue of their use, mode of action, chemistry, formulation toxicity safety, efficacy, marketability, or overal utility (Larew, 1992). Over 50 years approximately 2400 plant species reported to possess pest control properties (Grainge and Ahmed, 1988).

Neem has been found to have multifarious properties like antifeedant, repellent, toxicant, growth inhibitor, antiovipository etc (Fig. 12.10). The bioefficacy of neem oil (NO) and neem seed kernel powder (NSKP) against sucking pests have been observed in a field trial conducted in Darjeeling.

Fig. 12.10: Fruits of neem have properties like antifeedant, repellant, antiovipository etc.

NO and liquid soap at 1% concentration offered best control followed by NSKP 100 gm treatment. The average cost benefit ratio was 1: 2.13 (Bisen and Ghosh Hajra, 1995). Ghosh Hajra *et. al*. (1994a) also observed various plants exhibiting insecticidal activity which could be effectively utilized for the control of specific type of pests. They further noticed that the botanicals which are basically synthesized in the living system as a result of normal physiological process, naturally have very little or no harmful effect on life and environment as well as easily degradable. The antifeedant properties of the extracts derived from the aerial parts (leaves and stem) of *Artimisia vulgaris* Linn. (Fig. 12.11), *Urtica dioica* Linn. (Fig. 12.12), *Polygonum*

Fig. 12.11: *Artimisia vulgaris* – a common weed in tea plantations of Darjeeling – extracts of the aerial parts have antifeedant and repellant properties.

Fig. 12.12 : *Urtica dioica* – a common weed in tea plantations of Darjeeling – extracts of the aerial parts have antifeedant and repellant properties.

Fig. 12.13: *Polygonum runcinatum* – a common weed in tea plantations of Darjeeling – extracts of the aerial parts have antifeedant and repellant properties.

Fig. 12.14 : *Eupatorium glandulosum* — a common weed in tea plantations of Darjeeling – extracts of the aerial parts have antifeedant and repellant properties.

runcinetum Ham. (Fig. 12.13) and *Eupatorium glandulosum* Linn. (Fig. 12.14) which are at home in Darjeeling hills were evaluated in the laboratory against bunch caterpillar, *Andraca bipunctata*. The maximum antifeedant action was exhibited by the extract of *A. vulgaris* followed by *P. runcinatum*, *U. dioica* and *E. glandulosum* at 1 % concentration each (Bisen and Kumar, 1997).

The problem associated with much-hailed botanicals and biopesticides is their varying degrees of successes in the field situation. Actually while a sector of scientific community advocates in favour of them reporting their superiority in laboratory trials, others demonstrated their inefficiency in many-a-field trials.

F. CULTURAL CONTROL

This, probably is the most economical and widely applicable method of pest control. In tea, routine cultural operations such as mulching, plucking, pruning, manuring, regulation of shade and drainage which increase the vigour of the bush plays a predominant role in suppression of pests. Infestation by mites and thrips is seen more in tea fields devoid of shade. The heavily shaded and moist areas will be damaged more by tea mosquito bug. Leaf folding caterpillars during plucking can be manually removed. A few rounds of black plucking reduce the incidence of thrips, *Helopeltis* etc. Weeds serve as alternate hosts for many tea pests. *Mikania micrantha* HBK., *Melastoma malabathricum* Linn., *Eugenia jambolana* Lam., *Rosa sinensi, Fasminum scandens*are the alternate hosts of tea mosquito bug in north-east India. Effective weed control assumes greater significance in the management of tea mosquitoes. Further, the insect pests are also removed by plucking, skiffing and pruning. These operations are assumed to strongly affect the population of insect pests on new shoots. Balanced nutrition helps to improve vigour of plants to withstand pest attack. Application of high dose of potassium fertilizers is known to reduce the incidence of *Pratylenchus loosi*, lesion nematode (Gnanapragasam, 1982). The population of *Pratylenchus loosi* at certain extent decreased when castor oil and mahua cake added to the soil (Gnanapragasam, 1991). In Sri Lanka, *Glyricidia* has been used to trap swarming live wood-eating termites. In Japan, continuous application of fish manure in fields resulted in an increase in the populations of *Tetranychus kanzawai* and it was correlated to the high level of L-arginine in the leaves which increased the fecundity of the spider mite (Kaneko, 1981). The incidence of phytophagous mites on tea in India, Sri Lanka and Indonesia increased when copper fungicide applied repeatedly for the control of blister blight (Venkata Ram, 1966; Muraleedharan, 1991a; Cranham, 1966a; Oomen, 1982).

G. SEMIOCHEMICALS, INCLUDING PHEROMONES, ALLOMONES, KAIROMONES

Another approach for controlling pests is to modify their behaviour by exploiting their natural chemical communication or feeding cues. Semiochemicals are non-toxic behaviour modifying a mixture of chemical compounds released from exocrine glands of insects such as sex pheromones, used by insect females to attract their mates, and antifeedants which may be based on the compounds produced by plants. The pheromones, in general, do not find a place in pest control in tea though in Japan sex pheromones have been used for the management of the caterpillar pests of tea (Tamaki, 1980). However, in many moth species the females releases pheromones (odours) and in response males fly up wind to find them. For some species these compounds have been identified and synthesized. Such pheromones have been used mainly for monitoring populations, for disrupting mating, for mass trapping, and for lure and kill (Campion, 1984). Certain pheromones have been tested in the field and found useful for monitoring and mass trapping of *Homona magnanima*, *Caloptilia theivora* and *Adoxophyes* sp. (Kariya, 1977). Sex pheromones traps can be used for monitoring the population of *Caloptilia theivora* (Selvasundaram, 1990). Components of the sex pheromones of the smaller tea tortrix, *Adoxophyes* sp. have been identified and isolated (Tamaki *et. al.*, 1971, 1979). UV light traps and extracts of virgin females were equally attractive to males of *Adoxophyes orana* in the quantitative bioassay of the pheromones. Wind tunnel tests have proved the usefulness of disruption of pheromonal orientation in this species (Kainoh *et. al.*, 1984 and Hiyori *et. al.*, 1986). The synthetic sex pheromone of *Homona coffearia* when applied in plastic capsules and on rubber septa retains its attractiveness to males for fourteen days and more than one month respectively (Noguchi *et. al.*, 1981). The allelochemicals such as kairomones and allomones produced by plants and herbivores may help in enhancing natural enemy activity and may find a role in integrated pest management (IPM) in tea as well (Muraleedharan, 1993).

Synamones are volatiles emitted by host plants and triggered by feeding injury by pests. These chemical communication systems have been used in China and Sri Lanka for attracting parasitic wasps to parasitise geometrid caterpillar.

Insect growth regulators (hormones) can be effectively employed for adverse modification of the normal growth and behaviour of tea pests. Growth regulators like dimilin found to be biologically effective to hinder the normal growth of caterpillars and full grown larvae of bunch caterpillars fails to pupate (Chakravartee, 1994). The chitin inhibitor, diflubenzuron had been field tested for the management of *Cydia leucostoma* of tea (Muraleedharan and Varatharajan, 1985).

H. OIL, SOAP AND OTHER FATTY ACIDS

These are effective and have less impact on human health and wild life than synthetic organic chemicals. The spray of lemongrass (*Cymbopogon citratus* and *C. nardus*) oil and other vegetable oil had good control over some tea pests (Ghosh Hajra *et. al*., 1994a).

In some situation, implementation of some non-chemical control alternatives reported as available might be considered impractical, either because they are not perceived as cost-efficient or because they conflict with current production practices (Zalom and Strand 1990). However, keeping the significant limitations of chemical pest control through broad spectrum pesticides in view, there are welcome efforts to evolve an integrated pest management system.

12.12 Integrated Pest Management (IPM)

It comprises strategies aimed at minimising pest damage through the careful integration of available pest control technologies. IPM is not a new concept but has gained widespread scientific recognition in the past two decades. Agenda 21 of the United Nations Conference on Environment and Development (UNCED) at Rio de Janeiro in June, 1992 pinpointed the IPM as one of the requirements for promoting sustainable agriculture and rural development. However, it has yet to enjoy wide implementation. The term IPM is synonymous to integrated pest control (IPC), which has been defined by FAO panel of experts as:

"A pest management system that in the context of associated

environment and the population dynamics of the pest species, utilizes all suitable techniques and methods in as compatible a manner as possible and maintains pest populations at levels below those causing economic injury. It is not simply the juxtaposition or superimposition of two control techniques (such as chemical and biological controls) but integration of all suitable management techniques with natural regulating and limiting elements of the environment".

12.12.1 REQUIREMENTS OF IPM IMPLEMENTATION

Acceleration of IPM implementation in tea sector needs certain conditions to be met with and certain actions to be taken (NRI, 1992). These are:

- government support, in the form of policies that favour IPM, such as policies to reduce pesticide subsidies if any;
- government action, such as financial support to strengthen the research, technical and extension services required for IPM implementation;
- adequate institutional infrastructure, such as basic elements of a plant protection system incorporating research and development, diagnostic services, advisory services and training;
- scientists trained in IPM problem analysis and able to generate the locally-adopted technology for IPM and crop protection specialists and extensionist able to conduct large scale implementation programme;
- planters sensitized to the reliability of IPM and provided with incentives to adopt it;
- during the last decade, there have been numerous scientific and technological advances that have tremendous potential for refocussing and advancing IPM in right way in right time. One such advance is monitoring and crop-pest modelling. Planters need assistance in interpreting the pest monitoring data. Simple expert systems are to be designed and made available to them in different ways, including simple pegboards or charts, special booklets, radio and television programmes or more advanced aids like prediction models and computer based systems. Developing up-to-date database can be a key cardinal factor in successful IPM implementation. Training to the planters is the vital part of IPM implementation.

Components of IPM

The components are depicted below:

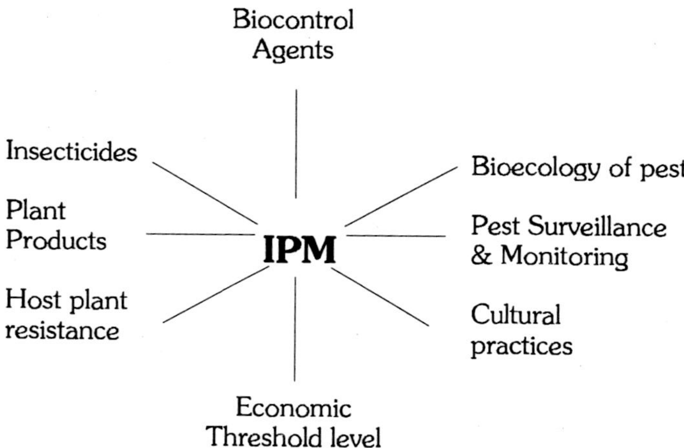

In IPM system when pesticide application is inevitable the following tips are important to keep the residue below MRL :

- Keep vigilance on pest incidence. Proper identification of pest, their developmental stages, site of attack, damage symptoms etc are necessary for selection of right pesticides. In case of combined infestation of red spider, pink and purple mites, recommended acaricides may be used. But combined infestation with scarlet mites, sulphur should be avoided as it has no effect on scarlet mite. However, recommendations issued by the Tea Research Institutes should be consulted for the right choice of pesticide and their dosage.
- Monitoring of the incidence and build up of pests, for example, tea mosquitoes, green fly, red spider, can show initial build up in the month of January-February in north-east India. If this initial build up is not controlled, repeated rounds of spraying during cropping season can not be avoided.
- Biological information—critical time for taking up control measures and economic threshold will have to be determined. Damage threshold limits have been established in India for mite pest (Banerjee, 1971b), in China for tea looper, tea Jassid, pink mite, leaf rollers, tussock moth

and the black aphid (Chen and Yin, 1988) and in Sri Lanka for the plant parasitic nematodes (Gnanapragasam and Herath, 1989).

- Identification of alternate hosts of pests and where possible use them as trap crop. Plant species which is preferred for oviposition by particular insect can be utilised as trap crop.
- Apply pesticides only when it is absolutely necessary.
- Proper supervision during spraying is important. Often, in absence of supervision, the operators may tend to over apply the pesticides at places and complete the job/task expeditiously. Such application may bound to carry residues in plucked leaves. Supervision is therefore must.
- Soil stirring during the cold weather reduces the incidence of caterpillars, thrips, termites, cockchafer grubs etc.
- Bush health and hygiene is prerequisite for good crop protection. It may be of little use to adopt control measures if, for example, the soil is not prepared well or soil fertility is low or unbalanced. Many crop pests affect primarily weakened plants. It is therefore, important to aim for a healthy bush by proper tillage, planting of proper clone at the right time, appropriate spacing and providing sufficient water/moisture and nutrients. A healthy bush can compensate pest damage by growing new shoots or spreading up the growth of undamaged parts.
- Alkaline washing of pruned bushes reduces pest incidence. The alkaline solution (washing soda 6 kg, quick lime 2-3 kg and water 100 litre) may be kept for overnight and strain by using cheese cloth. The strained solution may be applied to the frame by using a hand operated sprayer fitted with nozzle size 40.
- Use of pesticides on rotation. Synthetic pyrethroids should not be used repeatedly or at frequent intervals.
- Allow a minimum gap of eight to ten days between insecticide/acaricide application and plucking.
- Use of conventional sprayer preferable.

References

Agnihothrudu, V. 1961. A comparison of some techniques for the isolation of fungi from tea soils . *Mycopath. et Mycol. Applicata*, **16:** 235-242.

Agnihothrudu, V. and Muraleedharan, N. 1990. Pesticide residues in tea. *Planters' Chronicle*, **85:** 125-127.

Alkan, B. 1957. Teeshadlinge in der Turkei. *Zeitschrift fur Angewandte Entomologie*, **41(1 & 2):** 233-242.

Ananthakrishnan, N. R. 1963. Biology and control of *Scirtothrips bispinosus* (Bagnall). *Ann. Rep. UPASI Sci. Dept. 1962-63*, Appendix: 37-43.

Ananthakrishnan, N. R. 1964. Side effects of dieldrex spraying on tea. *Ann. Rep. UPASI Sci. Dept.* 1963-64, Appendix: 82-84.

Anonymous, 1974. Pests and diseases of the tea plant and their control (in Chinese). *Tea Research Institute, Chinese Academy of Agricultural Sciences*, Agricultural Press, Hangzhou, pp.166.

Anonymous, 1988. *Tea Encyclopaedia*, File J.3. Tocklai Experimental Station, Jorhat, Assam, Tea Research Association. India

Anonymous, 1992. *Tocklai Experimental Station 1911 to 1991*. Tea Research Association, Jorhat, India, 91-115.

Anonymous, 1994. Pests of tea in north-east India and their control. *Memorandum no. 27*. Tocklai Experimental Station, Jorhat, Assam, Tea Research Association, India, p.231.

Anonymous, 1995. *Annual Report 1994 - 1995*. Tea Research Foundation (Central Africa), Mimosa, p.55.

Anonymous, 1997. *71 st Annual Report.* UPASI Tea Scientific Department. Valparai, India, p.41.

Atwal, A. S. and Dhaliwal, G. S. 1997. *Agricultural pests of south Asia and their management*. Kalyani Publishers, Ludhiana, India. p. 486

Awasthy, R. C. and Venkatakrishnan, N. S. 1977. Benefit evaluation of Tocklai Recommendations III Control of red spider. *Two and a Bud*, **24 (2):** 37-38.

Banerjee, B. 1971a. Eclosion and oviposition rhythms of *Andraca bipunctata* Walker (Bombycidae: Lepidoptera) with a model for determining theoretical moth densities from eclosion rate. *Indian J. Entomol.*, 411-418.

Banerjee, B. 1971b. Threshold values in mite pest control. *Two and a Bud*, **18(1):** 20.

Banerjee, B. 1975. Growth of mounds and foraging territories in *Odontotermes redemanii* (Wasmann). *Insects Sociaux.*, **22:** 207-212.

Banerjee, B. 1976a. Pesticides and pesticide residues in tea. *Two and Bud*, **23(2):** 35-42.

Banerjee, B. 1976b. Development of the red crevice mite *Brevipalpus phoenicis* (Geijskes) on coffee and tea (Acarina: Tenuipalpidae). *Zeit. Angew. Ent.*, **80:** 342-346.

Banerjee, B. 1979. A factor analysis of the population fluctuations in *Andraca bipunctata* Walker (Lep: Bombycidae). *Bull.Ent. Res.*, **69:** 195-201.

Banerjee, B. 1981. An analysis of the effect of latitude, age and area on the number of arthropod pest species of tea. *J. Appl. Ecol.*, **18:** 339-342.

Banerjee, B. 1982. A strategy for the control of *Andraca bipunctata* Walker on tea. *Crop Proctection*, **1(1):** 115-119.

Banerjee, B. 1993. *Tea production and processing*. Oxford & IBH Publishing Co., Pvt. ltd., New Delhi, p.336.

Baptist, B. A. 1956. The leaf eating tortrix caterpillar (*Homona coffearia* Nietner) as a limiting factor in insecticide application in tea. *Tea Q.*, **27:** 28-35.

Barua, G. C. S. 1983. Fungi in biological control of tea pest and disease in north-east India. *Two and a Bud*, **30(1/2):** 5-7.

Benjamin, D. M. 1968. Insects and mites on tea in Africa and adjacent islands. *East Afr. Agric. For. J.*, **33:** 345-357.

Bhattacharya, A., Chowdhury, A., Somchowdhury, A. K., Pahari, A. K. and Roy, U. S. 1995. Studies on the residues, persistence and preharvest interval of cythion, durmet and ripcord in made tea of Darjeeling. *Pestology*, **21(2):** 28-36.

Bisen, J. S. and Ghosh Hajra, N. 1995. Bioefficacy of neem products in controlling certain pests of tea (*Camellia sinensis* L.). *Pestology*, **21(11):** 28-32.

Bisen, J. S. and Ghosh Hajra, N. 1997. Testing of delfin with some commonly used insecticides to control flushworm in young tea of Darjeeling. *Pestology*, **30(3):** 12–14.

Bisen, J. S. and Ghosh Hajra, N. 2000. Persistence and degradation of some insecticides in Darjeeling tea. *J. Plantation Crops*, **28(2):** 123-131.

Bisen, J. S. and Kumar, R. 1997. Studies on the antifeedant properties of some plant extracts against bunch caterpillar (*Andraca bipunctata*) on tea (*Camellia sinensis*). *Pestology*, **21(10):** 13-15.

Borthakur, M. C. 1986. Role of entomopathogenic bacteria for the control of tea looper caterpillar. *Two and a Bud*, **33:** 1-3.

Buchholz, H. 1988. Organochlorine pesticide residues in black tea. *Lebensmittelchem. Gerichtl. Chem.*, **42:** 140.

Calnaido, D. and Kanapathipillai, P. 1967. Tolerance and susceptibility of tea clones to shot hole borer (*Xyleborus fornicatus*) infestation in Sri Lank. *Tea Q.*, **40:** 47-52.

Campion, D. S. 1984. Survey of pheromones uses in pest control. *In:* Techniques in pheromone research (eds. H. E. Hummel and T. A. Miller), New york, Springer Verlag, 405-499.

Chakravartee, J. 1994. Towards minimizing pesticide residues in made tea. 1994. *Pestology*, **18(9):** 35-45.

Chakravartee, J. 1995. Tea pest management : Present status and research needs. *Two and a Bud*, **43(1):** 23-32.

Chandrasekaran, S. 1992. Insecticides act and scrutiny of pesticide containers. *In: Pesticides proper use and handling* (eds. A. Regupathy and K.Gopal Reddy), Tamil Nadu Agric. Univ. and IRWEC. India. pp.14-20.

Chen, H. T., Liew, T. L., Kao, M. C. and Hu, C. C. 1978. Ecological studies on the tea leaf hopper, *Empoasca formosana* Paoli and its control (In Chinese). *Bull. Taiwan Tea Res. Exp. Stn.*, **82:** 93-105.

Chen, X. and Yin K. 1988. Pest fauna of tea plant in China and its integrated management. *Proc. Intl. Tea Quality-Human Health Symposium*, November 4-9, 1987, Tea

Research Institute, Chinese Academy of Agricultural Sciences, Hangzhou, China, pp. 154–159.

Chen, Z. M. 1986. Use of pesticides in tea gardens in the Peoples Republic of China. *In: Precis of Proceedings of the Symposium: Tea - Important Aspects of Pesticide Usage* (held at the Society of Chemical Industry, London) (Bohea Ltd., Langport, Somerset, UK).

Chen, Z. M. and Chen, X. F. 1989a. An analysis of world tea pest fauna. *J. Tea Sci.*, **9(1):** 73-88.

Chen, Z. M. and Chen, X. F. 1989b. A world list of pathogens reported on tea plant (In Chinese). *J. Tea Sci.*, **9(1):** 73-88.

Chen, Z. M. and Wan, H. 1988. Factors affecting residues of pesticides in tea. *Pestic. Sci.*, **23:** 109-118.

Chen, Z. M., Wan, H., Wang, Y., Xue, Y. and Xia. H. 1987. Fate of pesticides in the ecosystem of tea garden. *Proc. Int. Symp.Tea - Quality - Human - Health*. November 4-9, 1987. Tea Research Institute, Chinese Academy of Agricultural Sciences, Hangzhou, China, pp.146-149

Cramer, H. H. 1967. Plant protection and world crop production. Bayer, Levekusen, pp. 524.

Cranham, J. E. 1966a. Insect and mite pests of tea in Ceylon and their control. *Monograph on tea production in Ceylon*. Tea Research Institute of Ceylon, Talawakelle. pp.122.

Cranham, J. E. 1966b. Tea pests and their control. *Ann. Rev. Entomol.* **11:** 491-514.

Dammermann, K. W. 1929. *The agricultural zoology of the Malay Archipelago*, J. H. Bussy Ltd., Amsterdam, p.473.

Danthanarayana, W. and Ranaweera, D. J. W. 1970. The red spider mite and scarlet mite of tea and their control. *Tea Q.*, **41:** 19-33.

Danthanarayana, W. and Ranaweera, D. J. W. 1971. The effect of rainfall and shade on the occurrence of three mite pests of Ceylon. *Ann. Appl. Biol.*, **70:** 1-12.

Das, G. M. 1957. *Enarmonia (Cydia) leucostoma* Meyr. (The flush worm). *Two Bud*, **4(3):** 3-4.

Das, G. M. 1959. Problems of pest control in tea. *Science and Culture*, **24:** 493-498

Das, G. M. 1965. *Pests of tea in north-east India and their control*. Tea Research Association, Tocklai Experimental Station, Cinnamara, Assam, p.115.

Das, G. M. and Das S. C. 1967. Effect of temperature and humidity on the development of the red spider mite *Oligonychus coffeae* (Nietner). *Bull. Ent. Res.*, **37:** 433-435.

Das, G. M. and Ganguli, R. N. 1961. Coccoids on tea in north east India. *Indian J. Entomol.*, 23: 245 - 256.

Das, G. M. and Sengupta, N. S. 1962. Biology and control of the purple mite, *Calacarus carinatus* (Green), a pest of tea in north-east India. *J. Zool. Soc. India*, **14:** 64-72.

Das, S. C. 1983. Control of thrips and cockchafer. *Tocklai Occassional Scientific Paper*, 12, pp.48.

Das, S. C. 1984. Resurgence of tea mosquito bug , *Helopeltis theivora* Waterhouse, a serious pest of tea. *Two and a Bud*, **31:** 36-39.

Dev, H. N. 1964. Preliminary studies on the biology of the Assam thrips, Scirtothrips dorsalis Hood, of tea. *Indian J. Entomol.*, **26:** 184-194.

Devnath, S. 1986. A new record of an entomogenous fungus *Paecilomyces tenuipes* (Peck) Samson on various Psychid pests of tea in Darjeeling. *Two and a Bud*, **33(1):** 24-25.

De Silva, M. P. and Thiemann, W. 1991. Organochlorine pesticide residues in different grades of Sri Lankan tea. Chem. *Mikrobiol. Technol. Lebensm.*, **13:** 89-94.

Dhaliwal, G. S. and Arora, R. 1996. *Principles of insect pest management*. National Agricultural Technology Information Centre, Ludhiana, India, p.374.

Dharmadi, A. and Sudjatmoko, F. X. B. 1981. Susceptibility of tea clones against flush worm , *Laspeyresia leucostoma* Meyr. (in Indonesian). *Warta BPTK*, **7(1/2):**155–164.

Dinham, B. 1993. The pesticide hazard: a global health and environmental audit. *The pesticides trust*, London.

Edwards, S. 1973. Persistent pesticides in the environment (end ed.) CRC Press, Cleveland, Ohio: 170.

Ekachai, S. 1992. Cabbage worse than opium poppies. *ILEIA Newslet.*, **1:** 21.

Dzhashi, V. S. 1970. [The nonspecialised pests of in the USSR and their control (in Russian)]. *Subtropi cheskie Kul'tury*, **6:** 174-187.

Fernandez, N., Sierra, M., Garcia, J. J., Diez, M. J. and Teran, M. T. 1993. Organochlorine pesticide residues in black tea, Camomile and linden. *Bull. Engiron. Contam. Toxicol.*, **50:** 479-485.

Fernando, E. F. W. 1967. Studies on the aetiology and biochemistry of *Oligonychus coffeae* Nietner (Acarina: Tetranychidae) the red spider mite of tea in Ceylon. *M. Sc. Thesis*, University of Peradeniya, Sri Lanka, pp.146.

Fertilizer and Pesticides Authority (FPA), 1988. FPA pesticide regulatory policies and implementing guidelines and procedures. Manila, Philippines.

Ghosh Hajra, N. and Kabir, S. E. 1993. Organic farming — an overview. *Indian Rev. Life Sci.*, **13:** 121-131.

Ghosh Hajra, N., Sarker, G. and Ghosh Hajra, A. 1994a. Environmental impact of pesticides and alternatives of pest control in tea (*Camellia sinensis* L.). *ARPC Monograph*, 1-30.

Ghosh Hajra, N., Kabir, S. E. and Bisen, J. S. 1994b. Botanical pest control — an alternative to minimise residue toxicity in Darjeeling tea. *Proc. Conf. Plant Protection & Human Health*, March 2, 1994, DPA, Darjeeling, India, pp.58-74.

Glover, P. M., Das, G. M. and Mukherjee, T. D. 1961. Pesticide residues and taint in tea. *Span*, **4(3):** 137-140.

Gnanapragasam, N. C. 1982. Effect of potassium fertilization and of soil temperature on the incidence and pathogenicity of the root lesion nematode, *Pratylenchus loosi* Loaf, on tea (*Camellia sinensis*), *Tea Q.*, **51(4):** 169-174.

Gnanapragasam, N. C. 1989. Limitation to growth and productivity of tea by plant parasitic nematodes. *Proc. Regional Tea Sci. Conf.*, (eds. P. Sivapalan and A. Kathiravetpillai), Tea Research Institute of Sri Lanka, Talawakalle, Sri Lanka, 123-132.

Gnanapragasam, N. C. 1991. Influence of soil amendments in reducing pathogenicity to tea by root lesion nematode, *Pratylenchus loosi*. *Proc. Intl. Symp. Tea Science.* Tea Expt. Stn., August 26 - 29, 1991, Shizuoka, Japan, 684-688.

Gnanapragasam, Nalini C. and Herath, U. B. 1989. Pathogenicity of the tea burrowing nematode *Radopholus similes* to young tea (susceptible clone TRI 2025) at different density. *Sri Lanka J. Tea Sci.*, **58(2):** 83–86.

Gope, B. 1991. Bio-ecological studies on the tea mosquito bug, *Helopeltis theivora* Waterhouse, in north-east India. *Ph.D. Thesis*, Dibrugarh University, Assam, India.

Grainge, M. and Ahmed, S. 1988. *Hand book of plants with pest-control properties.* John Willey & Sons, New York, p.470.

Green, E. E. 1890. Insect pests of the tea plant. Independent Press, Colombo, 103-104.

Hainsworth, E. 1952. *Tea pests and diseases and their control with special reference to north-east India*. W. Heffer & Sons Ltd., Cambridge, p.130.

Hiyori, T., Kainoh, Y. and Ninomyia, Y. 1986. Wind tunnel tests on the disruption of pheromonal orientation of the male smaller tea tortrix moth, *Adoxophyes* sp. (Lepidoptera : Tortricidae) I. Disruptive effect of sex pheromone components. *Appl. Ent. Zool.*, **21(1):**153-158.

Hock, W. K. 1987. Pesticide use: the need for proper protection, application and disposal. *In : Pesticides: Minimizing the Risks* (eds. N. N. Ragsdale and R. J. Kuhr). Developed from a symposium sponsored by the division of agrochemicals at the 191st meeting of the American Chemical Society, 13-8 April, 1986, New York, N. Y. Washington, D.C. : American Chemical Society.

Jayaratnam, J. 1990. Acute pesticide poisoning : a major global health problem. *WHO statistics Quarterly*, **43:** 139-143.

Joseph, C. P. D. 1991. Report of the advisory officer, Nilgiri - Wynaad. *Ann. Rep. UPASI Sci. Dept.*, UPASI Tea Research Institute, Valparai, India, **65:** 147.

Joseph, C. P. D. 1992. Disease and pest management in tea. *Proc. 30 th UPASI Sci. Conf.*, September 14, 1992, Coonoor, India, Bull. **45:** 84-100.

Kainoh, Y., Hiyori, T. and Ninomyia, Y. 1984. Conditions for wind tunnel test in studying pheromonal communication in the smaller tea tortrix moth, *Adoxophyes* sp. (Lepidoptera: Tortricidae). *App. Ent. Zool.*, **19(4):** 526-528.

Kalshoven, L. G. E. 1950. *De plagen van de cultuur - gewassen in Indonesie* (2 Vols.). W. Van Heeve, The Hague. p.1065.

Kandasamy, C. and Muraleedharan, N. 1983. The role of *Xyloborus fornicatus* in tea plantation ecosystem. *In: Insect Interrelations in Forest and Agroecosystems*. (eds. Sen Sarma, Sangal and Kulshreshta), D. A. V. College, Dehra Dun, India, 89-95.

Kaneko, T. 1981. Preference to oviposition of tea spider mite, *Tetranychus kanzawai* Kishida, with special reference to free L-arginine concentration in tea leaves (In Japanese). *Study of Tea*, **60:** 12-16.

Kaphalia, B. S., Hussain, M. M., Seth, T. D., Kumar, D. and Krishnamurthi, C. R. 1981. Organochlorine pesticide residues in some south Indian wild birds. *Pestic. Monit. J.*, **15:** 9-13.

Kariya, A. 1977. Control of tea pests with *Bacillus thuringiensis*. *JARQ*, **11(3):** 173-178.

Kawai, A. 1997. Prospect for integrated pest management in tea cultivation in Japan. *JARQ*, **31:** 213-217.

Kodomari, S. 1993. Microbial control of tea insect pests in Japan. *Proc. Int. Symp. Tea Science and Human Health*. Tea Research Association, January 11-14, Calcutta, India, 153-157.

Kodomari, S, and Horikawa, T. 1989. *Insect pests of tea and their control* (In Japanese). Japan Plant Protection Association, Tokyo, pp.172.

Kuang, H. and Zhao, J. 1988. Biology and population dynamics of the orange camellia rust mite. *J. Nanjing Agric. Univ.*, **11(2):** 48-53.

Loganathan, S. 1992. Studies on the pink mite, *Acaphylla theae* (Watt) (Acarina: Eriophyiodea: Eriophyidae) infesting tea in Southern India. *Ph.D. Thesis*, Bharathiar University, Coimbatore, India, pp.137.

Larew, H. 1992. Recent scientific advances in botanical pest control. *Proc. Final Workshop on Botanical Pest Control*. July 28 - 31, 1992. IRRI, Los Banos, Laguna, Philippines, 5-14.

Mahanta, P. K., Barua, S. K., Baruah, H. K. and Bora, L. C. 1992. Pigment profile as the quality criteria in various categories of black teas. *Proc. of the 31st Tocklai Conference.* Tocklai Experimental Station, Jorhat, India, 173-185.

Mahapatro, G. K. 1999. Pesticides : boon or bane ? Discussing its changed role in integrated pest management. *Pestology*, **23(6):** 30-36.

Minamikawa, J. and Osakabe, M. 1979. *Insect pests of tea and their control* (In Japanese). Japan Plant Protection Association, Tokyo, pp. 322.

Mkwaila, B. 1985. Red spider mite. *Quarterly Newsletter, Tea Research Foundation (Central Africa),* **77:** 7-9.

Morgan, D. P. 1982. Recognition and management of pesticide poisonings. *U. S. Environmental Protection agency*, 2nd edition, Office of Pesticide Programmes, Wasington, D. C.

Muraleedharan N. 1983. Tea entomology: an overview. *Occasional publication Entomology Research Institute*, Loyola college, Madras **4:** 1-20.

Muraleedharan, N. 1985. Short-hole borer management: present status and future strategies. *UPASI Tea Sci. Bull.* **39:** 51-52.

Muraleedharan, N. 1991a. *Pest management in tea.* UPASI Tea Research Institute, Valparai, p.130.

Muraleedharan, N. 1991b. Biology and control of the shot-hole borer of tea. *Planters' Chronicle*, **86(7):** 309-312.

Muraleedharan, N. 1991c. Entomology. *65th Ann. Rep.* UPASI Sci. Dept., UPASI Tea Research Institute, Valparai, India, 35-48.

Muraleedharan, N. 1992. Pest control in Asia. *In: Tea : Cultivation to Consumption* (eds. K. C. Willson and M. N. Clifford), Chapman & Hall, London, 375-412.

Muraleedharan, N, 1993. Ecology and management of tea pests. *In : Tea culture processing and marketing* (eds. M. J. Mulky and V. S. Sharma), Oxford & IBH Publishing Co. Pvt. Ltd., New Delhi, 129-144.

Muraleedharan, N. 1994. Pesticide residues in tea: problems and perspectives. *Planters' Chronicle*, **9:** 371-375.

Muraleedharan, N. and Chandrasekaran, R. 1981. Observations on the seasonal variations of *Acaphylla theae* (Watt) and *Calacarus carinatus* (Green) in a tea field at the Anamallais (south India). *Pestology*, **5(6):** 11-15.

Muraleedharan, N. and Chen, Z. M. 1997. Pests and diseases of tea and their management. *J. Plantation Crops*, **25(1):** 15-43.

Muraleedharan, N. and Radhakrishnan, B. 1989a. Recent studies on tea pest management in south India. *UPASI Tea Sci. Dept.Bull.*, **43:** 16 - 29.

Muraleedharan and Radhakrishnan, B. 1989b. Effect of midcycle chemical control of shot hole borer on tea yield. *Planters' Chronicle*, **84(3):** 97-98.

Muraleedharan, N. and Radhakrishnan, B. 1990. The shoot mealybug of tea. *Planters' Chronicle*, **85:** 533-534.

Muraleedharan, N., Radhakrishnan, B. and Devdas, V. 1988. Vertical distribution of three species of eriophyid mites on tea in south India. *Exp. Appl. Acarol.*, **4:** 359-364.

Muraleedharan, N., Radhakrishnan, B. and Selvasundaram, R. 1991. Observations on the life history and feeding rates of *Calleida nilgerensis* Straneo (Coleoptera: Carabidae) a predator of flushworm of tea. *J. Ecobiol.*, **3(3):** 189-192.

Muraleedharan, N. and Varatharajan, R. 1985. Flushworm control in tea with some new insecticides. *Pestology*, **9(7):** 5-8.

Muralidharan, S., Reghupathy, A. and Sundaramoorthy, T. 1992. Organochlorine residues in the eggs of selected colonial water birds breeding at Keoladeo National Park, Bharatpur, India, *N.H.R.I. Symposium Series 7*, Environment Canada, Saskatoon, 189-195.

Murthy, R. L. N. and Chandrasekaran, R. 1979. Effect of pest damage on quality of made tea. *Proc.* PLACROSYM II, June 26-29, 1979, UPASI, Valparai, India. 275-284.

Nag, D. D. 1983. Human safety in manufacture, handling and use of pesticides. *In: Course on toxicology of pesticides*, Regional Research Laboratory, Hyderabad, 220-229.

Nagayama, T., Kami, T., Kan, K., Lida, M., Tamura, Y.and Nishima, T. 1989. Residues of organophosphorus pesticides in commercial tea and their leaching into tea (In Japanese). *J. Pestic. Sci.*, **14(1):** 39-45.

Nimura, M., Saito, K., Nakano, K., Takahashi, M. and Takakura, Y. 1989. Analysis of pesticide residues in 'health foods' (tea leaves). *Fukushima- Ken eisei Kogai Kenkiusho Nenpo.*, (In Japanese) **6:** 44-48.

Noguchi, H., Tamaki, Y., Arai, S., Shimoda, M. and Ishikawa, I. 1981. Field evaluation of synthetic sex pheromone of the oriental tea tortrix moth, *Homona magnanima* Diakonoff (Lepidoptera: Tortricidae). *Jap. J. Appl. Ent. Zool.*, **25:** 170-175.

NRI. 1992. Integrated pest management in developing countries: experience and prospects. *Chatham: Natural Resources Institute*, United Kingdom, 1-77.

Oomen, P. A. 1981. The number of scarlet mites (*Brevipalpus phoenicis* Geijskes) that causes crop loss to tea. *BPTK*, **7(1 & 2):**125-129.

Oomen, P. A. 1982. Studies on population dynamics of the scarlet mite, *Brevipalpus phoenicis*, a pest of tea in Indonesia. *Mededelingen Landbouwhageschoolte, Wageningen*, **82(1):** 1-88.

Parthiban, M. and Muraleedharan, N. 1992. Symbiotic fungus of the shot-hole borer of tea. *J. Plantation Crops*, **20(1)**: 75-76.

Pathak, M. D. and Dhaliwal, G. S. 1993. Insect control. *In: Global aspects of food production* (eds. M. S. Swaminathan and S. K. Sinha). Natural resources and the environment series, 20, Tycooly Publishing Ltd., United Kingdom, 357-386.

Pingali, P. I., Marque, C. B. and Palis, F. G. 1992. Farmers health impact of long term pesticide exposure — a medical and economic analysis for the Philippines. *Paper* presented at a workshop on "Measuring the health and environmental effects of *pesticides"*. March 30 - April 3, 1992, Bellagio, Italy.

Porter, J. M., Parry, M. L. and Carter, T. R. 1991. The potential effects of climatic change on agricultural insect pest. *Agric. For. Meteorol.*, **57:** 221-240.

Prasad, S. 1992. Infestation of *Helopeltis*. Tea Times, Sandoz (India) Ltd., Bombay Radhakrishnan, B. 1989. Studies on the aphid, *Toxoptera aurantii* (Boyer de Fonscolombe) (Hemiptera: Aphididae) and its natural enemies in southern Indian tea plantations. *Ph.D. Thesis*. Bharathiar University, Coimbatore, India, pp154.

Radhakrishnan, B., Loganathan, S. and Muraleedharan, N. 1991. Bio-ecology of the pink mite, *Acaphylla theae* (Watt) (Eriophyidae : Acarina) infesting tea in southern India. *Proc.* PLACROSYM IX, December 5-7, 1990, Banglarore, India. 97-102.

Rajukannu, K., Balasubramanian, M. and Vasudevan, P. 1981. Residues of dicofol and tetradifon in tea leaves. *J. Plantation Crops*, **9:** 124-125.

Rao, G. N. 1970. Tea pests in southern India and their control. *Pesticides*, **4(1):** 71-79.

Rao, G. N. 1978. *Ann. Rep. UPASI Sci. Dept.*, UPASI Tea Research Institute, Valparai, India, 82-108.

Rao, G. N. and Murthy, R. L. N. 1976. Economics of tea pest control. *UPASI Sci. Dept. Bull.*, **33:** 88-100.

Rao, G. N. and Subramanian, H. 1968. Control of mites in tea results in increased yields. *Planters' Chronicle*, **63:** 412-413.

Rattan, P. S. 1984a. Incidence and losses caused by *Helopeltis schoutedeni* (mosquito bug) from 1980-1984. *Quarterly Newsletter, Tea Research Foundation (Central Africa)*, **75:** 6-12.

Rattan, P. S. 1984b. Effectiveness of Decis (deltamethrin) against *Helopeltis schoutedeni* (mosquito bug). *Quarterly Newsletter, Tea Research Foundation (Central Africa),* **73 :** 10-14.

Rattan, P. S. 1992a. Pest and disease control in Africa. *In: Tea: Cultivation to consumption* (eds. K. C. Willson and M. N. Clifford). Chapman & Hall, London, 331-352.

Rattan, P. S. 1992b. Economics of thrips control on tea. *Quarterly Newsletter, Tea Research Foundation (Central Africa),* **198:** 7-12.

Rattan, P. S. 1992c. Thrips (*Scirtothrips aurantii*), synthetic pyrethroid insecticides and alternatives. *Quarterly Newsletter, Tea Research Foundation (Central Africa),* **106:** 9-11.

Reddy, K. S.P., Chaudhury, T. C. and Ghosh Hajra, N. 1990. Infectivity of *Bacillus thuringiensis* var. *kurstaki* against bunch caterpillar. *Two and a Bud*, **37:** 15-16.

Regupathy, A. and Kuttalam, S. 1992. Monitoring of HCH and DDT residues in certain avian eggs in Tamil Nadu. *Pestic. Res. J.*, **4(2):** 149-153.

Sannigrahi, S. and Mukhopadhyay, A. 1993. Population dynamics of some tea pests infesting four young cultivars from Darjeeling plains. *J. Plantation Crops*, **21(Suppl.):** 195-201.

Selvasundaram, R. 1990. Studies on the leaf roller, *Caloptilia theivora* (Walsingham) (Lepidoptera: Gracillariadae) infesting tea in southern India. *Ph.D. Thesis*, Bharathiar University, Coimbatore, India, pp.118.

Selvasundaram, R. and Muraleedharan, N. 1989. Observations on the larval parasitoid of *Homona caffearia* Nietner (Tortricidae: Lepidoptera) in a tea field at the Anamallais. *J. Plantation Crops*, **16 (Suppl.):** 185-188.

Sen, A. R. and Chakraborty, R. P. 1964. Estimation of losses of crop from pests and diseases of tea from simple surveys. *Biometrics*, **20:** 792-504.

Shiao, S. N. 1975. An ecological study of the purple mite Calacarus carinatus (Green). *Plant Prot. Bull. Taiwan*, **18(2):** 183-198.

Singh, B. and Dhaliwal, G. S. 1994. Pollution hazards of insecticides in agriculture. *In: Management of agricultural pollution in India* (eds. G. S. Dhaliwal and B. D. Kansal). Commonwealth Publishers, New Delhi, India, pp. 25-26.

Singh, R. P. and Agnihothri, N. P. 1984. Residues of dicofol, endosulfan and malathion on tea, *Camellia sinensis* (L) O. Kuntze. *J. Ent. Res.*, **8(1):** 14-16.

Sivapalan, 1972. Nematode pests of tea. *In : Economic neematalogy* (ed. J. M. Webster), Academic Press, London, 285-311.

Sivapalan, P. 1977. Population dynamics of *Xyloborus fornicatus* Eichh. (Coleoptera : Scolytidae) in relation to yield trends in tea. *Bull. Ent. Res.*, **67(2):** 329-335.

Sonan, J. 1925. Insect pests of the tea plant in Formosa Part I. *Report of the Department of Agriculture and Government Research Institute*, Formosa, **12:** 1-132.

Subramaniam, Bala, 1995. *Tea in India*. Publication & Information Directorate and Wiley Eastern Limited, New Delhi, p.123.

Sudoi, V. 1990. In search for factors responsible for resistance in tea clones to red spider mite *Oligonychus coffeae* Nietner (Acarina : Tetranychidae) and possibility of their future use in screening programme. *Tea*, **13(1):** 62-67.

Surendra Mohan, M. 1989. Advisory service, Munnar. *63rd Ann. Rep*. UPASI Sci. Dep., UPASI Tea Research Institute, Valparai, India, 74.

Tamaki, Y. 1980. Insect sex pheromones and pest management. *Recent advances in Japan. Japan Pesticide Information*, **37:** 22-25.

Tamaki, Y., Noguchi, H., Sugie, H. and Sato, R. 1979. Minor components of the female sex-attractant pheromone of the smaller tea tortrix moth (Lepidoptera: Tortricidae) isolation and identification. *Appl. Ent. Zool.*, **14(1):** 101-113.

Tamaki, Y., Noguchi, H.and Yushima, T. 1971. Two sex pheromones of the smaller tea tortrix: isolation, identification and synthesis. *Appl. Ent. Zool.*, **6(3) :** 139-141.

Tatara, A. 1999. Determination of optimum spraying time for chemical control of Mulberry scale, *Pseudaulacaspis pentagona* (Targioni) (Hemiptera : Diaspididae) in tea fields. *JARQ*, **33:** 155-161.

Thiruganasuntharan, K. and Amarasinghe, L. D. 1990. Clonal susceptibility and population dynamics of tea spider mite, *Oligonychus coffeae* Nietner (Acarina: Tetranychidae) under laboratory conditions. *Sri Lanka J. Tea Sci.*, **59:** 9-15.

Thiruganasuntharan, K. and Calnaido, D. 1968. Observations on the tolerance and susceptibility of tea clones to the shot hole borer beetle, *Xyloborus fornicatus* infestation in Sri Lanka. *Tea Q.*, **40:** 47-52.

Thiruganasuntharan, K. and Jayachandran, I. C. R. 1989. Recent observations on resistance and susceptibility of tea clones to the shot hole borer beetle, *Xyloborus fornicatus* Eichh. (Coleoptera : Scolytidae). *Sri Lanka J. Tea Sci.*, **58:** 50-55.

Thiruganasuntharan, K., Perera, E. R. and Abeysekara, A. R. 1988. Preliminary observations on clonal susceptibility of tea clones to white grub, *Holotrichia disparilis* Arrow (Coleoptera: Scarabaeidae) attack. *Sri Lanka J. Tea Sci.*, **57(2):** 68-72.

Venkata Ram, C. S. 1966. New evidence on the increase of purple mite in tea with copper spraying. *Planters' Chronicle*, **61:** 209-210.

Visser, T. 1958. Report of the acting pathologist for the year 1957. *Bull. Tea Res. Inst. Ceylon*, **39:** 47-50.

Wadhwani, A. M. and Lal, I. J. 1972. Harmful effects of pesticides. Indian Council of Agricultural Research, New Delhi, India, 77-79.

Watt, G. and Mann, H. N. 1903. *The pests and blights of the tea plant*. Government printing press, Calcutta, India, XV, p. 429.

Wickremasinghe, R. L., Perera, B. P. M. and Perera, K. P. W. C. 1978. Spinasterol and moisture content as determining factors on infestation of *Camellia sinensis* by *Xyloborus fornicatus*. *Biochem. Syst. Ecol.*, **4:** 103-110.

Widayat, W. and Oomen, P. A. 1981. Scarlet mite population after pruning of tea. *Warta BPTK*, **7(1& 2):** 131-135.

World Health Organization (WHO), 1986. Early detection of occupational diseases. Geneva.

World Health Organization (WHO), 1990. Public health impact of pesticides used in agriculture. Geneva.

Yeh, C. C. and Huang, T. 1977. An ecological study on the tea eriophyid mite, *Acaphylla steinwadeni* Keifer (Acarina: Eriophyidae). *J. Agric. For.*, **24:** 73-95.

Zalom, F. G. and Strand, J. F. 1990. Alternatives to targeted pesticides: the DANR database. *California Agriculture*, **44(4):** 16-20.

Chapter 13

Diseases and management

13.1 Introduction

Tea like other agricultural crop is prone to attack by many diseases and some of which cause serious losses. About 12 fungal diseases were mentioned by Watt and Mann (1903) in their comprehensive report on tea diseases and later Butler (1918) discussed 17 diseases of tea. Petch (1923) described over 60 diseases of tea in his book, *"The diseases of the tea bush"*. In the early sixties, 209 genera of fungi comprising 385 species have been recorded on leaves, stems and roots of the tea all over the world (Agnihothrudu, 1964). Out of which 82 genera are reported to be parasitic and the rest are predominantly saprophytic. While Chen and Chen (1989) recorded a total of 507 fungi on tea after a thorough revision. The details of tea pathogens in India and their control have been discussed by different researchers (Sarmah, 1960; Satyanarayana, 1987; Banerjee, 1993). Informations on tea diseases have also been appeared in China (Chen and Chen, 1990), Japan (Hamaya, 1981), Sri Lanka (Arulpragasam, 1992), Africa (Rattan, 1992) and south-east Asia (Semangun, 1987). Most of the tea diseases are of fungal origin and only a few are caused by bacteria, viruses and algae.

13.2 Crop loss

The loss of crop due to disease mainly depends on pathogen and the geography of the area. Between 1948 to 1952, the annual loss caused by blister blight in India amounted to around 18 million kg of black tea before effective control measures could be developed and implemented (Venkata Ram, 1964). Crop loss as high as 35 % was estimated by Webster and Park (1956). Loos (1952) assessed 33 % loss in crop in Sri Lanka while 20-25 % loss was incurred in Indonesia (Schweitzer, 1950). In China, 40-45 % crop depression due to blister blight infestation was reported by Chen and Chen (1990). Blister blight inflicted enormous loss in yield and these losses are compounded because blister afflicted leaves produce flaky teas thereby affecting quality of tea (Banerjee, 1993; Sugha *et. al*., 1991). The yield in the second season crop in Japan decreased around 25 % when 500 leaves per square meter of canopy were infested by grey blight (Horikawa, 1986). Since sixties the area affected by grey blight had gradually increased in Japan and the loss in yield was around 10-20 % (Horikawa, 1986). In Taiwan, around 40% of the tea bushes were killed by twig die back and in south-east Asian countries root rot disease is responsible for formation of large scale vacancies in the field (Chen *et. al*., 1987; Muraleedharan and Chen, 1997). In Terai region of India, brown blight, a common foliar disease, is responsible for reducing yield and quality (Chakraborty *et. al*., 1998). However, the global loss due to diseases vary between 10 and 15% (Chen and Chen, 1990).

The important diseases of tea can be categorised as leaf, stem and root diseases (Table 13.1). However, it would not be possible to describe all known diseases in this book. Only common diseases are included.

13.3 Leaf diseases

Leaf diseases are most easily observed and have a more direct effect on the production of tea crop. The most important leaf disease in tea is blister blight caused by the fungus *Exobasidium vexans* Massee. It occurs in the area lying between 8^0S and 35^0S latitudes and between 75^0 and 138^0 E longitudes. This disease was first reported from upper Assam in 1868 and was endemic to Assam till 1908. It subsequently spread to the tea estates in Darjeeling. The first outbreak of blister blight was noticed in south India in 1947. The occurrence of the disease reported from Sri Lanka in 1947, Japan in 1922, Taiwan in 1920, Indonesia in 1949 and Cambodia in 1930. It occured in Bangladesh during 1964 and it is almost not in existence at present (Ali, 1990). Fortunately, blister blight disease is not occurring in Africa (Rattan, 1992). It is a seasonal disease and can spread

Table 13.1 : Major tea diseases.

Diseases	Country
A. Leaf diseases	
Asia	
Blister blight	China, Taiwan, Japan, India, Indonesia, Sri Lanka, Malaysia
Red rust	China, India, Sri Lanka, Indonesia, Malaysia
Bird's eye spot	China, Japan, India, Sri Lanka, Indonesia
Black rot	India, Indonesia, Sri Lanka
Anthracnose	China, Japan, Taiwan
Reticular blight	Japan, Taiwan
Africa	
Scab	Malawi, Tanzania, Mozambique, Uganda
Red rust	Central Africa, Uganda
Latin America	
Red rust	Peru
Scab	Brazil
B. Stem diseases	
Asia	
Branch canker	Japan, India, Sri Lanka, Indonesia, Taiwan
Thorny stem blight	India, Sri Lanka, Indonesia
Wood-rot	India, Sri Lanka
Collar canker	India, Sri Lanka
Die back	India, Sri Lanka
Africa	
Pink disease	Kenya, Burundi, Cameroon, Madagascar
Branch canker	Malawi, Kenya
Thorny stem blight	Malawi, Zimbabwe
Wood rot	Kenya
Collar canker	Kenya
Latin America	
Thorny stem blight	Columbia
Thread blight	Columbia
C. Root diseases	
Asia	
Brown root rot	China, India, Sri Lanka, Indonesia, Malaysia
Red root rot	China, India, Sri Lanka, Indonesia, Malaysia
Diplodia root rot	India, Sri Lanka, Indonesia, Malaysia
Black root rot	Japan, India, Indonesia, Malaysia
Root splitting	India, Sri Lanka, Indonesia
Africa	
Root splitting	Central Africa, Kenya, Malawi, Uganda, Zimbabwe
Black root rot	Central Africa
Diplodia root rot	Malawi
Pale brown root rot	Malawi

very rapidly by means of air borne spores which are produced in thousands from a single blister spot. It can only infect young and succulent leaves and stems of nursery, mature tea and teas recovering from pruning. It does not affect the mature leaves or stem. It is a multiple-cycle disease and pathogen completes its life cycle on tea itself in 11 to 28 days. No alternate hosts so far been reported. The basidiospores of the pathogen are sensitive to direct sunlight and require high humidity, cloudy weather, prolonged leaf wetness (11 to 13 hours) for germination of spore and occurrence of the disease. The main periods of outbreak of the disease in Assam and Malaysia is March to May and November to December, June to September in Darjeeling and September to November in Dooars. It occurs all round the year in south India but the incidence of disease is more pronounced between June and October in the estates that predominantly receive south-west monsoon rains, in the districts like Nilgiris which receives mostly north-east monsoon rains, the disease occurs between November and January. In Japan, peak occurrences are in June and September to October and in China, disease has two peaks, one in spring (March to April), and the other in autumn (September to October) (Muraleedharan and Chen, 1997). Aerobiological investigation in Darjeeling showed that the occurrence of blister blight spore in the atmosphere on the leaf surface was found to be highest in July, September and October (Dutta and Debnath, 1990).

The disease is first evident as pale yellow translucent spots on the first three young leaves within three to ten days after infection. The translucent spot gradually enlarge in size and seven to nine days later these spots become shiny, concave on the upper surface and white or pink powdery and convex on the underside where the spores are borne (Fig. 13.1). After sporulation the blisters turn brown and dry up. When several blisters develop on a single leaf, the affected leaf rolled irregularly and distorted. The young stem when affected by the disease become bent and distorted or girdled and may break off and die back (Arulpragasam, 1992).

Control: Normally 4-6 rounds copper fungicide containing 50 % metallic copper following removal of affected succulent growth is mostly afforded for protection against blister infection. It is sprayed at 625g / ha in 100-150 litres of water with a low volume motorised sprayer or at 0.125 % concentration (1 : 800) with a high volume hand sprayer. Two to three fortnightly rounds of hexaconazoles at 200 ml / ha with a low volume sprayer are also effective. Nickel chloride offered excellent eradicant control

Fig. 13.1: Blister blight (*Exobasidium vexans*) leaf disease of tea.

and a combination of copper fungicide and nickel chloride afforded superior disease control to either of them used alone. But nickel being a heavy metal has not been cleared by Environment Protection Agency (EPA), USA for use on food crops. It has been recently observed in a field trial at Darjeeling Tea Research Centre, Kurseong that 0.2 % Delan (Dithianon 75 % WP) gave a better control of blister blight than Baycor (Bitertanol 25 % WP) and Calixin (Tridemorph 80 %) (Anonymous, 1997). In south India, fortysix organic fungicides and their formulations were tested against blister blight and of them, those based on chlorothalonil (Daconil 2787) 560 g, dithianon (Delan) 750 g, pyracarbolid (Sicarol) 500 ml, and tridemorph (Calixin) 180

Box 13.1 : The cause of ill-health in plants

Living (biotic) agents		Non-living (abiotic) agents
Fungi:	*Extremes of weather:*	*Chemicals, principally:*
• *Bacteria and others single-celled organisms*	• *Drought*	• *Herbicides*
• *Viruses*	• *Sun scorch*	• *De-icing salt & blown sea salt*
• *Viroids*	• *Wind*	• *Fuel oils in the soil*
• *Biting insects*	• *Hail*	• *Fuel gases in the soil*
• *Sucking insects & mites*	• *Lightning*	• *Localised concentrations of toxic fumes in the air*
• *Nematodes*	• *Winter cold*	
• *Birds*	• *Spring & autumn frosts*	
• *Mammals including man*		
	Soil conditions:	*Others:*
	• *Compaction*	• *Factors associated with transplanting*
	• *Waterlogging*	• *Mechanical agents (eg. Strangulating tree ties, abrading tree ties, abrading stakes, impacting vehicles)*
	• *Mineral deficiencies*	• *Fire*
	• *Surface sealing*	

Source: Strouts and Winter (1994).

ml/ha offered good disease control. Pyracarbolid and tridemorph possessed long residual action and treatment interval with them could be extended upto 15 days. Chlorothalonil and dithianon on the other hand had to be sprayed at 7-10 day intervals for effective control (Chandra Mouli and Premkumar, 1989). Use of nonionic surfactants as spray adjuvants in blister blight spray especially during heavy monsoon is recommended. This enables retention of sprays on leaves, decrease run off of spray chemicals during rain, increase penetration of chemicals by lowering surface tension of the spray fluid, increasing area of contact and affecting membrane permeability (Baby and Premkumar, 1998).

In controlling blister blight, integration of chemical control with cultural practices would be advantageous. Black plucking which helps in removing the rudiments of disease shoots is still practiced in some tea estates. Nevertheless, the crop loss accruing from unselective plucking could be much higher than that caused by the pathogen. Therefore, it would only be prudent to control the disease by regular application of chemicals rather than resorting to black plucking (Hudson and Muraleedharan, 1998).

Assessment of fieldwise infection at every plucking round is very essential to keep track of the disease development to intensify the control measures whenever necessary. Further, the pluckers working in the affected areas are better not be allowed to pluck in the uninfested block unless the clothing is changed. When the disease appears on a small number of bushes in autumn, the affected leaves should be pruned, skiffed or slashed depending on the existing situation. Shade trees, if existed it should be optimum and thinned out in overshaded area.

Black rot disease caused by the fungi *Corticium theae* Bernard and *Corticium invisum* Petch is common in the plains of India, Indonesia and the southern part of China. Incidence of this disease was noticed in the tea areas of south India prior to incursion of blister blight. With the intensive spraying of fungicides for the control of blister blight, black rot became scarce and finally disappeared (Chandra Mouli, 1993). It generally affects the maintenance foliage of the mature bushes and is more prevalent on tea which has been cut across (pruned) without any cleaning. Skiffed and unpruned teas in general are highly susceptible to this disease (Sarmah, 1960). *C. invisum* produces irregular patches with a slightly raised wavy margin. The diseased patches of young affected leaves look black and slimy when wet (Sarmah, 1960). *C. theae* produces large greyish patches which sometimes even spreads over the entire surface of the leaf (Fig. 13.2). The adaxial surface of the affected leaves turn reddish-brown at the initial stage and the colour changes into a mixture of brown, yellowish-brown and grey at a later stage resembling some extent to the damage caused by sun-scorch. The abaxial surface of the affected leaves turn evenly brown or grey. The pathogen overwinters as sclerotia which is a resting stage thus carrying the disease from season to season. The sclerotium germinates during rainy season and produces disease symptoms. Dead leaves in both diseases remain attached to the stem held by the mycelial cords.

Control: Blanket fortnight spray of copper fungicide is effective. The abaxial surface of the affected bush and the frame should be thoroughly sprayed with the fungicide. Further, the spray of the organic fungicide viz. Delan (Dithianon 75 % WP) and systemic fungicide like Vitavex 75 WP suppress the development of sclerotia.

Grey blight is another leaf disease common in the field and nursery. The predominant species caused this disease are being *Pestalotia theae*

Fig. 13.2: Black rot (*Corticium theae*). Note suspended diseased leaf and fungal thread on stem [photograph from Sarmah (1960)].

Sawada in India and China (Semangun, 1987; Chen and Chen, 1990), *P. longiseta* Spegazzini in Japan (Hamaya, 1981) and *P. guepinii* (Desm.) Steyaert in Africa (Sutton, 1980). The diseased patches on the leaves turn light to dark brown with a greyish centre, circular to oval in outline, characterised by the concentric zonation from the centre to the edge (Fig. 13.3). Portion of the diseased patch on the lamina between the veins drop out leaving skeleton of vascular tissues. Recently, a number of nurseries in Malawi reported a rot of cuttings in the V.P. beds, both under misting and tunnels. The fungus responsible was *Pestalotia theae* Sawada. Typically the rot extended down from the cut surface of the stem but did not initially

Fig. 13.3 : Grey blight (*Pestalotia theae*).

affect the swelling at the base of the petiole which remained distinctly green. Eventually the whole cutting died (Rattan and Whittle, 1999). Lang'at *et.al.* (1998) investigated the factors responsible for resistance/ susceptibility of tea clones to infection by *P. theae* showed that the total polyphenol content of mature leaves influenced infection. Total polyphenol content of leaves of different clones and the size of lesions developed following infection by *P. theae* showed an inverse relationship. The water-soluble carbohydrate content and the pH of cell sap of mature leaves of clonal tea did not influence leaf infection by the pathogen. The brown blight caused by *Colletotrichum camelliae* Massee(= *Glomerella cingulata* (Stoneman) Spauld and Schrenk), also a common foliar disease in India, China, Sri Lanka, Africa and other tea growing regions. This disease develops from the leaf margins and spreads inwards. Entire leaf sometimes affected due to development of multiple infective patches. The edges of the patches are marked with a delicate concentric zonation. Initially, the colour of the affected upper surface of the leaf turns yellowish to chocolate brown and changing to grey gradually from the centre. Both the diseases are commonly found when tea plants are weakened due to waterlogging in the nursery beds and polythene sleeves. The other causes such as severe attack of red spider and Jassids, sun scorch, lack of shade, drought, overdose of chemical manure, lack of nitrogen, hail damage etc. For treating these ailments application of fungicide is not called for. Only predisposing cause(s) of attack should be ascertained and corrected.

The possible use of *Streptomyces roseosporus*, isolated from the grey blight spots which parasitized *P. longiseta* spores, against gray blight in Japan has been suggested (Horikawa, 1988). *Micrococcus luteus* isolated from the phyllosphere of tea was found to be highly antagonistic to *Glomerella cingulata* and growth of *G. cingulata* was also significantly inhibited in the culture (Chakraborty *et. al.*, 1998).

Anthracnose caused by *Colletrotrichum theae-sinensis* (Miyake) Yamomoto is an important leaf disease of tea in north-east Asia such as Japan, China and Taiwan and has not been reported from other countries. This disease is prevalent in hilly areas and occurs in rainy season. The germ tube of conidia penetrates the trichomes on the leaf, grows into the lumen of the trichome and invades the leaf tissue (Hamaya, 1981). There is a close relation between clonal resistance to anthracnose and a number of trichomes on abaxial surface of leaf. The incubation period of the disease is 20-30 days. It produces large reddish-brown lesions on leaves and most of the affected leaves fall off. If the extent of damage during autumn is severe then development of new shoots for the first crop in the following year will be hampered (Arulpragasam, 1992).

Control: Spray of copper fungicides, chlorothalonil and thiophanate methyl are effective in reducing the disease incidence.

Leaf spot (*Cercospora theae* Jan Breda), a disease mostly confined to the mature older leaves but young shoots also are occasionally affected. It produces the symtomatic reddish-brown circular spots on the lamina. Weak chinary bushes specially at Darjeeling gardens are comparatively more susceptible than healthy plants.

Control: It can be controlled by spraying two fortnight rounds of copper oxychloride at 1 : 400 dilution. The systemic fungicide i.e. Bavistin @ 200 gm / 200 litre also gives very good control.

Red rust caused by an intercellular parasitic alga (*Cephaleuros parasiticus* Karst.) attacks week plants of both young and old tea. The disease is common everywhere in the plains of north-east India and Kangra district of Himachal Pradesh. It is a major disease in Bangladesh (Ali, 1990). Despite leaves, it occurs on 1-2 years old branches. It is a secondary disease but may cause severe dieback of the stem. Older leaves are also attacked. The diseased leaves turn yellow, variegated and usually develop circular zones with purple margin. Leaves drop and the stems dieback giving the bush a thin and weakly appearance. Unsound cultural practices aggravate and encourage the disease. Improvement in health of the plant increases its resistance to the pathogen.

Cercosporella disease cause considerable leaf loss during rainy season is mostly restricted to Sri Lanka (Gadd, 1927) and south India (Subba Rao, 1945). The leaves of the young plants in the nurseries at high elevations in south India affected by this disease. *Calonectria theae* (Loos), the perfect stage of *Candelospora theae* (Petch) Wakefield ex Gadd (Syn. *Cercosporella theae* Petch) is the causal organism for this leaf disease. It produces small grey spots with a blackish margin on the leaves and in severe infections the bushes are defoliated. Spraying a 50 % wettable copper oxychloride fungicide formulation at 0.25 % concentration at weekly intervals will be useful in the nurseries.

Sooty mould (*Limacinula theae* Syd. & Butl.; *Capnodium footii* B & Desm.; *Meliola* sp.) on tea generally occurs in plants which are affected by mealy bugs or scale insects. The fungi growing on the sticky sugary secretion (honey dew) of scale insects form an entirely superficial coating on the leaf surface but disappear as soon as the scales are controlled (Banerjee, 1993). Though sooty mould damages the leaf slightly but it may interfere with photosynthesis by cutting off the light (Butler, 1918). Chemical treatment is hardly necessary since the fungi disappear soon after eradication of scale insects.

Red spot, a leaf disease in high elevation gardens of Darjeeling mostly occurs in unpruned chinary bushes during monsoon. It produces minute circular reddish spot, which causes a considerable deterioration of young tea leaves. The causal organism is not known (Jabbal, 1979).

In the VP nursery of Darjeeling hills, blackening of freshly planted tea cuttings is very common problem during monsoon. Sometimes high percentage of mortality was seen. *Fusarium oxysporum* Schlecht is associated with the same (Debnath and Barthakur, 1994). Copper oxychloride and Carbendazim (50 % WP), Tridemorph (80 % EC) at 1000 ppm have significant inhibitory influence on the growth of fungus.

Copper blight of tea caused by *Guignardia camellia* Butler occurs in the nursery in Himachal Pradesh. This disease is also observed in old bushes. High temperature coupled with high humidity is favourable for the growth of the pathogen. It infects plant after rainy season in Himachal Pradesh (Anonymous, 1994). This disease is also prevalent in China (Chen and Chen, 1982).

Bacterial shoot blight, *Pseudomonas syringae* pv. *theae* (Hori) Young, is an important bacterial disease in Japan. The pathogen infects tea shoots and remains prevalent after the typhoon season (Hamaya, 1981). The wounds on shoots caused due to typhoon favour entry of bacteria.

Tea white scab is another leaf disease caused by *Elsinoe leucospila* Bitan and Jenk in Japan (Fukuda and Takaya, 1978) and *Phyllosticta theaefolia* Hara in China threatens tea production in both the countries. Low temperature and high humidity are favourable for development of the disease. This disease produces minute circular spots on shoots which leads to defoliation. It imparts a bitter taste to the infusion (Muraleedharan and Chen, 1997). In China, white scab and brown blight diseases are controlled by combination spray of Benomyl and thiophanate (Chen and Chen, 1990).

13.4 Stem diseases

Pruning operation in tea increases the risk of diseases of stem and branches since it exposes the woody tissues to parasitic and saprophytic fungi. The fungal pathogens can easily enter through pruning cuts or tissue damaged by exposure to sunscorch. The pruning cuts also provide ideal surfaces for germination of spore. Stem diseases are important as they reduce the bush frame and sometimes kill bushes.

Thorny stem blight is a very common disease in Darjeeling and Dooars in north-east India. It is rarely found in the plains of Assam (Tunstall, 1947). It is known to occur in the up country districts of Sri Lanka and in

Malawi. The causal organism of this disease is *Tunstallia aculeata* (Petch) Agnihothrudu. The disease is also known as *Aglaospora* (= *A. aculeata* Petch) in Darjeeling. *T. aculeata* obtained sudden prominence in one of the districts of south India at high elevation (Joseph, 1992). The fungus completes its life cycle in 57 days (Chandra Mouli and Parthiban, 1992). The pathogen enters the stem through pruning cuts and wounds, then spreads downwards killing the branches until it reaches the collar and finally the roots, when the plant is killed. The disease is only recognised by fructifications which are small, black, raised, thorn like projections and are produced on dead branches (Fig. 13.4 and 13.5). These fructifications are arranged either in straight lines or in more or less concentric rings or sometimes scattered. Sometimes dull white stands of mycelia are seen when the bark is peeled off (Sarmah, 1960).

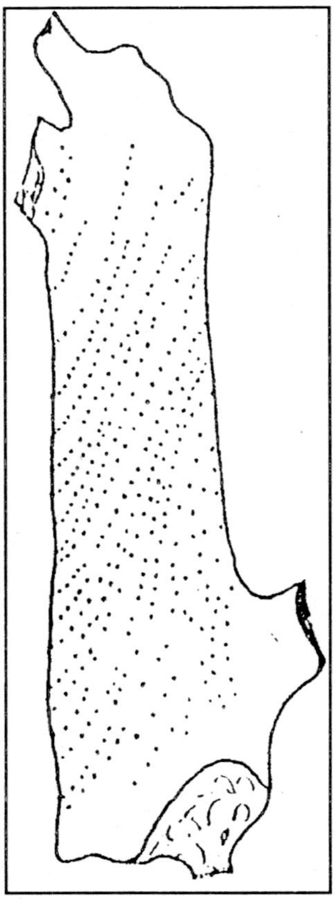

Fig. 13.4 : Thorny stem blight infected branch (diagrammatic).

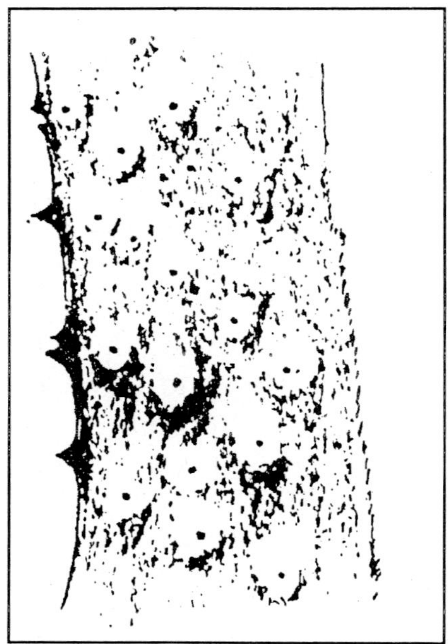

Fig. 13.5 : Fructifications (highly magnified) of thorny stem blight.

Control: The general health and vigour of the bushes should have to be improved. Further, it is advisable to excise the diseased stems right down to clean healthy wood. The cuts should be smoothed off and applied a protective fungicidal paste to prevent the entry of the fungus.

Branch canker caused by *Poria hypobrunnea* Petch is a wound parasite that gains entrance into the stem and then it slowly spreads downwards to the branches killing them until it reaches the main stem and finally the roots when the whole bush is killed. It is a stem-cum-root disease and debilitates, kills the bushes extremely slowly taking nearly 10 to 15 years for mature bushes and 2 to 3 years for young ones (Satyanarayana and Barua, 1983). The external symptoms are the presence of thin, irregular brown lines on the wood which turn yellowish and soft at the final stage of wood decay (Fig. 13.6). The grey corky encrustations on the collar and branches are the fructifications or spore bodies (Banerjee, 1993).

Control: Careful removal of all dead wood at each pruning and painting of the larger pruned surfaces with fungicidal paste (50 g fungicide in one litre water). Avoid injury to the plants during various cultural operations. Tea bush should be protected from sunscorch. Since maximum canker

Fig. 13.6: Branch canker (*Poria hypobrunnea*). Note brown, corky fructification on the stem [photograph from Sarmah (1960)].

development usually takes place under moisture stress therefore mulching, adequate shade etc would help in minimising the effect of disease.

Collar canker (*Phomopsis theae* Petch), is an important disease in south India recorded on a serious scale in pockets of young tea in the Anamallais, Central Travancore, High Range, the Nilgiris, and Singampatti (Venkata Ram, 1973). In Sri Lanka, the disease is mainly confined to the tea areas in the north-east monsoon zone (Shanmuganathan, 1965). It is also causing serious problem in the tea belt of Africa (Rattan, 1992). It has been reported that the clones SFS371 and MT12 have been withdrawn from the recommended list because of its susceptibility to this disease (Ellis and Nyirenda, 1995). *P. theae* primarily affects clonal tea and its development is influenced by low soil moisture and the concomitant low bark moisture. The worst attacks are found soon after prolonged drought. This is generally found to occur in the young clonal and replanted areas between two to eight years of age. Seedling tea is however rarely affected. Cankers are produced at the collar and on branches (Fig. 13.7). An injury on the collar under such conditions offers ready avenue of entrance of the fungus. Chlorosis, cesation of growth and profuse flowering are the common symptoms of the disease (Arunachalam, 1995). Application of fertilizer

Fig. 13.7 : Stem damaged by stem canker (*Phomopsis theae*).

close to the collar, surface watering during dry weather, deep planting, pegging or bending and time of prune have a marked effect on the development of this disease. Collar pruning helps the plants to recover from the disease (Fig. 13.8).

Wood rot caused by *Hypoxylon serpens* (Pers. ex Fr.) kick. has been reported from all tea growing districts of south India. It is also an important stem disease in Sri Lanka, Malaysia, Kenya, Malawi, Zimbabwe and Tanzania. In Africa, despite *H. serpens*, *Hypoxylon investiens* (Schw.) is also observed (Rattan, 1990). It is different from Tarry root rot of tea in

Fig. 13.8: Collar prune done and removed badly diseased branches affected by stem canker.

north-east India attributed to *Hypoxylon asarcodes* (Theiss.) Mill. *H. serpens* is said to enter the stem through moribund tissue following injury by sunscorch. Once the fungus gains entry into the stem it spreads gradually within the branches, ultimately reaching the main stem. Often only one branch or a portion of the bush which is affected by 'wood rot' dies; but when the infection girdles the main stem the entire bush is killed (Venkata Ram, 1970). Irregular, slightly raised, whitish-grey to dark patches (stromata) fructifications (Fig. 13.9) form on the affected wood (Onsando, 1985).

Control: Removal of dead and dying branches by rejuvenation pruning followed by painting of large pruning cuts with copper oxychloride can protect tea from this disease. In Kenya, painting with copper oxychloride in raw linseed oil is recommended (Anonymous, 1986). Application of increased levels of potash to generate sound primary frame helped significantly in removing the disease.

Stem canker (*Macrophoma theicola* Petch) disease is mostly

Fig. 13.9 : Typical fructifications of *Hypoxylon serpens* on wood rot affected stem.

prevalent in drought susceptible areas where soil is poor. The occurrence of the disease is very rare in Darjeeling (Sarmah, 1960). In Kangra Valley, Himachal Pradesh, this disease is observed after rainy season. *M. theicola* has been observed to cause twig die-back of mature tea in Taiwan (Arulpragasam, 1992). In general, tea bush affected by sun-scorch is prone to this disease. The diseased patches on the branches appear as slightly sunken lesions surrounded by a ring of callus growth (Sarmah, 1960). The affected branches are killed slowly by the invading fungus until the disease spreads to the collar when upper portion of the plant dies. In mild infestations the cankers are callused over completely within a few months, but the fungus may renew its growth forming concentric cankers under adverse conditions. Fructifications are produced on the dead bark during wet weather.

Control: Affected branches should be cut out to clean healthy wood. Plants should be protected from sunscorch and pruning during dry weather should be avoided.

Pink disease caused by *Corticium salmonicolor* Berk & Br. has been reported from India, Sri Lanka, Indonesia and southern China. This disease is frecuently encountered in north India and occasionally affects branches of tea in the wetter parts of Kenya, Burundi, Cameroon and Madagascar. The bark of the affected branch cracks and a pink encrustation travels down the branch on lower side. The affected branches lose leaves and die-back.

Control The remedy is to remove and burn the diseased branches following spray of copper oxychloride @ 0.5%. It is also effective in checking the spread of the disease (Flemal, 1986).

Die-back of branches in young tea recovering from pruning caused by *Leptothyrium theae* Petch known to occur in tea areas located at high altitudes (1500 m) of south India (Venkata Ram, 1960a). This fungus also causes stem cankers in tea (Arulpragasam, 1992). Gadd (1936) demonstrated that this fungus formed cankers on young bushes in Sri Lanka as well encircling the main stem and eventually dies. The fungus gains entry through pruning cuts and progressively invades the tender branches arising below the cut, causing the die-back of shoot in south India.

Control: Protection of pruning cuts by covering them thoroughly with oil based copper fungicide sprays immediately after pruning would be useful in preventing the disease (Joseph, 1992).

In north-east India, *Nectria cinnabarina* Fr. caused dieback of tea stem (Sarmah, 1960). They usually attack stems of mature tea wounded by hail, slight cuts, bruises etc produce cankers on branches which later dieback. Sometimes new shoots develop from the base of these stems may also be affected in a similar manner. During cold weather affected branches 10 cm below the point of dieback should be pruned. Improvement of general health of bush protect the plant from the disease.

Collar rot of tea seedlings in the nursery caused by *Rhizoctonia solani* Kuehn and its injury at the collar ranged from an initial disclouration and splitting of the bark, bark disintegration, wound development and callus formation. Finally, the entire stem was girdled, and a heavy callus growth developed below the injured part. It is suggested that young plants should not be planted deeper than nursery level and when mulching is restored to, some places around the collar of the plant is to be left uncovered. If wood-chip is used as mulching material, sufficient care should be taken to avoid injury to the collar tissues.

The other stem diseases viz. *Aschersonia* sp., velvet blight (*Septobasidium* spp.), jews-ear fungus (*Auricularia auricula* (Hooke) Underwood) etc though common in some areas but they do not cause any direct damage to tea (Sarmah, 1960).

13.5 Root diseases

Generally, root diseases are commonly observed wherever tea has been planted after clearing virgin forests or where shade trees have been uprooted without properly ring barking them. In China, root diseases mostly occur in south China, especially in the tea gardens reclaimed from primevial forests. The conducting tissues of the affected plant are damaged and translocation of nutrients, water are restricted which lead chlorosis, wilting, and subsequently death of either whole or part of the plant. Suddenly when a healthy plant or part of it dies with all its dried leaves intact, it is almost a case of root disease (Petch, 1923). In mature tea, more than 60 root rot diseases have been recorded (Chen and Chen, 1990). Tea root diseases can be broadly classed under primary and secondary root diseases. Primary root diseases are those in which the fungus attack itself is the direct cause of death of the bush. They generally spread through soil or by root contact and cause immediate crop loss as well as affects functional physiology of the tea bush. Secondary root diseases are those in which the fungus generally seen in association with the root is only of secondary importance, the main cause of bush decline being something else. In this case the fungus invades the root already weakened by other factors and perhaps accentuate the death of the infected plant (Venkataramani, 1954).

13.5.1 *Primary root diseases*

Charcoal stump rot caused by *Ustulina zonata* (Lev) Sacc. (= *U deusta* (Fr.) Petrak) is widespread in India including the hills of Darjeeling, Sri Lanka, Indonesia and Africa. It is of minor importance in Bangladesh. There are no visible symptoms on the surface of the affected root but if the bark near the collar is removed characteristic delicate fan-shaped patches of dull white, silky mycelium may be seen spreading over the diseased wood. The fungus develops fructifications abundantly at the collar or on the main stem which are irregularly ovoid, thin and plate-like with concentric furrows. They are charcoal black and brittle when old (Fig. 13.10). The spread of the disease is by root contact as well as by windborne spores. Sometimes it is seen that part of mature bush affected by this disease dies while other side remains

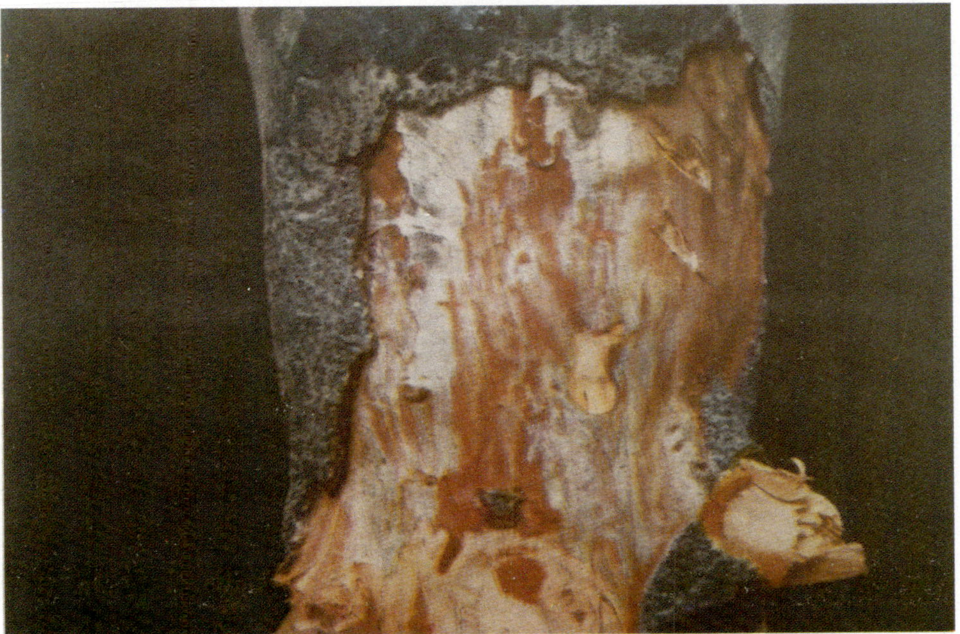

Fig. 13.10: Charcoal stump rot (*Ustulina zonata*). Note black crust of fructifications at the top, fan-like film of mycelium in the middle and black lines in the wood at the bottom of the photograph [photograph from Sarmah (1960)].

apparently healthy and functional.

Brown root rot (*Fomes noxius* Corner) is commonly occurring in hill and plains garden of India, Sri Lanka and Indonesia. Diseased roots are encrusted with earth and gravel bound together by typical tawny brown mycelium (Fig. 13.11). This crust can not easily be removed or washed off. When the infection on the roots is in advanced stage, the underside of the bark and the wood is marked by thin brown lines in a honey-comb pattern, known as reticulations. At this stage, the wood of the diseased root becomes soft, spongy and crumbling under slight pressure. Since infection spreads mostly through diseased root materials, dead tea bushes and stumps of the shade trees are potential sources of the disease.

Tarry root rot caused by *Hypoxylon asarcodes* (Theiss.) Mill. is found to occur in West Bengal, very rare in Assam and has not been reported from other countries of Asia. This disease is restricted to mature tea bushes. The fungus does not produce any external symptoms of damage on the affected roots. A smooth, hard, effused black encrustation develops on the

Fig. 13.11: Brown root rot (*Fomes noxius*). Note encrustation on the root, particularly at the collar, also brown reticulations in the piece of wood [photograph from Sarmah (1960)].

stem above the ground. Fructification of the fungus on the dead bark resembles dried patches of black paint. In an advance stage, the infection produces thin, irregular, black bands in the woody tissues. The affected bark does not easily peel off.

Red root rot is an important disease found in all soils but prevalent in Sri Lanka, Indonesia, Malaysia and southern China. In low elevation areas the predominant pathogen is *Ganoderma pseudoferreum* (Wak.) Stein. while at high elevation the main species is *Poria hypobrunnea* Petch. It is of minor importance in north-east India. In south and north India the species involved is *Poria hypolateritia* Berk. The young roots in the early stage of attack bear white, soft and somewhat fluffy mycelium (Fig. 13.12). At later stage of growth it becomes compact into smooth, thin, flat, dark red cords or sheets. The cords and sheets are white underneath. The affected root is invariably encrusted with soil and stones. Fructifications are either pink or fawn-coloured occasionally found on the main stem just above the collar The disease found to spread by means of mycelial strands that grow

Fig. 13.12: Red root rot (Poria hypolateritia). Note encrustation on the root surface on the left and white threads and films of fungus on a young tap root on the right [Photograph from Sarmah (1960)}.

out from infected stumps or roots (Arulpragasam, 1992).

Black root rot (*Rosellinia arcuata* Petch) disease though recorded from all tea districts, was found to be prominent in high altitude estates of Darjeeling, Sikkim (Satyanarayana *et. al.*, 1982), Nilgiris and High Range (Joseph, 1992). It occurs sporadically in Sri Lanka and has been found to attack both young and old tea (Arulpragasam, 1992). This is a serious disease in the tea estates of Indonesia located at high elevations. The disease is characterised by black wooly mycelial growth on the bark of the infected root and also at the collar region, and presence of star-shaped patches of white mycelium on the wood. Infection spreads through spores as well as direct contact with diseased materials. The disease spreads rapidly in soil rich in organic debris.

Root splitting disease caused by *Armillaria mellea* (Fr.) Vahl is widespread in Indonesia (Arulpragasam, 1992) and Africa (Rattan, 1992) but rare in India and not found in Sri Lanka. In India it is mostly confined to the hills of Darjeeling and Sikkim (Satyanarayana, *et. al.*, 1982). This disease was first recorded in south India in 1934 (Venkata Ram and Joseph, 1974a). The fungus invades the bark of the root and establishes thick white layer of mycelia between bark and the wood, which destroys the cambium of the root and the lower stem. The collar, tap root and lateral roots of affected plants often show longitudinal cracks on the bark and the wood which are usually filled with a compact fungal growth (Rattan, 1992). The fungus is known to remain viable for several years when it is colonised within the host roots and under favourable condition it spreads freely through the soil by means of the rhizomorphs.

Purple root rot (*Helicobasidium compactum* Boedign) is not very common disease in north-eastern India (Sarmah, 1960). Incidence of this disease is very localised in Africa (Rattan and Pawsey, 1981). It usually attacks young tea but it can kill old mature bushes. Diseased roots produce numerous round purplish mycelial cords which extend along the surface upto the collar region. A thick, velvety and purplish mycelial pad produced around the collar. The affected wood attains pinkish colour. Spores are wind-borne, produced on the velvety pad at the collar. Infection also spreads by contact with diseased materials.

It has been observed that the incidence of root rot disease in African and south-east Asian countries is comparatively more severe than north-east Asian countries (Muraleedharan and Chen, 1997). In south-east Asia, red root, brown root and charcoal stump root are predominant root rot diseases (Semangun, 1987) while in Africa the major root rots are root splitting diseases and *Cylindrocladium* root rot (Rattan, 1992). Several species of *Cylindrocladium* are reported to infect the roots of young and mature tea as well as cuttings and seedlings. The pock-marked appearance of the infected root is very typical disease (Venkataramani, 1954). They are mostly reported from India, Sri Lanka, Japan, Malawi and Mauritius. The predominant species are *C. parucum* Anderson, *C. nacrocarpum* Sherbaleott, *C. camelliae* Venkataramani and Venkata Ram, *C. scoparium* Morgan in India; *C. fluorideum* Sob. and Seym. in Japan; *C. parvum* Anders. in Mauritius and Malawi. The extent of their damage and disease etiology are not known in detail.

Xylaria root disease (*Xylaria* sp.) was identified in 1971 and sporadic in south India (Chandra Mouli, 1993). *Xylaria* has not been isolated from active lesions on tea roots neither from north east India nor from any other tea growing areas (Agnihothrudu, 1995). The disease is characterised by the development of typical black ribbon like compact strands of fungal hyphae on the surface of the affected roots. The growth of the mycelium is superficial and the effect of the fungal growth on the host tissue is confined more or less to the bark of the roots (Venkata Ram and Joseph, 1974b).

The young and mature tea roots infect by *Cylindrocarpon destructans* (Zins) Scholt in Japan, *C. tenue* Bugn. in India and Sri Lanka, and *C. lucidum* Booth in Kenya. The importance of *Cylidrocarpon* root disease in Kenya is next only to that of root splitting diseases (Muraleedharan and Chen, 1997).

White root rot caused by *Rosellinia necatrix* (Hartig) Berlese is the major root disease in tea plantations of Japan (Hamaya, 1981) where fagot wood has been buried prior to tea planting to improve drainage (Arulpragasam, 1992). White and black rots are very much localised in Africa (Rattan, 1992). The root rot of tea seedlings and violet root disease caused by *Sclerotium rolfsii* (Sacc.) Curzi and *Helicobasidium purpureum* Pat. respectively are the principal root diseases in southern China (Chen and Chen, 1990).

Box 13. 2 : How to send plant specimens for identification of diseases

Leaves and green shoots: These are best sent packed between sheets of blotting paper or newspaper.

Roots and stems: Before packing dry them partially to avoid development of moulds in transit. They may be sent wrapped up in a hessian cloth. Sawdust never be used during packing. Soil if sticking to the roots should not be washed off while sending root specimens.

Seeds and seedlings: Seeds should be packed dry in a tin container or wooden box. In case of seedlings, they should be sent after careful removal of entire root system from the soil. It would be better to send few seedlings packed in moist moss and a few others packed dry between folds of blotting paper. In absence of blotting paper, newspaper can be used.

When plants die out and it is not possible to discern which part is affected it is advisable to send the whole plant. Adequate and representative materials are required to facilitate free and thorough examination. When specimens are selected from more than one plant, the same should not be mixed up. Label them separately, making use of tie-on labels written with black-lead pencil. A record of the same should be retained with the senders.

Information required

1. *Seed jat / clone.*
2. *Age of plant and when last pruned ?*
3. *First appearance of the disease, symptoms and extent of damage.*
4. *Whether the disease occurs:*
   (a) *In nursery bed or in the field ?*
   (b) *In patches or they are scattered here and there. If affected in patches, do they correspond with any particular soil type ?*
   (c) *Is the field properly drained, or is it waterlogged at any time ?*
5. *Nature of the soil.*
6. *When dug out, was there any hard pan or rock below the root zone ?*
7. *Whether the area is affected by drought or hail or flooding or lightning that might have influenced the bushes concerned ?*
8. *Is the shade moderate, excessive or insufficient ?*
9. *What are the shade trees ? Are they free from disease ?*
10. *Has any spraying or dusting has been made to control the disease ? If so, furnish details.*
11. *Senders opinion of the cause of the trouble experienced.*

13.5.2 *Secondary root disease*

Violet root rot (*Sphaerostilbe repens* B. & Br.) is widespread in Assam but unknown in the hills (Sarmah, 1960). It is rarely found in other countries.

This disease is most common on stiff, clayey soils and pre-disposing factors of the disease are waterlogging and poor soil aeration. Surface roots of the affected plants are roughened due to enlargement of lenticels (because of waterlogging). The roots of the affected plants turn inky-black or light violet in colour and the freshly dug up roots have a characteristic sickly rancid odour. When the bark is peeled off the wood surface is found to be covered with thick, irregular purplish-black flattened strands of mycelial growth becomes visible.

Diplodia disease caused by *Botryodiplodia theobromae* (Pat.) is quite common in low and mid elevation gardens of Assam, south India and Sri Lanka. It can attack any part of the plant, young or old when the plant is debilitated due to other causes. When it occurs as root disease it does not produce any external mycelium or any other characteristic symptom by which it can be identified. The activity of the fungus increases with the depletion of starch reserves in the roots (Venkata Ram, 1960b) caused by continuous hard plucking and pruning the bushes after a rush crop (Arunachalam, 1995). The disease is more prevalent where the bushes suffer from repeated bouts of drought and particularly so in sandy soils. It is more in areas where the pruning of the bushes has not been done at the proper time of the year as well as where pruning cycles are short (Agnihothrudu, 1995) . In the advance stage of infection the surface of the roots generally turn black with clusters of hairy pycnidia. The wood has the characteristic black discoloration. However, the parasitism of the fungus has not been established.

Rhizoctonia bataticola (Tank) Butler also debilitates tea causing the affected root to become light due to gradual depletion of starch (Banerjee, 1993).

13.5.3 *Control of root diseases*

Root diseases, unlike other diseases, are difficult to control and it is far too late by the time their presence is detected. When disease is noticed immediate uprooting with all roots of the diseased bush (es) together with ring of apparently healthy bushes is to be done. When it involves into a large area, the diseased patches should be isolated by a trench of 45 cm wide and 100-120 cm deep all around until uprooting can be undertaken. Then the areas should be kept under thorough soil rehabilitation with

Guatemala or Napier or Manna grass at least for two years before replanting or refilling. The plants should be watered with systemic fungicides at the time of planting and subsequently at three to four months intervals for a period of one year. Though isolation of diseased patches by digging trench produced desirable results, was cumbersome and labour intensive, demanding frequent inspection to keep the trench clean. Improving the conditions of the soil is an important step towards the control of root diseases. Removal of stumps left in the soil will certainly eliminate the food base of secondary root rot pathogens. The operation of ring-barking carried out before felling the tree (shade tree) will deplete the starch reserves in the stump, thus making it unsuitable for the root rot pathogen. The ring barking is a method of killing big trees or shade trees. It consists of removing a ring of bark, about 30 cm wide at a height of about 1 m from the ground. While removing the bark, the cambium is also removed exposing the wood. Some toxic chemicals like 2, 4, 5 T mixed with diesel or kerosene (10 percent solution) may be applied to the ring.

Application of beneficial micro-organisms such as *Trichoderma viride* Pers. & Fr. and *T. harzianum* Pers. & Fr. showed inhibitory activity against *Poria hypobrunnea* (Barua *et. al.*, 1989).

Soil fumigation with Metham Sodium, Shell DD, 55 % Chloro Bromo Propane (CBP), formaline, methyl bromide, Telone etc is also helpful in preventing the spread of primary root pathogens. It has been observed that fumigation caused shifting in microbial balance with abrupt increase in the population of some fungi including *Trichoderma* sp. antagonistic to root disease. Further, the yield obtained in the first few years from fumigated plots were much higher compared to non-fumigated plots (Anonymous, 1992). Before taking up the fumigation, all dead / dying bushes should be cleared and burnt at the site. The soil should be levelled. Immense care is called for in fumigating living stands of tea because fumigation closer than 20 cm from the collar region is highly lethal to the plants (Satyanarayana *et. al.*, 1989). Secondary root diseases may be tackled by alleviating the pre-disposing factors responsible for their incidence.

Sometimes, the root of the cuttings when started growing, occasionally parasitised by the species of *Pythium* and *Fusarium*. They invade the tissues and the infected roots degenerate rapidly. Application of Mancozeb (Dithiocarbamates) or Captan (Phthalimides) @ 0.3 % (30 grams in 10 litres of water) is recommended.

13.6 Plantation level strategies for disease management

Like other crops, sound agronomic and cultural practices, planting of tolerant and resistant varieties, operational disease prediction system in addition to sanitation enable the disease to keep below the economic threshold. Vanderplank (1963) defines sanitation as a process that reduces or completely eliminates or excludes the initial inoculum from creating epidemics. He has shown mathematically that severity levels with or without sanitation, under identical situations, remain proportional. According to him sanitation is unlikely to offer any gain when the apparent rate of infection is high.

When varieties with a common genetic base occupy large areas of plantation, there is no space and time obstacle against the epidemic. But planting varieties differing disease resistance will result in varietal diversity at the plantation level. It would also minimise the risk of spread of disease from one block to another.

In general, it has been seen that a healthy plant grown under optimum conditions, has more resistance to diseases than a debilitated one. Many root rot pathogen infect tea plants only when they are weak. However, this is not always the case as can be seen in blister blight leaf disease where it has been seen on healthy plants. On the otherhand, recovery from such attacks is quicker in healthy plants.

13.6.1 *Management by cultural methods*

Pruning cuts are suitable entry points for fungi. Tea bush sanitation at the time of pruning helps to remove snags and dead wood and in many cases the source of inoculum of the pathogens with prunings. The non-phytotoxic good wound dressing having fungicidal property on the pruned cuts will reduce infection. Pruning and adequate spacing are important in disease management. They allow sufficient circulation of air within the canopy and rapid drying of surface of tea bushes, therefore spread of infection is restricted. Further, these operations facilitate penetration of fungicide within the canopy where target sites are often located. Deep ploughing was found useful in controlling anthracnose, brown blight and grey blight where the causal pathogen overwintered on the diseased fallen leaves. Ploughing promoted decay of fallen leaves, leading to death of pathogen (Muraleedharan and Chen, 1997).

13.6.2 *Management by chemical application*

When breeding for resistance, biological control and cultural methods fails to control plant diseases there is one option of disease management left—the chemical control. One of the main limitations imposed on chemical control of plant disease is the appearance of fungicide resistant strains (Dekker, 1977). Fungicidal resistance is a stable herited adjustment of the organism to external toxic or inhibitory substances. Continuous usage of a particular chemical against a pathogen is liable to cause such a situation. In Japan, Benzimidazole, tolerant strains of the pathogens of brown blight and grey blight (*Pestatoliopsis longiseta*) were discovered (Horikawa, 1988). In plantation crops where many protective sprays are given, fungicidal resistance of the pathogen should also be monitored and spray schedules be adjusted accordingly (Nagarajan, 1993).

Soil fumigants used for the control of root diseases leaves no residue on tea leaves since they are highly volatile compounds and evaporate rapidly from the soil (Muraleedharan and Chen, 1997). But sufficient care should be taken before using systemic fungicides. Systemic fungicides viz. thiophanate, thiophanate-methyl, benomyl, chlorothalonil and tridemorph are characterised by low vapour pressure and their degradation is rather low. The safety interval is generally more than 10 days. Degradation is relatively low during manufacturing process and the extraction rate during brewing ranged from medium to high (Muraleedharan and Chen, 1997). Since the surface area per unit weight of tea leaf is large and the interval between fungicide application and harvest is short hence presence of applied fungicide on the leaf surface at the same application rate is probably be higher than other crops. Therefore, the dose and timing for the foliar application of fungicide are important in respect of residues in the processed tea. Arulpragasam (1992) opined that in most of the black tea producing countries in Asia, tea is harvested at frequent intervals and hence it must be ensured that fungicide application should be made atleast three to four days prior to the harvest. Keeping the limitations for disease control through chemical fungicides there are welcome efforts to evolve integrated control approach i.e. blending of all the available control techniques to minimise the economic damage caused by diseases, with a minimum disturbance to the common environment will be the best proposition. In fact, Integrated Disease Management was proposed in the mid seventies (Apple, 1977).

However, at plantation level it means that resistant varieties, sanitation, changed agronomy and disease monitoring for operational disease management needs to be done in addition to applying timely sprays with appropriate dose of chemicals.

References

Agnihothrudu, V. 1964. A world list of fungi reported on tea. *J. Madras Univ.* Section B., **35:** 155-271.

Agnihothrudu, V. 1995. Some controversial diseases of the tea bush. Proc. *Intl. Symp. Tea Quality and Human Health.*, Nov. 7-10, 1995, Sanghai, China, 392-403.

Ali, M. A. 1990. Pest spectrum of Bangladesh tea and its management. *Proc. Conf. Tea Research Global Perspective.* January 11-12, 1990, Tea Research Association, India, 112-118.

Anonymous, 1986. *Tea Growers Handbook.* Tea Research Foundation of Kenya, Kericho, Kenya.

Anonymous, 1992. *Tocklai Experimental Station 1911 to 1991.* Tea Reseach Association, Jorhat, India, pp.123.

Anonymous, 1994. *Annual Progress Report of Tea Experiment Station for the year 1993-1994.* Department of Horticulture, Himachal Pradesh Krishi Vishvavidyalaya, Palampur, Himachal Pradesh, India, pp. 32.

Anonymous, 1997. Annual Scientific Report 1996-1997. Darjeeling Tea Research Centre, Tea Board, Kurseong, India, pp. 17.

Apple, J. L. 1977. The theory of disease management. *In: Plant Ddisaease* (eds. J. G. Horsfall and E. B. Cowling). Academic Press, New York.

Arulpragasam, P. V. 1992. Disease control in Asia. *In: Tea: Cultivation to consumption* (eds. K. C. Willson and M. N. Clifford). Chapman and Hall, London, 353-374.

Arunachalam, K. 1995. *A Hand Book on Indian Tea.* Arunachalam Associates (Plantation consultants), The Nilgiris, Tamil Nadu, India, p. 271.

Baby, U. I. and Premkumar, R. 1998. Tea blister blight management: importance of surfactants. *Planters' Chronicle,* **93(1):** 21-23.

Banerjee, B. 1993. Tea Production and Processing. Oxford & IBH Publishing Co., Pvt. Ltd., New Delhi, India, p. 336.

Barua, K. C., Borthakur, B. K., Datta, P. and Begum, R. 1989. Biocontrol of *Poria* stem disease of tea (*Camellia sinensis* L.) O. Kuntze caused by *Poria hypobrunnea* Petch. *In: Abstracts of the papers presented at the National Symposium on Recent Trends in Plant Disease Control,* North Eastern Hill Univ., Nagaland, India. pp.226.

Butler, E. J. 1918. Fungi and diseases in plants. Agricultural Research Institute, Pusa, India.

Chakraborty, U., Das, G., Das, S. K. and Chakraborty, B. N. 1998. Evaluation of *Micrococcus luteus* as a potential biocontrol agent of brown blight disease of tea caused by *Glomerela cingulata. J. Plantation Crops,* **26(1):** 50 - 57.

Chandra Mouli, B. 1993. Management of tea diseases—strategies and prospects. *In: Tea culture, processing and marketing* (eds. M. J. Mulky and V. S. Sharma). Oxford & IBH Publishing Co. Pvt. Ltd., New Delhi, 120-128.

Chandra Mouli, B. and Parthiban, M. 1992. Thorny stem blight disease of tea. *Newsletter,* UPASI Tea Research Institute, **1(2):** 6-7.

Chandra Mouli, B. and Premkumar, R. 1989. Control of tea blister blight (*Exobasidium vexans* Massee) with organic fungicides. *J. Plantation Crops* (Suppl.), **16:**249-253.

Chen, Z. M. and Chen, X. F. 1989. A world list of pathogens reported on tea plant (In Chinese). *J. Tea Sci*., **9(1):** 73-88.

Chen, Z. M. and Chen, X. F. 1982. Diseases of tea and their control in the Peoples Republic of China. *Plant Disease*, **66:** 961-965.

Chen, Z. M. and Chen, X. F. 1990. *The Diagnosis of Tea Diseases and Their Control* (In Chinese). Shanghai Scientific and Technical Publishers, Shanghai, p. 275.

Chen, J. S., Thseng, F. M. and Ko, W. H. 1987. Die back disease of tea in Taiwan. *Ann. Phytopath. Soc. Japan*, **53:** 198-203.

Debnath, S. and Barthakur, B. K. 1994. Mortality of tea cuttings infected by *Fusarium oxysporum* Schlecht. *Two and a Bud*, **41(2):** 44-46.

Dekker, J. 1977. Resistance. *In: Systemic Fungicides* (ed. R. W. Marsh). Longman, London, NewYork, 176-197.

Dutta, B. K. and Debnath, S. K. 1990. Disease forecasting as an aid to control blister blight of tea in Darjeeling. *Proc. Int. Conf. R & D in Tea*. January, 11-12, 1990, Calcutta, 107-109.

Ellis, R. and Nyirenda, H. E. 1995. A successful plant improvement programme on tea (*Camellia sinensis*). *Expl. Agric*., **31:** 307–323.

Flemal, J. 1986. La culture du theier au Burundi. *Publications du Service Agricole*, 8, Institut des Sciences Agronomiques du Burundi.

Fukuda, T. and Takaya, S. 1978. Invasion process of *Elsinoe leucospila*, the causal fungus of the white scab of tea and recognition of the hyphae in diseased tissue. *Study of Tea*, **54:** 11-20.

Gadd, C. H. 1927. *Report of the Mycologist*. Tea Research Institute of Ceylon, p. 2.

Gadd, C. H. 1936. Report of the Mycologist for 1935. *Bull. Tea Res. Inst. Ceylon*, **13:** 29-31.

Hamaya, E. 1981. Diseases of tea plant in Japan and their control. *Rev. Plant Prot. Res*., **14:** 96-111.

Horikawa, T. 1986. Yield loss of new tea shoots due to grey blight caused by *Pestalotia longiseta* Spegazzini. *Bull. Shizuoka Tea Exp. Stn*., 12: 1 - 8.

Horikawa, T. 1988. Occurrence of tea grey blight and tea shoot blight caused by *Pestalotia longiseta* Spegazzini and control. *In: Proc. Int. Symp. Recent Development in tea Production*. Taiwan Tea Experimental Station, Taiwan, 181-195.

Hudson, J. B. and Muraleedharan, N. 1998. Impact of cultural operations on blister blight control. *Planters' Chronicle*, **93(8):** 360-361.

Jabbal, M. S. 1979. Plant protection and weed control. *Proc. Seminar on 'Operation 15 quintals'*. Tea Research Association, Area Scientific Committee, Darjeeling, 1-29.

Joseph, C. P. D. 1992. Disease and pest management in tea. *Proc. Thirtieth UPASI Sci.*

Conf., Sept. 14, 1992, Coonoor, Tamil Nadu, India, 84-100.

Lang'at, J. K., Otieno, W. and Musau, J. M. 1998. Evaluation of some Kenyan tea (*Camellia sinensis*) clones for resistance/susceptibility to *Pestalotiopsis theae* (Sawada) as influenced by some chemical attributes of mature green leaf. *Tea*, **19(1):** 6-10.

Loos, C. A. 1952. Report of the pathological division for 1952. *Ann. Rep. Tea Res. Inst. Ceylon*, Talawakelle, 40-49.

Muraleedharan, N. and Chen, Z. M. 1997. Pests and diseases of tea and their management. *J. Plantation Crops*, **25(1):** 15-43.

Nagarajan, S. 1993. Disease problems and their management. *In: Global Aspects of Food Production*. Natural resources and the environment series; 20 (eds. M. S. Swaminathan and S. K. Sinha), Tycooly Publishing Ltd., U.K. 387-415.

Onsando, J. M. 1985. Wood rot disease of tea (*Hypoxylon serpens*), a review. *Tea*, **6(2):** 39-42.

Petch, T. 1923. *The Diseases of the Tea Bush*. Macmillan & Co., London, p. 220.

Rattan, P. S. 1990. Incidence of *Hypoxylon* wood rot and thorny stem blight in Malawi and Zimbabwe. *Tea Research Foundation (Central Africa) Quarterly Newsletter*, **97:** 13-14.

Rattan, P. S. 1992. Pest and disease control in Africa. *In: Tea: Cultivation to consumption* (eds. K. C. Willson and M. N. Clifford), Chapman & Hall, London, 331-352.

Rattan P. S. and Pawsey, R. G. 1981. Death of tea in Malawi caused by *Pseudophaeolus baudonii*. *Tropical Pest Management*, **27(2):** 225-229.

Rattan, P. S. and Whittle, A. M. 1999. Infection of vegetatively propagated cuttings by the grey blight fungus. *Tea Research Foundation (Central Africa) Quarterly Newsletter*, **136:** 7-9.

Sarmah, K. C. 1960. Diseases of tea and associated crops in north-east India. *Indian Tea Association Memorandum 27*, Tocklai Experimental Station, India, pp. 61.

Satyanarayana, G. 1987. Tea diseases of importance in India and their control. *Rev. Trop. Plant Path.*, **4:** 89-106

Satyanarayana, G. and Barua, G. C. S. 1983. Trends in disease control with special reference to tea in north east India. *Tocklai Occasional Scientific Paper* no. **12:** 38–45.

Satyanarayana, G., Barua, G. C. S., Barua, K. C. and Chakravorty, P. C. 1989. 'Durofume' a new soil fumigant in control of root rot diseases in tea. *Two and a Bud*, **30(1/ 2):** 24-26.

Satyanarayana, G., Padmanaban, R., Barua, G. C. S., Baruah, K. C. and Debnath, S. K. 1982. *Armillaria mellae*- A primary root-rot disease of tea, *C. sinensis* (L) O. Kuntze in N. E. India. *Two and a Bud*, **29(1):** 18-21.

Schweitzer, J. 1950. De situatie inzake de blister blight ziekte van de thee op Sumatra's oostkust. *Bergcultures*, **19:** 207-211.

Semangun, K. 1987. Tea diseases in south-east Asia. *Rev. Trop. Plant Path.*, **4:** 49-87.

Shanmuganathan, N. 1965. Collar and branch canker of young tea caused by *Phomopsis theae* Petch. *Tea Q.*, **36:** 14-21.

Stouts, R. G. and Winter, T. G. 1994. Diagnosis of ill-health in trees. *In: Research for*

Amenity Trees no. 2. Forestry Commission, Dept. of the Environment. HMSO, London, 286 pp.

Subba Rao, M. K. 1945. *Report of the mycologist 1943-44*. UPASI Tea Scientific Deptt., India.

Sugha, S. K., Singh, B. M., Sharma, D. K. and Sharma, K. L. 1991. Effect of blister blight on tea quality. *J. Plantation Crops*, **19(1):** 58-60.

Sutton, B. C. 1980. *The Coelomycetes* (Fungi Imperfecti with pycnidia, acervuli and stromata). Commonwealth Mycological Institute, Kew, U. K., p. 696.

Tunstall, A. C. 1947. Notes on stem disease of tea in north-east India. *Indian Tea Association Memorandum* **16**, Tocklai Experimental Station, India,

Venderplank, J. E. 1963. *Plant Disease : Epidemics and Control*. Academic Press, New York, p. 349.

Venkata Ram, C. S. 1960a. Report of the Plant Pathologist (1959-60). *Annual Report of the United Planters Association of South India*, Tea Section, 19-32.

Venkata Ram, C. S. 1960b. A reassessment of the *Diplodia* root disease of tea plants. UPASI *Tea Sci. Dept. Bull.*, **19:** 18-28.

Venkata Ram, C. S. 1964. Plant protection problems in tea—blister blight control in southern India. *Plant Prot. Bull.*, **16:** 1-12.

Venkata Ram, C. S. 1970. Diseases as limiting factor in tea crop production—wood rot and zinc deficiency in southern India. UPASI *Tea Sci. Dept. Bull.*, **28:** 6-8.

Venkata Ram, C. S. 1973. Influence of genetic factors and cultural practices on *Phomopsis* canker disease of tea. UPASI *Tea Sci. Dept. Bull.*, **30:** 5-17.

Venkata Ram, C. S. and Joseph, C. P. D. 1974a. Development in the occurrence and control of root disease in tea. 2. Soil fumigation against *Armillaria mellea* and the status of *Ustulina zonata* as a root pathogen. UPASI *Tea Sci. Dept. Bull.*, **31:** 11-18.

Venkata Ram, C. S. and Joseph, C. P. D. 1974b. Development in the occurrence and control of root disease in tea. 1. A new root disease incited by a species of *Xylaria*. UPASI *Tea Sci. Dept. Bull.*, **31**: 6-10.

Venkataramani, K. S. 1954. The principles of tea root disease control. UPASI *Tea Sci. Dept. Bull.*, **12:** 12-17.

Watt, G. and Mann, H. H. 1903. *The Pests and Blights of the Tea Plant*. Government Printing Press, Calcutta, India, p. 429.

Webster, B. N. and Park, P. O. 1956. Development in blister control. 1. Introduction to the 1955 series of blister blight control experiments. *Tea Q.*, **27:** 3-6.

Chapter **14**

Organic tea farming

14.1 Introduction

In attempts to increase the food production to feed ever increasing population, modern farming systems have played significant role in revolutionizing agriculture. However, these farming systems involve intensive use of chemical fertilizers and pesticides. Chemical fertilizers nourish plants and microbes but also jeopardized environment through nitrate poisoning and exterminating the beneficial soil microflora and microfauna by adversely altering the chemical and physical structures of the soil. Consistent use of pesticides, mostly soil applied persistent herbicides, pollute ground water resulting in adverse effects on fragile ecosystem.

Tea grown in widespread monoculture, is attacked by many diseases and insect pests. Weed infestation is also high due to absence of secondary tillage operation in tea plantations. This apart, time for degradation of the applied agrochemicals is too short as the tea crop harvested at weekly intervals. Consequently, threat from residual effects of pesticides applied is also high especially so with processed tea. Based on a survey report the consumption of pesticides in north-east Indian tea estate has been reported between 4.47 to 7.67 kg/litre ha^{-1} which is higher than other field crops.

Having become conscious of such over-dependence on synthetic chemicals and the associated degradation in environmental quality, some ecologists and environmentalists have taken to the other extreme to demand agricultural produce farmed in accordance with a specialized system that is totally independent of any form of synthetic chemical influences. This has led to the emergence of a movement towards specified farming concept based on the traditional farming philosophy which is popularly known as 'organic farming' slogans like 'back to nature'. People and groups committed to these concepts began to spread this message all over the world. In addition, there was growing health consciousness in the population of the world, especially among those of the developed world, due to the growing realization that the accumulation of high toxic chemicals in their system through the food chain could not only be damaging to them but also to the future generations. Hence, there was a more to purchase items of food which were organically grown. The organic farming system has already become popular in the developed countries, it is only gaining ground in most of the developing countries. The demand for organically grown products is increasing worldwide and the concept of organic farming is gaining ground with Indian farmers too. Today, more than 130 countries produce certified organic foods and beverages of which USA, Canada and many European countries play a leading role. The organic products from developing countries are tea, coffee, sugar, soya, rice, bananas, mangoes and cocoa. Today, Europe imports such products from India, Morocco, Togo, Egypt, Brazil, Columbia, Bolivia, Mexico, Nicaragua (Harkaly, 1990). It is also assumed that about 2% of the internationally traded agricultural produce are organically farmed. The corresponding figure for organic tea was 1735 tonnes during 1996.

14.2 Definition and concept

Organic farming is defined as 'production system which avoids or largely excludes the use of synthetically compounded fertilizers, pesticides, growth regulators, and livestock feed additives. To the maximum extent feasible, organic farming systems rely upon crop rotations, crop residues, animal manures, legumes, green manures, off-farm organic wastes, mechanical cultivation mineral-bearing rocks, and aspects of biological pest control to maintain soil productivity, to supply plant nutrients and to control insects, weeds and other pests' (USDA, 1980). The concept of the soil as a living system that develops the activities of beneficial organisms is central to this definition. Further, keeping the sustainability of natural resources and environment in view, Dahama (1999) defined organic farming as 'production system which favours maximum use of organic material (crop residues, animal excreta, legumes, on and off-farm organic wastes, growth regulators, bio-pesticides etc) and discourages use of synthetically produced agro-inputs, for maintaining soil productivity and fertility and pest management under conditions of sustainable natural resources and healthy environment'.

Organic farming does not imply the simple replacement of synthetic fertilizers and other chemical inputs with organic inputs and biologically active formulations. Instead, it envisages a comprehensive management approach to improve the health and the underlying productivity of the soil. In a healthy soil, the biotic and abiotic components covering organic matter including soil life, mineral particles, soil air and soil water exist in a state of dynamic equilibrium and regulate the ecosystem processes in mutual harmony by complementing and supplementing each other. When the soil is in good health, the population of soil fauna and flora multiplies rapidly which in turn will sustain the biochemical processes of dissolution and synthesis at a high pitch. This state of soil life and the associated organic transformations will enhance the regenerative capacity of the soil and make it resilient to absorb the effects of climatic vicissitudes and occasional failures in agronomic management (Thampan, 1995).

Mohan and Bachi (1999) however stated that the modern concept of organic agriculture is not a return to some past golden age, but rather a networking of scientific advances with many of the traditional practices. The principles governing organic farming are more scientific than even the principles followed in modern agriculture. It is a knowledge intensive, labour oriented and complex system integrating several organic recycling processes. A high degree of motivation, a firm commitment and innovative ability seem to make successful organic farmers.

Tea is comparatively a new entrant in the field of organic products. Organic tea production involves assisting nature to grow a healthy bush, and conserve natural enemies of insect pests, diseases and weeds to minimize crop losses caused by them. This is done by following all modern agronomic practices but completely eliminating synthetic pesticides and fertilizers. The change towards a sustainable farming concept in this perennial plantation tree crop cannot be easily achieved as much as what may possibly be achieved with annual crops, on a rotation system of cultivation. A far greater effort and specialized skills are necessary to bring about such a change and yet manage such systems at economic levels (Sivapalan, 1993).

14.3 Organic tea Gardens

Organic tea was first taken up in Sri Lanka in 1985. Since then progress has been made with only four tea estates and several neighbouring smallholdings (Modder *et. al.*, 1997). In Darjeeling and south India it was in 1885 and 1991 respectively. Needwood tea estate of Sri Lanka converted to organic during 1986. In Darjeeling, the conventional tea estates such as Makaibari, Monteveiot, Mullootar, Singell, Seeyok, Dooteriah, Pussimbing, Chamong, Selim Hill, Selimbong, Ambootia, Samabeong etc are converted to organic estates. In south India, Oothu which is fully clonal tea estate, certified during 1991 as organic tea estate. In Dooars, sections of Dalgaon and Pathurjhora tea estates are converted to organic. Rembong, Banaspaty and one division of Sankar and Sewpur tea estates of Assam are converted to organic tea estate. Luponde tea estate located in the Livingstonia mountains of Tanzania exported organically grown tea in 1989 in the UK market under the 'Natureland' brand name. There is only one organic tea estate in Nepal. Despite Tanzania, China and Sechelle were also in the list of organic tea producing countries but discontinued production at present. The production of organic tea is around 1750 tonnes (Table 14.1).

Table 14.1 : World area and production of organic tea in 1996

Country	T. E. (no.)	Area (ha)	Production (tonnes)	Export (tonnes)
India	15	2996	1415	1300
Sri Lanka	4	560	315	300
Nepal	1	100	20	20
China*	2	135	53	
Seychelle*	1	55	50	
Tanzania*	2	182	100	

*Discontinued

Source: Darjeeling Planters' Association, IFOAM Directory of Organic Agriculture, World Wide, 1996; Modder et. al. (1997).

14.4 Certification of organically grown produce

Tea qualifies as 'organic' when it is grown by using ecofriendly techniques and processed in accordance with the principles of organic farming. Further, a sound system of certification and labelling of the produce by a competent agency is essential. The tea estate where production is organised as per accepted standards has to be supervised and inspected at frequent intervals and at different stages of production before certification in order to ensure quality and authenticity. The cetification agency has to adopt very reliable methods such as soil tests, leaf analysis, water tests, quality tests besides taking note of the natural qualitative indicators so as to ensure credibility of the system and to prevent fraudulent labelling of the produce.

While certification gives reliability to a product as organic, the certification itself would need a guarantee. This is provided through a system of accreditation. International Federation of Organic Agriculture Movement (IFOAM) which has been in the forefront of promoting, monitoring and certifying organically farmed produce.

All the organizations who have gardens growing organic tea in India have recently formed an association called "Indian Bio-Organic Tea Association", to further their interests and also to disseminate information about the benefits of organic tea, both in terms of healthier alternative and a balanced environment.

14.5 Organic techniques

Though there has been a series of scientific and policy conferences on this issue, the organic farming methods are yet to get approval of most agricultural scientists. However, a few isolated examples of enterprising planters, private institutions stand testimony to the utility of safe and productive organic farming methods. To qualify a crop as "organic" active use of following environment friendly techniques can be made.

(a) terracing, contour planting, soil covers and mulching to prevent soil erosion,

(b) increase of organic matter by using legumes, shade tree lopping etc and reduction of dependency on chemical fertilizers,

(c) soil activation by correcting the pH,

(d) use of clones or seedlings resistant to pests and diseases,

(e) regulation of the microclimate and improvement of the ecological diversity to control pests and diseases,

(f) replenishment of nutrients removed by using a mineral balance calculation,

(g) using weeds as a (living) mulch to prevent soil moisture evaporation and using them in compost preparation,

(h) planting of a variety of selected trees to provide cattle fodder, improve the soil, supply green leaf manure and serve as a wind break, and

(i) installation of biogas plants which provide methane gas for use as fuel and organic manure in the form of slurry which is comparatively better in quality and cheaper source of fertilization.

14.5.1 *Weed management*

Weeds could be controlled by adopting eco-friendly practices which are as follows:

i. Mulching

The single most important routine in the organic process is mulching. Advantage of mulching in the form of protecting soil from the beating effect of heavy rainfall, suppression of weeds, surface runoff, preservation of the fertile top soil from erosion, multiplication of the bacterial population of the soil ensuring better plant nutrient availability and conservation of soil moisture is well established. In tea fields, legumes such as *Crotalaria, Mimosa, Tephrosia, Sesbania, Guatemala* (*Tripsacum laxum* Nash) and Mana grass (*Cymbopogon confertiflorus*), water hyacinth, *Citronella* clippings etc. could be used for mulching (see also Chapter 5 and section 5.8.5). Rice straw is one of the cheapest and most available source of organic material found in Asia. But more critical research on the impact of mulch on weed, insect, insect pathogen and nematode pests, and the mechanisms like allelopathy, light exclusion, antagonism etc should go on.

ii. Cover cropping

Leguminous cover crops such as *Mimosa invisa* Mart., *Stylosanthes guyanensis* SW var. *gracilis* (HBK) Vog., *Calapogonium mucunoides* Desv., *Centrosema pubescens* Benth., *Pueraria phaseoloides* (Roxb.) Benth., may be sown before rains. *Mimosa invisa* and *Desmodium* sp. in combination with Guatemala (*Tripsacum laxum*) would be ideal. It suppresses weed growth, improves soil structure and fertility, and helpful in keeping down the soil temperature during summer months. Cover crop can be grown in between the tea rows and in walking rows. Cover cropping with *Mimosa invisa* at Looksan tea estate in north-east India during 1993–

94 reduced the cost of weed control by 15% over manual weed control besides addition of organic matter to the soil (Singh *et. al.*, 1999). Non-leguminous cover crops have also been used with success. *Polygonum, Oxalis* are of considerable value in south India, Sri Lanka and east Africa. In Kenya, Oats (*Avena sativa* L) and buckwheat (*Fagopyrum esculentum* Moench.) are widely used as cover crop for higher and lower altitudes respectively. Crotalaria is also widely used as nurse crop planted between rows of young tea plants. In former USSR, the green crops like Soyabean or Lupines for a period of one year or so are planted.

iii. Weed killing and weed suppressing plants

The crop, weed and tree residues constitute the major source of organic matter in the soil. They also release numerous organic and inorganic compounds in the soil. These compounds (allelochemicals) are generally inhibitory to the growth of other crops/trees and in some instances to their own plants (autotoxicity) depending on their residue amount, length of decomposition period and type of residue. A list of these plants which could be grown in organic garden has been presented by Singh *et. al.* (1999).

14.5.2 *Soil pH*

It has significant influence on the availability of plant nutrient as well as microbial activity. Tea grows well within a range of 4.5 to 5.5. Any change of pH beyond this limit affect yield adversely. For correction of pH, see Chapter 5 and section 5.5.3.

14.5.3 *Soil erosion*

This has been discussed in the Chapter 8 and section 8.5.

14.5.4 *Planting material*

Clonal materials which have very high requirements of fertilizers are generally found susceptible to different stresses, may not be ideal for planting in organic tea fields. Low input responsive clones or seed stocks having wide adaptability with tolerant to different stresses may be considered. Polyclonal or boclonal varieties with high tolerance to stresses having modest yield potential may also be considered.

14.5.5 *Pest management*

In order to tackle the pests in organic gardens one should thorough the life cycle of the attacking pests. Neem-based pesticides produced from neem karnel extracts could be used to repel pests in organic farming. Spraying of diluted cow urine and extracts of marigold leaves were frequently practiced. Natural predators are brought into action like ladybird, beetles and others, which devour the pests but leave the fragile eco-system undisturbed. This has also been discussed in Chapter 12 and section 12.10.

14.5.6 *Nutrient management*

Though there are several doubts in the minds of not only the farmers but the scientists too, as to whether it is possible to supply the minimum required nutrients to crops through organic sources alone and even it is possible how are we going to mobilize that much of organic matter. Growing leguminous crops (shade trees, cover crops, green crops etc) could be a good source for the supply of nutrients. Nutrients can also be supplemented using oil cakes, farm yard manure, lopping of green manures, recycling of

Fig. 14.1 : Compost pits: compost ready for application in the field (Photograph from Monteviot tea estate, Darjeeling by the permission of the Manager).

pruning litters and shade tree droppings. Figure 14.1 shown compost ready for application.

i. Biofertilizer

Use of bio-fertilizer fortified with tea suitable bio-inoculants makes following contributions in organic tea fields. Bio-fertilizer has no side-effects rather all positive impacts help tea growth and better quality.

i. It could give 20–50 kg nitrogen and 30–50 kg phosphate in tea for instant uptake.

ii. It liberates growth promoting hormones, vitamins and helps to improve soil-fertility.

iii. It suppresses incidence of pathogens, secret antibiotic to control diseases.

iv. It increases tea yield by fixing-N and solubilising P. It reduces depletion of soil-nutrient and provides sustainability to farming system etc.

v. It is pollution free and improves soil's physical properties, and soil health for better quality-flavour,

(A) VERMICULTURE

Now-a-days, improvement of soil through vermiculture is also becoming popular and is an important part of organic farming to hasten the decomposition of organic residues. It is the technique by which organic wastes, manure, sewage sludge and household wastes can be processed by indigenous species of earthworms (Fig. 14.2) into vermicompost. Vermicompost is accepted as humus bio-fertilizer, soil fertility booster, soil activator, soil conditioner, with required plant nutrients, humus, organic carbon, vitamins, enzymes, growth hormones, and beneficial microbes like nitrogen fixing, phosphate solubilising denitrifying and decomposing bacteria (Ghosh Hajra, 1994). It is estimated that 1000 tonnes of moist organic matter can be converted by earthworms into 300 tonnes of compost. It is also estimated that 1800 worms which is an ideal population for one square meter can feed on 80 tonnes of humus per year. The average nutrient content of vermicompost is N -1.6%; P_2O_5 -2.2%; K_2O -0.67%; Ca -0.99% and Mg -0.15%. Preparation of vermicompost (Fig. 14.3) and vermiculture ecotechnology have been discussed by (Gaur and Sadasivam, 1993; Bhawalkar, 1995).

Fig. 14.2 : Earthoworms – friends of the farmers – the effective biomanagers. Worms collected from the soils of the tea estate of Darjeeling hill.

Fig. 14.3 : Vermicompost bins (40'x 10' x 3' – capacity 10 mt) in north Tukvar tea estate, Darjeeling shaded with nylon net to protect from bright sunlight, rain and also to provide suitable environment for growth and multiplication of the earthworms.

Vermicompost, excreta or castings of earthworms can be mixed in nursery soil while filling up of polysleeves for good rooting of the cuttings. This can also be mixed with the soil at the time of infilling and replanting for easy establishment of plants. The compost can also be used in mature tea fields for enrichment of poor soil organic matter and enhancement of the productivity of the soil. During replanting or new planting of tea, vermicasting produced by the burrowing earthworms may be used. Compost @ 5 tonnes/ha may be applied in presence of sufficient moisture.

14.6 Transitional period

The biggest challenge we have at present is how to change from conventional farming to organic farming without any loss of yield during the transition period as well as in the future. Tea, being a perennial crop, is likely to take longer time than annual crops in switching over to organic farming. It is believed that a minimum of three years is required as transitional period to convert a conventional high input plantation to organic tea allowing time for chemicals to disperse (Singh *et. al.*, 1999). A study of the ecological farmers in Tamil Nadu and Karnataka, India conducted by the Agriculture-Man-Ecology Programme (AMEP) at Pondicherry revealed that the switchover from conventional farming to organic agriculture should be gradual. If field of low fertilizer—use with less application of plant protection chemicals are to be converted, it took three to four years. This duration of time could be curtailed to three months by harnessing and seeding the soil with vermicastings (Bhawalkar and Bhawalkar, 1993). In case of farm where high dose of fertilizers and insecticides were used, the transition phase extends to 7 or 8 years. During the interim years, the farmers could build up a sufficient organic base in their fields and improve productivity of the soil. If the switchover is rapid, the yields will fall dramatically and the farmers get demoralized.

Hadfield (1991) gathered experimental results from many sources indicated that without normal fertilization (about 100 kg nitrogen plus potash and phosphate per hectare), yields of tea decline until a 'steady state' is reached and this is between 500–600 kg ha^{-1} depending on the fertility level of the original soil. Tea yielding 2000 kg will take about 5 years to decline to 1000 kg, and it may be another 5 to 6 years before it reaches 600 kg. Hadfield (1991) had shown in Figure 14.4 what is likely to happen to tea yielding 2000, 1500 and 1000 kg ha^{-1} in Assam, West Bengal in

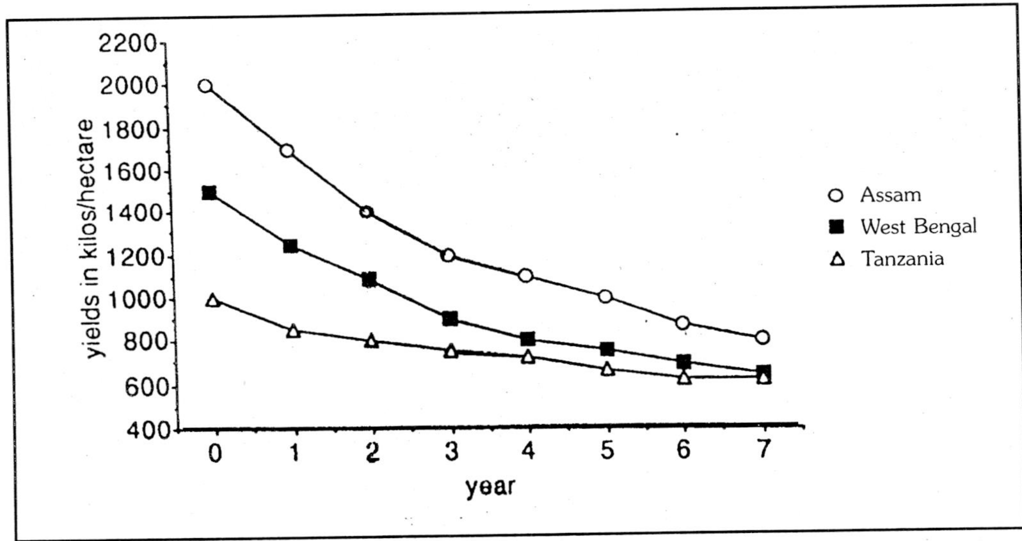

Fig. 14.4 : Declination of yearly yield when application in fertilizer is cessed [Source: Hadfield, (1991)].

India and Tanzania respectively if manuring is cessed. It could be argued from Figure 14.4 that it is those planters who are the most efficient producers of conventional tea who stand to lose the most by converting to organic

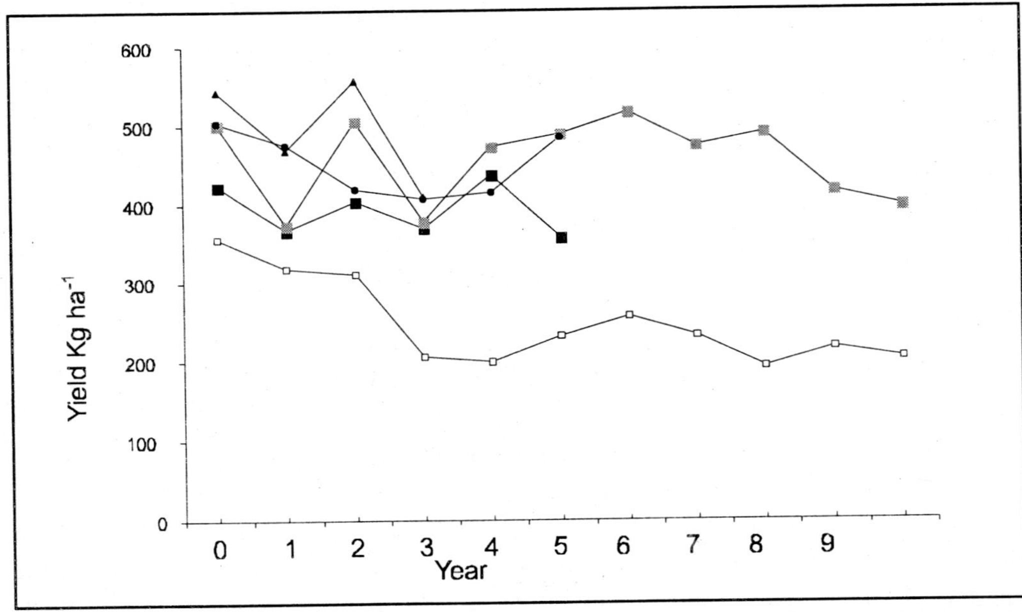

Fig. 14.5 : Yield trend of few Darjeeling tea estates after switching over to organic garden (0' year is year of conversion).

methods of production unless they can maintain a reasonable level of yield combined with a premium sufficiently high to justify the conversion. In Darjeeling hills, the yield trends of few organic tea estates after switching over are presented in Figure 14.5

Mohan and Bachi (1999) reported when commercial tea estates switched over to organic farming the usual tea fields lost 10% crop in the first year, 20, 30 and 40% in second, third and fourth year respectively and stabilized to gain 20% in fifth year as a consequence of recycling organic matter and the resurgence of effective natural enemies of the crop pests. But Werf (1990) reported that in south India tea yield decreases of 21 to 33% in the first year of transition from conventional (240 to 300 kg N/ha/year) to organic cultivation.

Mackay (1990) laid out a rotational trial with an objective to establish the yield loss associated with converting to organic production and observed that yield of clover, winter wheat, oats, potatoes, spring barley and swedes has been reduced at least 50% when converted to organic production, provided disproportionate quantities of manure are not applied. These lower yields could be compensated when one takes into consideration reduced cost of chemical inputs and premiums for organic produce.

Box 14.1 : Organic farming

- Organic farming or organic agriculture is one of the alternatives for sustaining production without sacrificing the environment and ecology. The principles governing organic farming are more scientific than even the principles followed in modern agriculture.
- Organic farming is still very marginal and occupies between 1% and 10% of Europe's agricultural area. Ecofarming accounts for less than 1% of agriculture in France, the UK and Spain, in the Netherlands around 1%, Sweden, Finland and Denmark between 5 and 7% and Austria and Switzerland 10% (Kieft, 1999).
- Organic farming has been perceived differently by different people in different countries.
- In the developed countries, it may be meant for preventing contamination by poisonous chemicals. In India, the major problem is the low productivity of the soils, because of its low organic matter content. Even the efficiency of the uptake of fertilizer depends on the organic matter content of the soil.
- The statement has often been made by critics that 'Natural farming works fine, but there are not enough organic materials to go around'. On the contrary, there is more organic wastes around us than we know what to do with. The handling and disposal of organic wastes is a great national problem, and many millions of tonnes of potential humus are going to wastes or are creating environmental problems.

But no drastic reduction in yield was observed in an experiment laid out at Tea Research Association, Tocklai on the comparisons of changeover of conventional plots to organic tea in a 25-year-old clonal tea block. However, only marginal differences between the comparative plots manured with organics (manure, FYM and mulching @ 3, 10 and 10 t/ha respectively) and inorganics was observed. This indicate switch over to organic farming does not drastically reduce yield as is widely observed. However, the declination of yield towards the steady state can be thwarted if the initial yield had been significantly higher and if the required infrastructural base for organic farming is established well within the transition phase (Sivapalan, 1993). Singh *et. al.* (1999) stated that the yield losses could be avoided / minimised if new extension or uprooted area is adequately rehabilitated prior to planting. The withdrawal of synthetic fertilizers and pesticides which account for 20–25% of the total cost of production lowers the unit cost (Anonymous, 1995).

14.7 Future of organic tea

With a little lower crop productivity, organically grown tea command a premium of 30–40 % over conventionally produced tea. Further, organic tea requires around 20% lower input costs by way of not applying chemical fertilizers and pesticides. The demand for certified organic tea is growing at a very rapid rate all over the world, with Europe, Japan, and the U.S. leading the way. Organic tea is a value added product and most organic tea producers have developed direct marketing strategies and partnerships with leading companies worldwide, thereby bypassing the traditional channels of the tea trade such as auctions. Organic tea producers could seize the emerging opportunity of this niche in the world market.

In the global movement towards conservation and nurturing of the environment, respect for the ecosystem and concern for the well-being of all living creatures, the organic way of growing tea will surely achieve more popularity.

References

Anonymous, 1995. Financial performance of tea companies1973-74 to 1986-87. *Occassional paper*, 1, Indian Tea Association, Calcutta.

Bhawalkar, U. S. 1995. Vermiculture Ecotechnology. *Bhawalkar Earthworm Research Institute*, Pune, India, p. 332.

Bhawalkar, U. S. and Bhawalkar, V. U. 1993. Vermiculture biotechnology. In *Organics in Soil Health and Crop Production* (ed. P. K. Thampan). Peekay Tree Crops Development Foundation, Cochin, India. 69 – 85.

Dahama, A. K. 1999. *Organic Farming for Sustainable Agriculture.* Agro Botanica, Bikaner, India, p. 294

Gaur, A. and Sadasivam, K. V. 1993. Theory and practical considerations of composting organic wastes. In *Organics in Soil Health and Crop Production* (ed. P. K. Thampan). Peekay Tree Crops Development Foundation, Cochin, India, 1 - 22.

Ghosh Hajra, N. 1994. Vermiculture and vermicasting in tea cultivation. *Assam Rev. & Tea News*, 83(5): 9 – 12.

Hadfield, W. 1991. Production problems during conversion. *Proc First Intl. Conf. Organic Tea and Coffee.* June 2 – 4, 1991, Frauenfeld, Switzerland, IFOAM, 1 – 4.

Harkaly, A. 1990. Promote local consumption of organic export crops. *Ecology and Farming*, 1: 18

Kieft, Henk. 1999. Biodiversity in Dutch Agriculture. COMPAS Newsletter, 2, October 1999, Page 26 - 28.

Mackay, J. M. 1990. Organic production systems: A means of reducing chemical inputs. *J. Sci. Food Agric.*, 53: 112 – 117.

Modder, W. W. D., Kathiravetpillai, A. and Sivaram, B. Organic tea: a Sri Lankan case study. *Proc. 2nd Intl. IFOAM Conf.*, April 28 – 29, 1997, Calcutta, India, 101 – 106.

Mohan, Brij and Bachi, Rainer. 1999. Organic tea. In *Global Advancement in Tea Science* (ed. N. Jain). Aravali Books International, New Delhi, India, 813 – 816.

Singh, I. D., Bhattacharya, A. K. and Bera, B. 1999. Organic tea cultivation an ecofarming approach. *Assam Rev. & Tea News*, 88(9): 5 – 15.

Sivapalan, P. 1993. Organic farming of tea as an aid to maintaining environmental quality. *Proc. Intl. Symp. on Tea Science and Human Health*. TRA, Calcutta, January 11-14, India. 300 – 302.

Thampan, P. K. 1995. Perspectives on organic agriculture. In *Organic Agriculture* (ed. P. K. Thampan). Peekay Tree Crops Development Foundation, Cochin, India, 1 - 37.

Werf, Erik J. van der. 1990. Organic tea cultivation at Singampatti Group of B.B.T.C., India. ETC – Foundation, Leusden, The Netherland.

Glossary

Abscission Refers to the separation of a leaf, fruit or other part from the body of a plant. It is controlled by growth hormone, especially by abscisic acid.

Absorption Refers to the movement of fluid or dissolved substances across a cell membrane. In plants, mineral salts and water get absorbed by the roots from soil.

Acaricide A chemical used for controlling mite pests.

Acid soil A soil with a pH value <7.0. Usually applied to surface layer or root zone, but may be used to characterize any horizon.

Adaxial Facing away from the stem (Dorsal surface of a leaf).

Adsorption The attraction of ions or compounds to the surface of a solid. Soil colloids adsorb large amounts of ions and water.

Aeration, soil The process by which air in the soil is replaced by air from the atmosphere. In a well-aerated soil, the soil air is very similar in composition to the atmosphere above the soil. Poorly aerated soils usually contain more carbon dioxide and correspondingly less oxygen than the atmosphere above the soil.

Allelopathy The release by a plant of a chemical that influences the growth of other plants.

Ammonification The biochemical process whereby ammoniacal nitrogen is released from nitrogen–containing organic compounds.

Anatomy The branch of biology which deals with the study of internal structure of an animal or a plant or of any of its parts.

Anion	Negetively charged ion; ion that during electrolysis is attracted to the anode.
Anthocyanin	Water soluble pigments, found in certain plants. Impart red, purple, or blue colouring to flower, fruits, and autumn leaves.
Apical bud	The bud in the apex of an extending shoot.
Aroma	A fragrant smell usually derived from the dry leaf or infused leaf. The concept of tea flavour can be broadly divided into sensations of taste and aroma. Non-volatile components are generally responsible for the test whilst the more volatile components comprise the aroma. Aroma is considered to have the most impact on the flavour of food products.
ATP	Adenosine- triphosphate, a nucleotide, occurring in plant and animal cells is a major energy source for vital processes. Energy is released when ATP is hydrolysed to ADP (Adenosine diphosphate).
Available nutrient	That portion of any element or compound in the soil that can be readily absorbed and assimilated by growing plants.
Axil	The angle between a leaf and the stem.
Axillary shoot	A shoot which develops from an axillary bud.
Banjhi	This is an Assamese word meaning sterile. The dormant phase of the cyclical growth pattern of tea. Also applicable to the dormant bud at the apex of a dormant shoot.
Bari	It is an Assamese word denoting comparatively small piece of highland for growing tea plants producing tea seeds.
Bheel	An Assamese word. Low lying area in and around the tea plantation, very common in north east India.

Bheel soil — A type of tea soil in Cachar district of Assam (Surma Valley). The soils of these areas are rich peats, black, and very rich in organic matter and nitrogen.

Bheti — It actually means a raised bed of earth. Term used for the ball or cylinder of soil which encloses the roots of a plant when transplanting.

Biodegradable — Capable of being decomposed by natural biological processes.

Biofertilizer — Fertilizer containing minerals combined with microorganisms, the action of which renders the minerals available to plants.

Biological control — The control of pests using other organisms, usually their natural predators, parasites or diseases.

Biological fixation — Formation of organic compounds containing nitrogen by the living microorganisms.

Biomass — The weight of living organism or all organisms in an ecosystem or in a habitat.

Bio-tea — Also called as organic tea. See organic tea.

Breaking back — The operation of removing thick or knotted stems to a level below the plucking table.

Breathers — The stems which left unpruned on the heavy or medium prune bushes. Breathers are normally prevents the bushes from carbohydrate starvation during recovery from pruning. Breathers are removed when new stems are sufficiently large and vigorous. Also called as 'lungs' or 'kickers'.

Broadcasting — A manuring method when manure is applied broadcast all over the area without leaving any unmanured patches.

Broad spectrum pesticide A pesticide which is effective against more than one pest at a time.

Bud (1) An outgrowth on a stem or branch, often enclosed in protective scales, comprising a shortened stem and immature leaves or floral parts. (2) The apex of an elongated tea shoot wrapped tightly in a foliage leaf which is not different from expanded leaf except it is unwrapped.

Bud break Resumption of growth by the dormant buds.

Bush (1) Refers to the tea plant under cultivation. The plants during cultivation are trained to form a low spreading bush by a succession of prunes from early years. (2) A low and thick shrub without a distinct trunk.

Calcium A silvery, moderately hard metallic element, which is a basic component, shells, and teeth.

Cambium The plant tissue with fast dividing cells and contribute to the increase in girth of stem and root in dicots.

Cation A positively charged ion.

Cation exchange capacity The sum total of exchangeable cations that a soil can adsorb. Sometimes called 'total-exchange capacity'.

Caterpillar Wormlike, often brightly coloured, hairy or spiny larva of a butterfly or moth, having many legs and biting jaws.

Cell The basic structural and functional unit of plants and animals.

Cell wall The rigid outermost layer of a plant cell, consisting of cellulose and other polysaccharides.

Chloroform test A simple test for assessing fermentation ability.

Chlorophyll Green pigments found in plants that trap energy from sunlight for use in photosynthesis.

Chloroplast Plastid containing chlorophyll in photosynthetic plants.

Chlorosis Abnormal condition in plants, characterized by deficiency in green pigment due to lack of light, mineral deficiency, or genetic disorders.

Chromosome A structural unit in the nucleus which carries the genes in a linear constant order; it preserves its individuality from one cell generation to the next and its typically constant in number in any species.

Clone A group of plants originating by vegetative propagation from a single plant.

Collar A line of junction which is present between a root and its stem.

Compost Organic residues, or a mixture of organic residues and soil, that have been piled, moistened, and allowed to undergo biological decomposition. Mineral fertilizers are sometimes added. Often called "artificial manure" or "synthetic manure" if produced primarily from plant residues.

Contour An imaginary line connecting points of equal elevation on the surface of the soil. A contour terrace is laid out on a sloping soil at right angles to the direction of the slope and nearly level throughout its course.

Conversion efficiency The proportion of solar radiation intercepted by leaves which is converted for production of dry matter by photosynthesis.

Cover crop A low-growing, often spreading, often annual, crop grown all over a field before planting. This protects the soil against erosion and may provide some shelter for young tea plants.

Creep — The rise of plucking table due to continuous pluckings during the season which ultimately results in loss of crop.

Crop factor — The ratio of the water loss by transpiration from a crop to the water loss from an open water surface under the same meteorological conditions. For many crops providing complete coverage of the soil surface by vegetation the crop factor is close to 0.85.

Cross pollination — It is the transfer of pollen from an anther on one plant to a stigma in a flower on a different plant.

Crow's feet — A knot formed in the plucking table where a number of new flushes have formed from the same height after successive plucking.

Cuticle — The protective layer of cutin covering the epidermis of plants.

Cultural control — Adoption of ordinary field practices in such a way that the pests are either eliminated or reduced in population.

Day degree — The average number of degrees by which the atmospheric temperature is above the minimum critical temperature for tea growth (12-13^{0}C) but below the maximum (ca. 30^{0}C) multiplied by the duration (in days) of the period over which the average is taken.

Dicotyledon — Division of angiosperms, characterized by a pair of embryonic seed leaves that appear at germination.

Dormancy — See Banjhi.

Eelworm — Very small worms abundant in many soils and important because some of them attack and destroy plant roots.

Epidermis Outer, protective layer of cells covering of a plant or plant part.

Epiphyte A fungus or bacterium existing on the surface of a plant or plant organ without causing infection.

Erosion (1)The removal of soil and other materials by moving water, wind or by other geological agents, including such processes as gravitational creep. (2) Detachment and movement of soil or rock by water, wind, ice, or gravity.

Essential elements A chemical element required for the normal growth and development of plants.

Evaporation The change of state of water from solid and liquid to the vapour and its diffusion into the atmosphere is referred to as evaporation. It plays a major role in the redistribution of thermal energy between the earth and the atmosphere, and is an essential part of the hydrological cycle.

Evapotranspiration It is the combined loss of water from vegetation – both as evaporation from soil and transpiration from plants. Both the processes are basically the same and involve a change of state – from liquid to vapour. When water is adequately available at the site of transformation (i.e. soil or plant surfaces) the rate of evapotranspiration is primarily controlled by meteorological factors, like solar radiation, wind, temperature and the evaporating power of the atmosphere.

Extension planting Planting tea in new areas.

Feeder roots The fine, fibrous, white or cream coloured rootlets of the root system, responsible for the absorption of water and nutrients from the soil.

Fertile soil A soil containing all essential elements in available forms and in amounts necessary for maximum growth and yield.

Field capacity	The maximum level of water which the soil can hold in or attached to the soil particles.
Fish leaves	The appendages that is unfold after the janams while producing a flush. It is also known as 'Gol-pat' in Assam. It is deep green with prominent venation and often partially serrated without apex and photosynthetically very efficient.
Flower	Reproductive structure of an angiosperm having male or female organs enveloped within petals or sepals.
Flush	(1) The young tea shoots which form the major part of the harvest. (2) The amount of growth produced by a growing bud in between banjhi stages.
Fumigant	A toxic chemical which is volatile at room temperature, used normally against pests of stored products, soil pests and pathogens.
Guard cells	Kidney-shaped cells on either side of stoma which control the opening and closing of it.
Habitat	Type of environment in which an organism or biological population normally lives or occurs.
Haploid	Having a single set (genome) of chromosomes in a cell or an individual; the reduced number (n), as in a gamete.
Harvest index	The fraction of the total dry matter produced in a plant which can be harvested.
Harvesting	It is called as plucking in most countries. In China, it is known as picking. Harvesting is done at frequent intervals in India and other tea producing countries. In China, it is done 3-4 times a year and 3 times in Japan. Each harvesting period lasting for about two weeks.
Herb	Angiosperm with soft fleshy stem.
Herbicide	A class of pesticides that destroys or reduces the negetive effects of weeds.

Humus That more or less stable fraction of the soil organic matter remaining after the major portion of added plant and animal residues have decomposed. Usually it is in dark in colour.

Hybrid Offspring of genetically dissimilar parents or stock.

Infection Invasion of the body by pathogenic micro-organisms.

Insect Invertebrate animal. Adult stage has three pairs of legs, body with three major divisions, and usually two pairs of wings.

Internode A section or part between two successive nodes.

Ions Atoms, group of atoms, or compounds that are electrically charged as a result of the loss of electrons (cations) or the gain of electrons (anions).

Irradiation In genetics and plant breeding, exposing seed, pollen, or other plant parts to x-rays or other radiations to increase mutation rates.

Larva Early form of an organism which is very unlike the parent and undergoes great changes to become an adult.

Leaching The removal of materials in solution from the soil by percolating waters.

Leaf Usually green, flattened structure of vascular plants functioning as a principal organ of photosynthesis and transpiration.

Leaf stock The stalk by which leaf is attached to a stem; a petiole.

Legume Any member of the family leguminosae, a pulse family which produce legumes. A pod, that splits into two halves with the seeds attached to the lower edge of one of the halves e.g. pea, bean.

Macronutrient Element required in relatively large proportion for growth e.g. carbon, hydrogen, oxygen, nitrogen etc.

Mesophyll The cells inside the leaves containing chloroplasts located between the upper and lower epidermis.

Micronutrient A chemical element necessary in only extremely small amounts (<50 ppm in the plant) for the growth of the plants

Miticide A chemical used to control, destroy mites.

Mother bush Bush selected or planted for its desirable characteristics and allowed to be used for vegetative propagation. The plants so produced form a clone.

Natural farming A system of agriculture devised by Masanobu Fukuoka that seeks to follow Nature by minimizing human interference: no mechanical cultivation, no synthetic fertilizers or prepared compost, no weeding by tillage or herbicides, no dependence of chemicals.

Nematode See eelworm.

Nucleus plot The multiplication plot or area maintained separately for the mother bushes of the clones.

Nutrient Any substance which promotes growth or provides energy for physiological processes.

Organic tea Tea qualifies as 'organic' when it is grown by using ecofriendly techniques and processed in accordance with the principles of organic farming. No legislation on organic standards has been passed but the standards of IFOAM in general are accepted.

Ovicide A chemical for killing eggs.

Palisade parenchyma The closely packed mesophyll cells in leaves towards the upper epidermis containing chloroplasts.

Parasite Any organism that grows, feed, and is sheltered on or in a different organism while contributing nothing to the survival of its host.

Perennial A plant that lives for three or more years and which normally flowers and fruits at least in its second and subsequent years.

Persistent A term applied to chemicals that remain active for long period after its application.

Pest An organism causing damage to man's, crops and animals.

pH A quantitative expression for acidity or alkalinity of a solution. It is the negetive logarithm of the hydrogen ion concentration. Higher values are alkaline and lower are acidic.

Phenotype (1) Physical or external appearance of an organism as contrasted with its genetic constitution (genotype). (2) A group of organisms with similar physical or external makeup.

Pheromones Chemical used by insects to communicate each other; the odours emitted by species of insects to communicate between males and females of the same species.

Phloem The nutrient-conducting tissue of vascular plants, consisting of sieve tubes, companion cell and phloem parenchyma.

Photolysis The breaking down of water into H^+ and OH^- ions in the presence of light.

Photosynthesis The physiological process by which chlorophyll-containing cells in plants convert incident light to chemical energy and prepare food in the form

	of carbohydrates from carbon dioxide and water, with the simultaneous released of oxygen as bye-product.
Phyllochron	The time interval between unfolding of successive leaves on a growing shoot. It is also called as leaf period.
Picking	See harvesting.
Pith	Soft, sponge like substance in the centre of stems and branches of most vascular plants.
Pollination	Transfer of pollen from the anther to stigma.
Pollution	An undesirable and excessive addition of substance to water, air or land which adversely affects the natural quality of the environment.
Polyploid	An organism with more than two sets (genomes) of chromosomes in its body cells.
Progeny selection	Selection based on progeny performance.
Respiration	The metabolic process by which an organism assimilates oxygen, oxidizes organic substances in the cells, with the release of energy, carbon dioxide and water vapour.
Resurgence	The increase in a pest population after it is freed from natural controls, most commonly following the application of a pesticide that destroys its natural enemies.
Run-off	Rainfall or other water that flows across the soil surface and does not infiltrate into the soil.
Scion	Twig containing bud from a woody plant used in grafting.
Self pollination	Transfer of pollen from an anther to the stigma of the same flower or another flower on the same plant, or within a clone.
Spongy parenchyma	The mesophyll cells towards the lower epidermis with large intercellular spaces and cavities.

Starch	A carbohydrate, found chiefly in the seeds, fruits, tubers, roots, and stem pith of plants.
Stem	Ascending axis of plant. Bears leaves, flowers and buds.
Stock	A plant or stem onto which a graft is made.
Thermal time	See day degree.
Weed	A plant in a place where it is not wanted by humans.
Wilt	To become limps or flaccid; droop.
Xylem	The supporting and water conducting tissue of vascular plants.

Subject Index

N

O

P

R

I

J

K

S

T

U

V

W

X

Y

Z